Spring Security

Effectively secure your web apps, RESTful services, cloud apps, and microservice architectures

Badr Nasslahsen

Spring Security

Group Product Manager: Kaustubh Manglurkar

Publishing Product Manager: Vaideeshwari Muralikrishnan

Book Project Manager: Arul Viveaun S

Senior Editor: Nathanya Dias

Technical Editor: K Bimala Singha

Copy Editor: Safis Editing

Indexer: Tejal Soni

Production Designer: Jyoti Kadam

DevRel Marketing Coordinators: Anamika Singh and Nivedita Pandey

First published: May 2010

Second edition: December 2012

Third edition: November 2017

Fourth edition: June 2024

Production reference: 1310524

Published by Packt Publishing Ltd.

Grosvenor House

11 St Paul's Square

Birmingham

B3 1RB, UK

ISBN 978-1-83546-050-4

www.packtpub.com

Contributors

About the author

Badr Nasslahsen is a lead security and cloud architect with over 17 years of experience. He holds an executive master's degree from Ecole Centrale Paris and an engineering degree from Telecom SudParis. He is an Oracle Certified Java SE 11 Professional, CISSP, TOGAF, CKA, and Scrum master.

Badr has extensive experience in public cloud providers: AWS, Azure, GCP, Oracle, and IBM. He is also the author of the `springdoc-openapi` project.

About the reviewer

Zainab Alayande is an accomplished software engineer renowned for her proficiency in Java development, particularly within the Spring Framework ecosystem. Her expertise spans various aspects of Spring, including Spring Security. Zainab has leveraged her skills to architect and implement comprehensive enterprise solutions. She has a strong passion for Spring Security, evidenced by her extensive writing on the topic and her dedication to educating fellow developers on its intricacies.

Prior to her current venture in Lagos, Nigeria, where she cofounded a start-up aimed at decentralizing logistics, Zainab gained valuable experience at Convexity Technology and Semicolon Africa.

Table of Contents

Part 2: Authentication Techniques

5

Authentication with Spring Data   117

6

LDAP Directory Services   141

7

Remember-me Services   183

8

Client Certificate Authentication with TLS   209

Part 3: Exploring OAuth 2 and SAML 2

9

Opening up to OAuth 2   233

10

SAML 2 Support   263

Part 4: Enhancing Authorization Mechanisms

11

12

13

Part 5: Advanced Security Features and Deployment Optimization

14

15

Additional Spring Security Features   393

16

Migration to Spring Security 6   425

17

Microservice Security with OAuth 2 and JSON Web Tokens   467

18

Single Sign-On with the Central Authentication Service   489

19

Build GraalVM Native Images   519

Preface

Knowing that experienced hackers are itching to test your skills makes security one of the most difficult and high-pressured concerns of creating an application. The complexity of properly securing an application is compounded when you must also integrate this factor with existing code, new technologies, and other frameworks. Use this book to easily secure your Java application with the tried and trusted **Spring Security** framework, a powerful and highly customizable authentication and access-control framework.

The book starts by integrating a variety of authentication mechanisms. It then demonstrates how to properly restrict access to your application. It also covers tips on integrating with some of the more popular web frameworks. An example of how Spring Security defends against session fixation, moves into concurrency control, and how you can utilize session management for administrative functions is also included.

It concludes with advanced security scenarios for RESTful web services and microservices, detailing the issues surrounding stateless authentication, and demonstrates a concise, step-by-step approach to solving those issues.

Who this book is for

If you are a Java Web and/or RESTful web service developer and have a basic understanding of creating Java 17/21, Java Web and/or RESTful web service applications, XML, and the Spring Security framework, this book is for you.

What this book covers

Chapter 1, *Anatomy of an Unsafe Application*, covers a hypothetical security audit of our calendar application, illustrating common issues that can be resolved through the proper application of Spring Security. You will learn about some basic security terminology and review some prerequisites for getting the sample application up and running.

Chapter 2, *Getting Started with Spring Security*, demonstrates the `"Hello World"` installation of Spring Security. After that, the chapter walks you through some of the most common customizations of Spring Security.

Chapter 3, Custom Authentication, incrementally explains Spring Security's authentication architecture by customizing key pieces of the authentication infrastructure to address real-world problems. Through these customizations, you will gain an understanding of how Spring Security authentication works and how you can integrate with existing and new authentication mechanisms.

Chapter 4, JDBC-based Authentication, covers authenticating against a database using Spring Security's built-in **Java Database Connectivity (JDBC)** support. We then discuss how we can secure our passwords using Spring Security's new cryptography module.

Chapter 5, Authentication with Spring Data, looks at the Spring Data project, and how to leverage **Jakarta Persistence (JPA)** to perform authentication against a relational database. We will also explore how to perform authentication against a document database using MongoDB.

Chapter 6, LDAP Directory Services, will review the **Lightweight Directory Access Protocol (LDAP)** and learn how it can be integrated into a Spring-Security-enabled application to provide authentication, authorization, and user information services to interested constituents.

Chapter 7, Remember-me Services, demonstrates the use of the remember-me feature in Spring Security and how to configure it. We also explore additional considerations to bear in mind when using it. We'll add the ability for an application to remember a user even after their session has expired and the browser is closed.

Chapter 8, Client Certificate Authentication with TLS, demonstrates that, although username and password authentication is extremely common, as we discussed in *Chapter 1, Anatomy of an Unsafe Application*, and in *Chapter 2, Getting Started with Spring Security*, forms of authentication exist that allow users to present different types of credentials. Spring Security caters to these requirements as well. In this chapter, we'll move beyond form-based authentication to explore authentication using trusted client-side certificates.

Chapter 9, Opening up to OAuth 2, explains that OAuth 2 is a very popular form of trusted identity management that allows users to manage their identity through a single trusted provider. This convenient feature provides users with the security of storing their password and personal information with the trusted OAuth 2 provider, optionally disclosing personal information upon request. Additionally, the OAuth-2-enabled website offers the confidence that the users providing OAuth 2 credentials are who they say they are.

Chapter 10, SAML 2 Support, will deep dive into the world of **Security Assertion Markup Language (SAML 2.0)** support and how it can be seamlessly integrated into their Spring Security applications. SAML 2.0 is an XML-based standard for exchanging authentication and authorization data between **identity providers (IdPs)** and **service providers (SPs)**.

Chapter 11, Fine-Grained Access Control, will first examine two ways to implement fine-grained authorization—authorization that may affect portions of a page of the application. Next, we will look at Spring Security's approach to securing the business tier through method annotation and the use of interface-based proxies to accomplish **Aspect-Oriented Programming (AOP)**. Then, we will review an interesting capability of annotation-based security that allows for role-based filtering on collections of data. Last, we will look at how class-based proxies differ from interface-based proxies.

Chapter 12, Access Control Lists, will address the complex topic of **Access Control Lists (ACLs)**, which can provide a rich model of domain object instance-level authorization. Spring Security ships with a robust, but complicated, ACL module that can serve the needs of small- to medium-sized implementations reasonably well.

Chapter 13, Custom Authorization, will include some custom implementations for Spring Security's key authorization APIs. Once we have done this, we will use the understanding of the custom implementations to understand how Spring Security's authorization architecture works.

Chapter 14, Session Management, discusses how Spring Security manages and secures user sessions. The chapter starts by explaining session fixation attacks and how Spring Security defends against them. It then discusses how you can manage logged-in users and restrict the number of concurrent sessions a single user has. Finally, we describe how Spring Security associates a user with HttpSession and how to customize this behavior.

Chapter 15, Additional Spring Security Features, covers other Spring Security features, including common security vulnerabilities such as **cross-site scripting (XSS)**, **cross-site request forgery (CSRF)**, **synchronizer tokens**, and **clickjacking**, and how to protect against them.

Chapter 16, Migration to Spring Security 6, provides a migration path from Spring Security 5, including notable configuration changes, class and package migrations, and important new features, including Java 17 support and new authentication mechanisms with OAuth 2.1.

It also highlights the new features that can be found in Spring Security 6.1 and provides references to examples of the features in the book.

Chapter 17, Microservice Security with OAuth 2 and JSON Web Tokens, looks at microservices-based architectures and how OAuth 2 with **JSON Web Tokens (JWT)** plays a role in securing microservices in a Spring-based application.

Chapter 18, Single Sign-On with the Central Authentication Service, shows how integrating with a **Central Authentication Service (CAS)** can provide single sign-on and single logout support to your Spring-Security-enabled applications.

Chapter 19, Build GraalVM Native Images, looks at Spring Security 6 support for building native images using GraalVM. This can be a great way to improve the performance and security of your Spring Security applications.

To get the most out of this book

The primary method for integrating with the sample code is providing **Gradle-** and **Maven**-compatible projects. Since many **Integrated Development Environments (IDEs).** have rich integration with Gradle and Maven, users should be able to import the code into any IDE that supports either Gradle or Maven. As many developers use Gradle and Maven, we felt this was the most straightforward method of packaging the examples. Whatever development environment you are familiar with, hopefully, you will find a way to work through the examples in this book.

Many IDEs provide Gradle or Maven tooling that can automatically download the Spring and **Spring Security 6** Javadoc and source code for you. However, there may be times when this is not possible. In such cases, you'll want to download the full releases of both **Spring 6** and Spring Security 6. The Javadoc and source code are top notch. If you get confused or want more information, the samples can provide an additional level of support or reassurance for your learning.

To run the sample application, you will need an IDE such as **IntelliJ IDEA** or **Eclipse** and build it with **Gradle** or **Maven**, which don't have strict hardware requirements. However, these are some general recommendations to ensure a smooth development experience:

1. System Requirements:

    * A modern computer with at least 4GB of RAM (8GB or more is recommended).

    * A multi-core processor for faster build and development.

2. Operating System:

    * Spring applications can be developed on Windows, macOS, or Linux. Choose the one that you are most comfortable with.

3. Disk Space:

    * You will need disk space for your project files, dependencies, and any databases you might use. At least 10GB of free disk space is advisable.

4. Network Connection:

    * A stable internet connection may be needed for downloading dependencies, plugins, and libraries during project setup.

Software/hardware covered in the book	Operating system requirements
IntelliJ IDEA and Eclipse are both popular choices for Spring development	Windows, macOS, or Linux
JDK versions: 17 or 21	
Spring- Security 6.	
Spring- Boot 3.	
Thymeleaf 6.	

If you are using the digital version of this book, we advise you to type the code yourself or access the code from the book's GitHub repository (a link is available in the next section). Doing so will help you avoid any potential errors related to the copying and pasting of code.

From *Chapter 3*, *Custom Authentication* onwards, the book shifts its emphasis to delve deeper into Spring Security, particularly in conjunction with the Spring Boot framework.

Download the example code files

You can download the example code files for this book from GitHub at `https://github.com/PacktPublishing/Spring-Security-Fourth-Edition/`. If there's an update to the code, it will be updated in the GitHub repository.

We also have other code bundles from our rich catalog of books and videos available at `https://github.com/PacktPublishing/`. Check them out!

Code in Action

The Code in Action videos for this book can be viewed at `https://packt.link/Om1ow`.

Conventions used

There are a number of text conventions used throughout this book.

`Code in text`: Indicates code words in text, database table names, folder names, filenames, file extensions, pathnames, dummy URLs, user input, and Twitter handles. Here is an example: "We are loading the `calendar.ldif` file from `classpath`, and using it to populate the LDAP server."

A block of code is set as follows:

```
//src/main/java/com/packtpub/springsecurity/configuration/
SecurityConfig.java
@Bean
public SecurityFilterChain filterChain(HttpSecurity http,
        PersistentTokenRepository persistentTokenRepository,
RememberMeServices rememberMeServices) throws Exception {
    http.authorizeHttpRequests( authz -> authz
```

```
                    .requestMatchers("/webjars/**").permitAll()
    ...

    // Remember Me
    http.rememberMe(httpSecurityRememberMeConfigurer ->
httpSecurityRememberMeConfigurer
            .key("jbcpCalendar")
            .rememberMeServices(rememberMeServices)
            .tokenRepository(persistentTokenRepository));

    return http.build();
}
```

The first line, `//src/main/java/com/packtpub/springsecurity/configuration/ SecurityConfig.java`, indicates the location of the file to be modified.

Any command-line input or output is written as follows:

```
X-Content-Security-Policy: default-src 'self' X-WebKit-CSP: default-
src 'self'
```

Bold: Indicates a new term, an important word, or words that you see onscreen. For instance, words in menus or dialog boxes appear in **bold**. Here is an example: "Right-click on **World** and select **Search**."

> **Tips or important notes**
> Appear like this.

Get in touch

Feedback from our readers is always welcome.

General feedback: If you have questions about any aspect of this book, email us at customercare@ packtpub.com and mention the book title in the subject of your message.

Errata: Although we have taken every care to ensure the accuracy of our content, mistakes do happen. If you have found a mistake in this book, we would be grateful if you would report this to us. Please visit www.packtpub.com/support/errata and fill in the form.

Piracy: If you come across any illegal copies of our works in any form on the internet, we would be grateful if you would provide us with the location address or website name. Please contact us at copyright@packtpub.com with a link to the material.

If you are interested in becoming an author: If there is a topic that you have expertise in and you are interested in either writing or contributing to a book, please visit authors.packtpub.com.

Share Your Thoughts

Once you've read *Spring Security*, we'd love to hear your thoughts! Scan the QR code below to go straight to the Amazon review page for this book and share your feedback.

https://packt.link/r/1-835-46050-X

Your review is important to us and the tech community and will help us make sure we're delivering excellent quality content.

Download a free PDF copy of this book

Thanks for purchasing this book!

Do you like to read on the go but are unable to carry your print books everywhere?

Is your eBook purchase not compatible with the device of your choice?

Don't worry, now with every Packt book you get a DRM-free PDF version of that book at no cost.

Read anywhere, any place, on any device. Search, copy, and paste code from your favorite technical books directly into your application.

The perks don't stop there, you can get exclusive access to discounts, newsletters, and great free content in your inbox daily

Follow these simple steps to get the benefits:

1. Scan the QR code or visit the link below:

https://packt.link/free-ebook/9781835460504

2. Submit your proof of purchase
3. That's it! We'll send your free PDF and other benefits to your email directly

Part 1: Fundamentals of Application Security

This part delves into the foundational aspects of application security, laying the groundwork for understanding potential vulnerabilities. We embark on a comprehensive exploration of application security using Spring Security. This part introduces you to the process of conducting a security audit on a hypothetical calendar application. Through this audit, we uncover common security vulnerabilities and lay the groundwork for implementing robust security measures.

Building upon this foundation, this part guides you through the installation and configuration of Spring Security. We start with a basic `"Hello World"` example, gradually customizing Spring Security to suit the specific needs of our application.

We will also delve deeper into the authentication process within Spring Security. By customizing key components of the authentication infrastructure, we address real-world authentication challenges and gain a comprehensive understanding of Spring Security's authentication mechanisms. Through practical examples and hands-on exercises, we learn how to integrate custom authentication solutions seamlessly into our applications.

This part has the following chapters:

- *Chapter 1, Anatomy of an Unsafe Application*
- *Chapter 2, Getting Started with Spring Security*
- *Chapter 3, Custom Authentication*

1

Anatomy of an Unsafe Application

Security is arguably one of the most critical architectural components of any web-based application written in the *21st* century. In an era where malware, criminals, and rogue employees are always present and actively testing software for **exploits**, smart and comprehensive use of security is a key element of any project for which you'll be responsible.

This book is written to follow a pattern of development that, we feel, provides a useful premise for tackling a complex subject—taking a web-based application with a Spring 6 foundation, and understanding the core concepts and strategies for securing it with Spring Security 6. We complement this approach by providing sample code for each chapter in the form of complete web applications.

In this chapter, we will delve into an example scenario to highlight several prevalent security vulnerabilities. Our journey will commence by examining the fundamental principles of secure coding. We will then shift our focus to explore common vulnerabilities in the next chapters, such as SQL injection, **Cross-Site Scripting (XSS)**, and **Cross-Site Request Forgery (CSRF)**.

Whether you're already using Spring Security or are interested in taking your basic use of the software to the next level of complexity, you'll find something to help you in this book.

During the course of this chapter, we will cover the following topics:

- Exploring software architecture styles
- Understanding security audit
- Addressing the security audit findings

It is recommended to have a basic understanding of the Spring framework before delving deeper into the subsequent sections.

By the end of this chapter, you will know how an application can be vulnerable to attack, and you will have mastered the core security mechanisms that can be used to protect an application.

If you are already familiar with basic security terminology, you may skip to *Chapter 2, Getting Started with Spring Security*, where we start using the basic functionality of the framework.

Exploring software architecture styles

Numerous businesses are acquiring computational capabilities from online cloud service platforms and embracing a primary reliance on the cloud for the development of most applications. This shift has prompted a transformation in the design of applications.

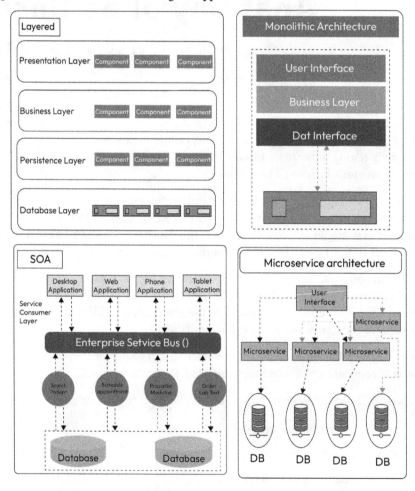

Figure 1.1 – Monolithic versus layered versus SOA versus microservices

The selection of the most suitable application architecture is contingent upon your specific business requirements. We will explore four architectural options designed to facilitate digital transformation and tailored to meet general business needs

Monolithic architecture

A traditional architecture where the entire application is constructed as a unified and closely integrated entity.

While it is easy to develop and deploy initially, scaling and maintaining it can pose challenges as the project expands.

N-Tier architecture (layered architecture)

N-tier architecture, also known as a hierarchical structure with distinct layers, refers to a design approach for software systems that organizes the application into multiple layers, typically four: *Presentation*, *Business*, *Persistence*, and *Data*. This architectural model is commonly employed in enterprise applications to enhance maintainability by compartmentalizing and promoting modular development. Each layer has specific responsibilities.

The **Model-View-Controller** (**MVC**) software design pattern separates an application into three interconnected components: Model (data and business logic), View (user interface), and Controller (handles user input and updates the Model and View accordingly).

This segmentation facilitates scalability, ease of maintenance, and flexibility in adapting to changing business requirements.

SOA

The **service-oriented pattern**, also known as **Service-Oriented Architecture** (**SOA**), is an architectural style that structures a software application as a collection of loosely coupled and independently deployable services.

Before SOA came into use in the late 1990s, connecting an application to services housed in another system was a complex process involving point-to-point integration.

Microservices architecture

Microservices derive from SOA, but SOA is different from microservices.

This architecture involves fragmenting the application into small, autonomous services that communicate via APIs. It provides scalability, flexibility, and simplified maintenance, but introduces challenges in handling distributed systems complexities.

> **Important note**
>
> Historically, the emphasis was on functionality and statefulness, but now, the majority of consumer-oriented applications are transitioning to **Software-as-a-Service** (**SaaS**) and digital platforms. Presently, the focus in application design has shifted toward enhancing user experience, embracing statelessness, and prioritizing agility.

Choosing between traditional web applications and Single-Page Applications

There are two primary approaches to constructing web applications in the contemporary landscape: conventional web applications, which execute most of the application logic on the server, and **single-page applications** (**SPAs**), which handle most of the user interface logic in a web browser and primarily communicate with the web server through web APIs. An alternative hybrid approach is feasible, wherein one or more feature-rich SPA-like sub-applications are hosted within a larger traditional web application.

Select a traditional web application under the following circumstances:

- Your application's client-side requirements are straightforward or even limited to read-only functionality
- Your application needs to operate in browsers lacking JavaScript support
- Your application is publicly accessible and benefits from search engine visibility and referrals

Choose a SPA under the following circumstances:

- Your application requires a sophisticated user interface with numerous features
- Your development team is well-versed in JavaScript, Angular, ReactJS, VueJS, TypeScript, or WebAssembly
- Your application already exposes an API for other clients, whether internal or public

The enhancements in user experience facilitated by the SPA approach should be carefully considered against these factors.

In the next sections of the book, we will use a traditional Spring MVC application as an example to illustrate various security principles.

> **Important note**
>
> It's important to note that these security principles apply to all the architectural styles discussed in this chapter.

Understanding security audit

It's early in the morning at your job as a software developer for the **Jim Bob Circle Pants Online Calendar** (JBCPCalendar.com), and you're halfway through your first cup of coffee when you get the following email from your supervisor:

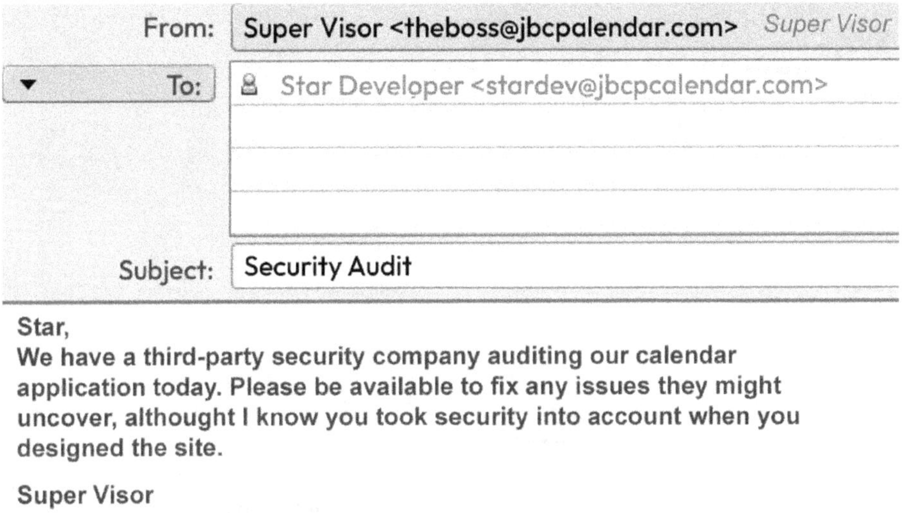

From:	Super Visor <theboss@jbcpalendar.com> *Super Visor*
To:	Star Developer <stardev@jbcpcalendar.com>
Subject:	Security Audit

Star,
We have a third-party security company auditing our calendar application today. Please be available to fix any issues they might uncover, although I know you took security into account when you designed the site.

Super Visor

Figure 1.2 – The email from the supervisor

What? You didn't think about security when you designed the application? In fact, at this point, you are not even sure what a **security audit** is. Sounds like you'll have a lot to learn from the security auditors! Later in this chapter, we will review what an audit is, along with the results of an audit. First, let's spend a bit of time examining the application that's under review.

Exploring the example application

Although we'll be working through a contrived scenario as we progress through this book, the design of the application and the changes that we'll make to it are drawn from the real-world usage of Spring-based applications. The calendar application allows users to create and view events:

myCalendar
Welcome
All Events
My Events
Create Event
H2
Event Information
Attendee Email
Event Date/Time (yyyy-MM-dd HH:mm)
Summary
Description
Create
Auto Populate Form

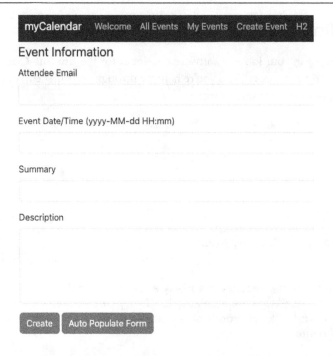

Figure 1.3 – The calendar application event information

After entering the details for a new event, you will be presented with the following screenshot:

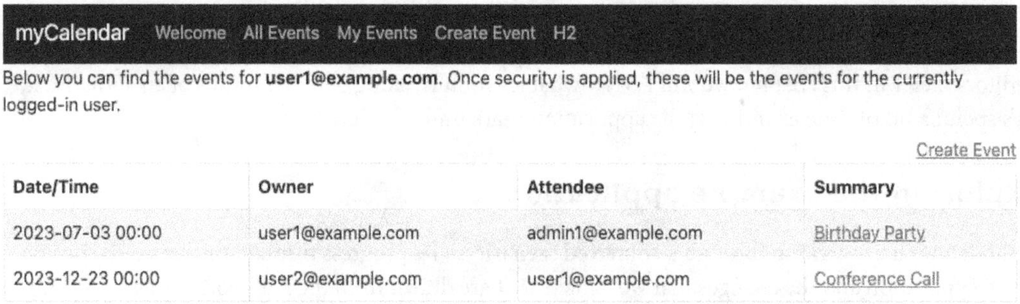

myCalendar
Welcome
All Events
My Events
Create Event
H2
Below you can find the events for user1@example.com. Once security is applied, these will be the events for the currently logged-in user.
Create Event
Date/Time
Owner
Attendee
Summary
2023-07-03 00:00
user1@example.com
admin1@example.com
Birthday Party
2023-12-23 00:00
user2@example.com
user1@example.com
Conference Call

Figure 1.4 – The calendar application summary

The application is designed to be simple to allow us to focus on the important aspects of security and not get tied up in the details of **object-relational mapping** (**ORM**) and complex UI techniques. We expect you to refer to other supplementary materials in the *Appendix*, *Additional Reference Material* (in the *Supplementary materials* section of this book) to cover some of the baseline functionality that is provided as part of the example code.

The code is written in Spring and Spring Security 6, but it would be relatively easy to adapt many of the examples to other versions of Spring Security. Refer to the discussion about the detailed changes between Spring Security 4 and 6 in *Chapter 16, Migration to Spring Security 6*, for assistance in translating the examples to the Spring Security 6 syntax.

> **Important note**
>
> Please don't use this application as a baseline to build a real online calendar application. It has been purposely structured to be simple and to focus on the concepts and configurations that we illustrate in this book.

In the next section, we will explore the application architecture.

The JBCP calendar application architecture

The web application follows a standard three-tier architecture consisting of a web, service, and data access layer, as indicated in the following diagram:

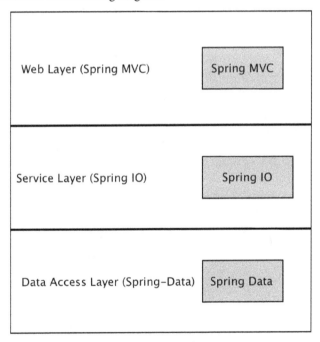

Figure 1.5 – JBCP calendar application architecture

You can find additional material about MVC architectures in the *Supplementary materials* section of the *Appendix, Additional Reference Material*.

The web layer encapsulates MVC code and functionality. In this example application, we will use the Spring MVC framework, but we could just as easily use **Spring Web Flow (SWF)**, **Apache Struts**, or even a Spring-friendly web stack, such as **Apache Wicket**.

In a typical web application that uses Spring Security, the web layer is where much of the configuration and augmentation of code takes place. For example, the `EventsController` class is utilized to convert an HTTP request into the process of storing an event in the database. If you haven't had a lot of experience with web applications and Spring MVC specifically, it would be wise to review the baseline code closely and make sure you understand it before we move on to more complex subjects. Again, we've tried to make the website as simple as possible, and the construction of a calendar application is used just to provide a sensible title and light structure to the site.

> **Important note**
>
> You can find detailed instructions on setting up the example application within the *Appendix, Additional Reference Material*.

The service layer encapsulates the business logic for the application. In our application, we use `DefaultCalendarService` as a very light facade over the data access layer to illustrate particular points about securing application service methods. The service layer is also used to operate on both Spring Security APIs and our Calendar APIs within a single method call. We will discuss this in greater detail in *Chapter 3, Custom Authentication*.

In a typical web application, the service layer incorporates business rule validation, composition and decomposition of business objects, and cross-cutting concerns such as auditing.

The data access layer encapsulates the code responsible for manipulating the contents of database tables. In many Spring applications, this is where you would see an ORM, such as Hibernate or JPA. The data access layer exposes an object-based API to the service layer. In our application, we use basic JDBC functionality to achieve persistence in the in-memory H2 database. For example, `JdbcEventDao` is used to save event objects to the database.

In a typical web application, a more comprehensive data access solution would be utilized. As ORM, and more generally data access, tends to be confusing for some developers, this is an area we have chosen to simplify as much as possible for clarity.

In the next section, we will review the audit results.

Reviewing the audit results

Let's return to our email and see how the audit is progressing. Uh-oh… the results don't look good:

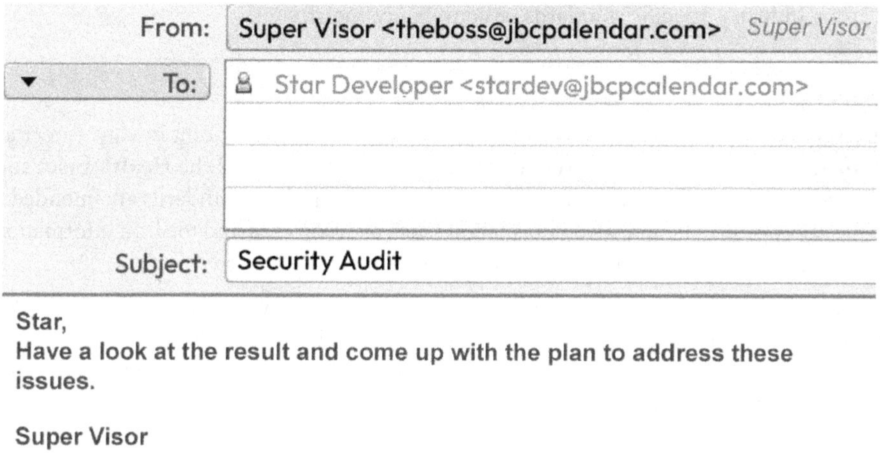

From:	Super Visor <theboss@jbcpalendar.com> *Super Visor*
To:	Star Developer <stardev@jbcpcalendar.com>
Subject:	Security Audit

Star,
Have a look at the result and come up with the plan to address these issues.

Super Visor

Figure 1.6 – The email of the audit results

APPLICATION AUDIT RESULTS: This application exhibits the following insecure behavior:

- Inadvertent privilege escalation due to a lack of URL protection and general authentication
- Inappropriate or non-existent use of authorization
- Missing database credential security
- Personally identifiable or sensitive information is easily accessible or unencrypted
- Insecure transport-level protection due to lack of SSL encryption

The risk level is high. We recommend that this application should be taken offline until these issues can be resolved.

Ouch! This result looks bad for our company. We'd better work to resolve these issues as quickly as possible.

Third-party security specialists are often hired by companies (or their partners or customers) to audit the effectiveness of their software security through a combination of white hat hacking, source code review, and formal or informal conversations with application developers and architects.

White hat hacking or **ethical hacking** is done by professionals who are hired to instruct companies on how to protect themselves better, rather than with the intent to be malicious.

Typically, the goal of **security audits** is to provide management or clients with the assurance that basic secure development practices have been followed to ensure the integrity and safety of the customer's data and system functions. Depending on the industry the software is targeted at, the auditor may also test it using industry-specific standards or compliance metrics.

> **Important note**
>
> Two specific security standards that you're likely to run into at some point in your career are the **Payment Card Industry Data Security Standard (PCI DSS)** and the **Health Insurance Privacy and Accountability Act (HIPAA)** privacy rules. Both these standards are intended to ensure the safety of specific sensitive information (such as credit card and medical information) through a combination of process and software controls.
>
> Many other industries and countries have similar rules about sensitive or **Personally Identifiable Information (PII)**. Failure to follow these standards is not only bad practice but also something that could expose you or your company to significant liability (not to mention bad press) in the event of a security breach.

Receiving the results of a **security audit** can be an eye-opening experience. Following through with the required software improvements can be the perfect opportunity for self-education and software improvement and can allow you to implement practices and policies that lead to secure software.

In the next section, we will review the auditor's findings and come up with a plan to address them in detail.

Addressing the security audit findings

In this section, we will meticulously examine the outcomes of our security audit, shedding light on the vulnerabilities and areas of concern within our application's security landscape. We'll dissect the audit results and embark on a journey to explore various effective strategies and patterns to secure and mitigate these identified risks. This chapter serves as a roadmap for enhancing the robustness of our application's security, ensuring it stays resilient against potential threats and vulnerabilities.

Authentication

Authentication is one of two key security concepts that you must internalize when developing secure applications (the other being authorization). Authentication identifies who is attempting to request a resource. You may be familiar with authentication in your daily online and offline life, in very different contexts, as follows:

- **Credential-based authentication**: When you log in to your web-based email account, you most likely provide your username and password. The email provider matches your username with a known user in its database and verifies that your password matches what they have on record. These credentials are what the email system uses to validate that you are a valid user of the system. First, we'll use this type of authentication to secure sensitive areas of the JBCP calendar application. Technically speaking, the email system can check credentials not only in the database but anywhere, for example, a corporate directory server such as **Microsoft Active Directory**. Several examples of this type of integration are covered throughout this book.

- **Two-factor authentication**: When you withdraw money from your bank's ATM, you swipe your ID card and enter your personal identification number before you are allowed to retrieve cash or conduct other transactions. This type of authentication is similar to username and password authentication, except that the username is encoded on the card's magnetic strip. The combination of the physical card and user-entered PIN allows the bank to ensure that you have access to the account. The combination of a password and a physical device (your plastic ATM card) is a ubiquitous form of two-factor authentication. In a professional, security-conscious environment, it's common to see these types of devices in regular use for access to highly secure systems, especially when dealing with finance information or PII. A hardware device, such as **RSA SecurID**, combines a time-based hardware device with server-based authentication software, making the environment extremely difficult to compromise.

- **Hardware authentication**: When you start your car in the morning, you slip your metal key into the ignition and turn it to get the car started. Although it may not feel similar to the other two examples, the correct match of the bumps on the key and the tumblers in the ignition switch function as a form of hardware authentication.

There are literally dozens of forms of authentication that can be applied to the problem of software and hardware security, each with their own pros and cons. We'll review some of these methods as they apply to Spring Security throughout the first half of this book. Our application lacks any type of authentication, which is why the audit included the risk of inadvertent privilege escalation.

Typically, a software system will be divided into two high-level realms, such as unauthenticated (or anonymous) and authenticated, as shown in the following screenshot:

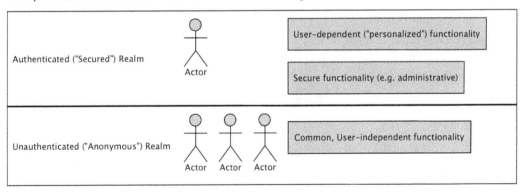

Figure 1.7 – High-level realms in a software system

Application functionality in the anonymous realm is the functionality that is independent of a user's identity (think of a welcome page for an online application).

Anonymous areas do not do the following:

- Require a user to log in to the system or otherwise identify themselves to be usable
- Display sensitive information, such as names, addresses, credit cards, and orders
- Provide functionality to manipulate the overall state of the system or its data

Unauthenticated areas of the system are intended for use by everyone, even by users whom we haven't specifically identified yet. However, it may be that additional functionality appears to identify users in these areas (for example, the ubiquitous `Welcome {First Name}` text). The selective displaying of content to authenticated users is fully supported through the use of the Spring Security tag library and is covered in *Chapter 11, Fine-Grained Access Control.*

We'll resolve this finding and implement form-based authentication using the automatic configuration capability of Spring Security in *Chapter 2, Getting Started with Spring Security.* Afterward, we will explore various other means of performing authentication (which usually revolves around system integration with enterprise or other external authentication stores).

In the next section, we will explore **authorization**.

Authorization

Authorization is the second of two core security concepts that are crucial in implementing and understanding application security. Authorization uses the information that was validated during authentication to determine whether access should be granted to a particular resource. Built around the authorization model for the application, authorization partitions the application functionality and data so that the availability of these items can be controlled by matching the combination of privileges, functionality, and data to users. Our application's failure at this point of the audit indicates that the application's functionality isn't restricted by the user role. Imagine if you were running an e-commerce site and the ability to view, cancel, or modify orders and customer information was available to any user of the site!

Authorization typically involves the following two separate aspects that combine to describe the accessibility of the secured system:

- The first is the mapping of an authenticated principal to one or more authorities (often called **roles**). For example, a casual user of your website might be viewed as having visitor authority, while a site administrator might be assigned administrative authority.
- The second is the assignment of authority checks to secured resources in the system. This is typically done when the system is developed, either through an explicit declaration in code or through configuration parameters. For example, the screen that allows for the viewing of other users' events should be made available only to those users with administrative authority.

> **Important note**
>
> A secured resource may be any aspect of the system that should be conditionally available based on the authority of the user.

Secured resources of a web-based application could be individual web pages, entire portions of the website, or portions of individual pages. Conversely, secured business resources might be method calls on classes or individual business objects.

You might imagine an authority check that would examine the principal (the identity of the current user or system interacting with the application), look up its user account, and determine whether the principal is, in fact, an administrator. If this authority check determines that the principal who is attempting to access the secured area is, in fact, an administrator, then the request will succeed. If, however, the principal does not have sufficient authority, the request should be denied.

Let's take a closer look at an example of a particular secured resource, the **All Events** page. The **All Events** page requires administrative access (after all, we don't want regular users viewing other users events), and as such, looks for a certain level of authority in the principal accessing it.

If we think about how a decision might be made when a site administrator attempts to access the protected resource, we'd imagine that the examination of the actual authority versus the required authority might be expressed concisely in terms of the set theory. We might then choose to represent this decision as a **Venn** diagram for the administrative user:

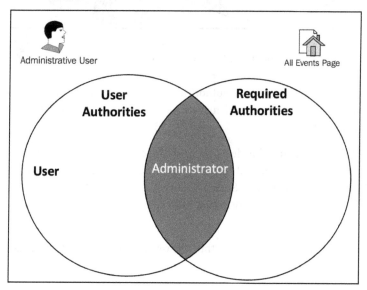

Figure 1.8 – Venn diagram for the administrative user

There is an intersection between **User Authorities** (users and administrators) and **Required Authorities** (administrators) for the page, so the user is provided with access.

Contrast this with an unauthorized user, as follows:

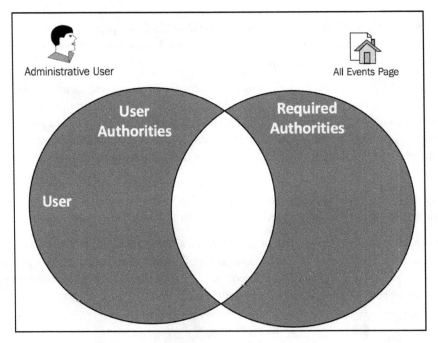

Administrative User   All Events Page

Figure 1.9 – Venn diagram for the visiting (unauthorized) user

We have a Symmetric difference. The sets of authorities are disjointed and have no common elements. So, the user is denied access to the page. Thus, we have demonstrated the basic principle of the authorization of access to resources.

In reality, there's real code making this decision, with the consequence that the user is granted or denied access to the requested protected resource. We'll address the basic authorization problem with the authorization infrastructure of Spring Security in *Chapter 2, Getting Started with Spring Security*, followed by more advanced authorization in *Chapter 12, Access Control Lists*, and *Chapter 13, Custom Authorization*.

Now that we have covered the concept of authorization, we will explore how it can be applied to database security.

Database credential security

In Spring terminology, database credentials typically refer to the information required to establish a connection between a Spring application and a database. These credentials include the following:

- Username: The username or user ID associated with the database account that the Spring application uses to connect.
- Password: The corresponding password for the specified username, providing authentication for access to the database.
- Database URL: The URL that specifies the location and details of the database. It includes information such as the host, port, and database name.

Through the examination of the application source code and configuration files, the auditors noted that user passwords were stored in plain text in the configuration files, making it very easy for a malicious user with access to the server to gain access to the application.

As the application contains personal and financial data, a rogue user being able to access any data could expose the company to identity theft or tampering. Protecting access to the credentials used to access the application should be a top priority for us, and an important first step is ensuring that one point of failure in security does not compromise the entire system.

We'll examine the configuration of database access layers in Spring Security for credential storage, which requires **JDBC connectivity**, in *Chapter 4, JDBC-based Authentication*. In the same chapter, we'll also look at built-in techniques to increase the security of passwords stored in the database.

After covering the database credential security, we will explore the sensitive information audit finding.

Sensitive information

Personally identifiable or sensitive information is easily accessible or unencrypted. The auditors noted that some significant and sensitive pieces of data were completely unencrypted or masked anywhere in the system. Fortunately, there are some simple design patterns and tools that allow us to protect this information securely, with annotation-based **Aspect-Oriented Programming** (AOP) support in Spring Security.

Transport-level protection

There is insecure transport-level protection due to the lack of SSL encryption.

While, in the real world, it's unthinkable that an online application containing private information would operate without SSL protection, unfortunately, the JBCP calendar is in just this situation. SSL protection ensures that communication between the browser client and the web application server is secure against many kinds of tampering and snooping.

In the *HTTPS setup in Tomcat* section, in the *Appendix, Additional Reference Material*, we'll review the basic options for using transport-level security as part of the definition of the secured structure of the application.

Using Spring Security 6 to address security concerns

Spring Security 6 provides a wealth of resources that allow many common security practices to be declared or configured in a straightforward manner. In the coming chapters, we'll apply a combination of source code and application configuration changes to address all of the concerns raised by the security auditors (and more) to give ourselves the confidence that our calendar application is secure.

With Spring Security 6, we'll be able to make the following changes to improve our application's security:

- Segment users of the system into user classes
- Assign levels of authorization to user roles
- Assign user roles to user classes
- Apply authentication rules globally across application resources
- Apply authorization rules at all levels of the application architecture
- Prevent common types of attacks intended to manipulate or steal a user's session

Spring Security exists to fill a gap in the universe of Java third-party libraries, much as the Spring Framework originally did when it was first introduced. Standards such as **Java Authentication and Authorization Service (JAAS)** or **Jakarta EE Security** do offer some ways of performing some of the same authentication and authorization functions, but Spring Security is a winner because it includes everything you need to implement a top-to-bottom application security solution concisely and sensibly.

Additionally, Spring Security appeals to many because it offers out-of-the-box integration with many common enterprise authentication systems, so it's adaptable to most situations with little effort (beyond configuration) on the part of the developer. It's in wide use because there's really no other mainstream framework quite like it!

Technical requirements

We have endeavored to make the application as easy to run as possible by focusing on some basic tools and technologies that almost every Spring developer would have on their development machine. Nevertheless, we have provided the *Getting started* section as supplementary information in the *Getting started with the JBCP calendar sample code* section in the *Appendix, Additional Reference Material*.

The primary method for integrating with the sample code is providing projects that are compatible with **Gradle** and **Maven**. Since many IDEs have rich integration with Gradle and Maven, users should be able to import the code into any IDE that supports either Gradle or Maven. As many developers use Gradle and Maven, we felt this was the most straightforward method of packaging the examples. Whatever development environment you are familiar with, hopefully, you will find a way to work through the examples in this book.

Many IDEs provide Gradle or Maven tooling that can automatically download the Spring and Spring Security 6 Javadoc and source code for you. However, there may be times when this is not possible. In such cases, you'll want to download the full releases of both Spring 6 and Spring Security 6. The Javadoc and source code are top-notch. If you get confused or want more information, the samples can provide an additional level of support or reassurance for your learning. Visit the *Supplementary materials* section in the *Appendix, Additional Reference Material*, to find additional information about Gradle and Maven, including running the samples, and obtaining the source code and Javadoc.

To run the sample application, you will need an **Integrated Development Environments** (**IDEs**) such as IntelliJ IDEA or Eclipse and build it with Gradle or Maven, which don't have strict hardware requirements. However, these are some general recommendations to ensure a smooth development experience:

- System requirements:

    - A modern computer with at least 4GB of RAM (8GB or more is recommended)

    - A multi-core processor for faster build and development

- Operating system: Spring applications can be developed on Windows, macOS, or Linux. Choose the one that you are most comfortable with.

- IDE: IntelliJ IDEA and Eclipse are both popular choices for Spring development. Make sure your IDE is up to date.

- **Java Development Kit** (**JDK**): Spring applications require a minimum of Java 17. Install the latest JDK version compatible with your IDE.

    Upon the publication of this book, all the code has been validated with Java 21, the most recent **Long-Term Support** (**LTS**) version.

- Disk space: You will need disk space for your project files, dependencies, and any databases you might use. At least 10 GB of free disk space is advisable.

- Network connection: A stable internet connection may be needed for downloading dependencies, plugins, and libraries during project setup.

> **Important note**
>
> You can now look at the sample code of the JBCP calendar application: `chapter01.00-calendar`.
>
> From *Chapter 3, Custom Authentication* onwards, we shift our emphasis to delve deeper into `spring-security`, particularly in conjunction with the `spring-boot` framework.

If `Maven` is your choice, run the following command with Maven from the project directory:

```
./mvnw package cargo:run
```

If you are using `Gradle`, run the following command from the project directory:

```
./gradlew tomcatRun
```

Then open a browser at `http://localhost:8080/`.

Summary

In this chapter, we have reviewed the common points of risk in an unsecured web application and the basic architecture of our example application. We began by scrutinizing the audit results, highlighting the areas of concern and potential vulnerabilities. The chapter then branched into key security concepts, including **authentication**, **authorization**, and **database credential security**. We also discussed the strategies for securing the application based on the spring framework.

In the next chapter, we'll explore how to get **Spring Security** set up quickly and get a basic understanding of how it works.

2

Getting Started with Spring Security

In this chapter, we'll apply a minimal Spring Security configuration to start addressing our first finding—inadvertent privilege escalation due to a lack of URL protection and general authentication from the security audit discussed in *Chapter 1, Anatomy of an Unsafe Application*. We will then build on the basic configuration to provide a customized experience for our users. This chapter is intended to get you up and running with Spring Security and to provide a foundation for any other security-related tasks you will need to perform.

During the course of this chapter, we will cover the following topics:

- Implementing a basic level of security on the **JBCP Calendar** application, using the automatic configuration option in Spring Security

- Learning how to customize both the login and logout experience

- Configuring Spring Security to restrict access differently, depending on the URL

- Leveraging the expression-based access controls of Spring Security

- Conditionally displaying basic information about the logged-in user using the **JavaServer Pages** (**JSP**) library in Spring Security

- Determining the user's default location after login, based on their role

In each chapter of the book, you are instructed to begin their exploration from a designated project `chapterX.00-calendar`, where X denotes the chapter number. During this chapter, the focus is on using an application built on Spring Security within the Spring Framework, excluding the use of Spring Boot.

The primary goal is to help readers become acquainted with the integration of Spring Security in the context of the Spring Framework.

From *Chapter 3, Custom Authentication* onwards, the book shifts its emphasis to delve deeper into Spring Security, particularly in conjunction with the Spring Boot framework.

This chapter's code in action link is here: `https://packt.link/fZ96T`.

Hello Spring Security

Although Spring Security can be extremely difficult to configure, the creators of the product have been thoughtful and have provided us with a very simple mechanism to enable much of the software's functionality with a strong baseline. From this baseline, additional configuration will allow for a fine level of detailed control over the security behavior of the application.

We'll start with our unsecured calendar application from *Chapter 1, Anatomy of an Unsafe Application*, and turn it into a site that's secured with a rudimentary username and password authentication. This authentication serves merely to illustrate the steps involved in enabling Spring Security for our web application; you'll see that there are some obvious flaws in this approach that will lead us to make further configuration refinements.

Importing the sample application

We encourage you to import the `chapter02.00-calendar` project into your **Integrated Development Environments (IDEs)** and follow along by obtaining the source code from this chapter, as described in the *Getting started with JBCP calendar sample code* section in the *Appendix, Additional Reference Material*.

For each chapter, you will find multiple revisions of the code that represent checkpoints within the book. This makes it easy to compare your work to the correct answers as you go. At the beginning of each chapter, we will import the first revision of that chapter as a starting point. For example, in this chapter, we start with `chapter02.00-calendar`, and the first checkpoint will be `chapter02.01-calendar`. In *Chapter 3, Custom Authentication*, we will start with `chapter03.00-calendar`, and the first checkpoint will be `chapter03.01-calendar`. There are additional details in the *Getting started with JBCP calendar sample code* section in the *Appendix, Additional Reference Material*, so be sure to refer to it for details.

Updating your dependencies

The first step is to update the project's dependencies to include the necessary Spring Security JAR files. Update the `build.gradle` Gradle file (from the sample application you imported previously) to include the Spring Security JAR files that we will use in the following few sections.

> **Important note**
> Throughout the book, we will be demonstrating how to provide the required dependencies using Gradle. The `build.gradle` file is located in the root of the project and represents all that is needed to build the project (including the project's dependencies). Remember that Gradle will download the transitive dependencies for each listed dependency. So, if you are using another mechanism to manage dependencies, ensure that you also include the transitive dependencies. When managing the dependencies manually, it is useful to know that the Spring Security reference includes a list of its transitive dependencies. A link to the Spring Security reference can be found in the *Supplementary materials* section in the *Appendix, Additional Reference Material*.

Let's take a look at the following code snippet:

```
// build.gradle
dependencies {
    implementation 'org.springframework.security:spring-security-config'
    implementation 'org.springframework.security:spring-security-web'
    ...
}
```

This code provides the necessary dependencies for basic Spring Security features in a Spring Framework project. However, depending on your specific requirements, you might need to add additional configurations as we will explore in the upcoming sections.

Using Spring 6 and Spring Security 6

Spring 6 is used consistently. Our sample applications provide an example of the former option, which means that no additional work is required by you.

In the following code, we present an example fragment of what is added to the `build.gradle` Gradle file to utilize the dependency management feature of Gradle; this ensures that the correct Spring version is used throughout the entire application. We are going to leverage the Spring **Bill Of Materials (BOM)** dependency, which will ensure that all the dependency versions imported by the BOM will work together correctly:

```
// build.gradle
dependencyManagement {
    imports {
        mavenBom 'org.springframework:spring-framework-bom:6.1.4'
        mavenBom 'org.springframework.security:spring-security-bom:6.2.2'
    }
```

```
}
dependencies {
    ...
}
```

> **Important note**
>
> If you are using IntelliJ IDEA, any time you update the build.gradle file, ensure you
> right-click on the project, navigate to **Gradle | Reload Gradle Project…**, and select **OK** to
> update all the dependencies.

For more information about how Gradle handles transitive dependencies, as well as the BOM, refer
to the Gradle documentation, which is listed in the *Supplementary materials* section, in the *Appendix,
Additional Reference Material*.

Implementing a Spring Security configuration

The next step in the configuration process is to create a Java configuration file representing all of the
Spring Security components that are required to cover standard web requests.

Create a new Java file in the src/main/java/com/packtpub/springsecurity/
configuration/ directory with the name SecurityConfig.java, and the following
content. Among other things, the following file demonstrates user login requirements for every page
in our application, provides a login page, authenticates the user, and requires the logged-in user to
be associated with a role called USER for every URL element:

```java
//src/main/java/com/packtpub/springsecurity/configuration/
SecurityConfig.java
@Configuration
@EnableWebSecurity
public class SecurityConfig {
    @Bean
    public InMemoryUserDetailsManager userDetailsService() {
        UserDetails user = User.withDefaultPasswordEncoder()
                .username("user1@example.com")
                .password("user1")
                .roles("USER")
                .build();

        return new InMemoryUserDetailsManager(user);
    }
    @Bean
    public SecurityFilterChain filterChain(HttpSecurity http) throws
Exception {
```

```
        http
                .authorizeHttpRequests( authz -> authz
                    .requestMatchers("/**").hasRole("USER")
                    .anyRequest().authenticated()
                )
                .formLogin(withDefaults())
                .csrf(AbstractHttpConfigurer::disable);
        // For H2 Console
        http.headers(headers -> headers.
 frameOptions(FrameOptionsConfig::disable));
        return http.build();
    }
}
```

> **Important note**
>
> If you are using IntelliJ IDEA, you can easily review `SecurityFilterChain` by pressing *F3*.
> Remember that the next checkpoint (`chapter02.01-calendar`) has a working solution,
> so the file can be copied from there as well.

This is the only Spring Security configuration required to get our web application secured with a
minimal standard configuration. This style of configuration, using a Spring-security-specific Java
configuration, is known as **Java Configuration**.

Let's take a minute to break this configuration apart so we can get a high-level idea of what is happening.
In the `filterChain (HttpSecurity)` method, the `HttpSecurity` object creates a `Servlet
Filter`, which ensures that the currently logged-in user is associated with the appropriate role. In
this instance, the filter will ensure that the user is associated with ROLE_USER. It is important to
understand that the name of the role is arbitrary. Later, we will create a user with ROLE_ADMIN and
will allow this user to have access to additional URLs that our current user does not have access to.

In the `userDetailsService ()` method, the `InMemoryUserDetailsManager` object is
how Spring Security authenticates the user. In this instance, we utilize an in-memory data store to
compare a username and password.

Our example and explanation of what is happening are a bit contrived. An in-memory authentication
store would not work in a production environment. However, it allows us to get things up and
running quickly. We will incrementally improve our understanding of Spring Security as we update
our application to use production quality security throughout this book.

> **Important note**
>
> General support for **Java Configuration** was added to Spring Framework in Spring 3.1. Since Spring Security 3.2 release, there has been Spring Security Java Configuration support, which enables users to easily configure Spring Security without the use of any XML. If you are familiar with *Chapter 6, LDAP Directory Services*, and the Spring Security documentation, then you should find quite a few similarities between it and Security **Java Configuration** support.

Updating your web configuration

The next steps involve a series of updates to the `WebAppInitializer.java` file. Some of the steps have already been performed because the application was already using **Spring MVC**. However, we will go over these requirements to ensure that more fundamental Spring requirements are understood, if you are using Spring Security in an application that is not Spring-enabled.

The ContextLoaderListener class

The first step of updating the `WebAppInitializer.java` file is to use `jakarta.servlet.ServletContainerInitializer`, which is the preferred approach to Servlet 4.0+ initialization. Spring MVC provides the `o.s.w.WebApplicationInitializer` interface, which leverages this mechanism. In Spring MVC, the preferred approach is to extend `o.s.w.servlet.support.AbstractAnnotationConfigDispatcherServletInitializer`. The `WebApplicationInitializer` class is polymorphically `o.s.w.context.AbstractContextLoaderInitializer` and uses the abstract `createRootApplicationContext()` method to create a root `ApplicationContext`, then delegates it to `ContextLoaderListener`, which is registered in the `ServletContext` instance, as shown in the following code snippet:

```
//src/main/java/com/packtpub/springsecurity/configuration/JavaConfig.
java

@Configuration
@Import({ SecurityConfig.class, DataSourceConfig.class })
@ComponentScan(basePackages =
        {
                «com.packtpub.springsecurity.configuration»,
                «com.packtpub.springsecurity.dataaccess»,
                «com.packtpub.springsecurity.domain»,
                «com.packtpub.springsecurity.service»
        }
)
public class JavaConfig {}
```

The updated configuration will now load the `SecurityConfig.class` from the `classpath` of the WAR file.

ContextLoaderListener versus DispatcherServlet

The `o.s.web.servlet.DispatcherServlet` interface specifies configuration classes to be loaded on their own using the `getServletConfigClasses()` method:

```
//src/main/java/com/packtpub/springsecurity/web/configuration/
WebAppInitializer
public class WebAppInitializer extends
AbstractAnnotationConfigDispatcherServletInitializer {

    @Override
    protected Class<?>[] getRootConfigClasses() {
        return null;
    }

    @Override
    protected Class<?>[] getServletConfigClasses() {
        return new Class[] { JavaConfig.class, WebMvcConfig.class };
    }
. . .
```

The `DispatcherServlet` class generates an `o.s.context.ApplicationContext` within the Spring framework, functioning as a child of the root `ApplicationContext` interface.

Generally, components specific to Spring MVC are initialized within the `ApplicationContext` interface of `DispatcherServlet`.

On the other hand, the remaining components are loaded through `ContextLoaderListener`.

It's crucial to understand that beans in a child `ApplicationContext`, like the ones generated by `DispatcherServlet`, can reference beans from the parent `ApplicationContext`, established by `ContextLoaderListener`.

However, the parent `ApplicationContext` interface cannot make references to beans within the child `ApplicationContext`.

This is illustrated in the following diagram, in which **Child Beans** can refer to **Root Beans**, but **Root Beans** cannot refer to **Child Beans**:

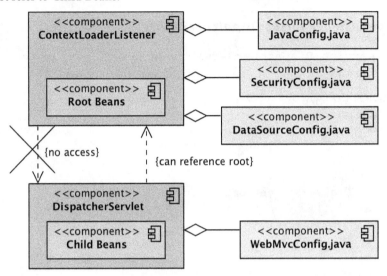

Figure 2.1 – ContextLoaderListener and DispatcherServlet relationships

As in most use cases of Spring Security, we do not need Spring Security to refer to any of the MVC-declared beans. Therefore, we have decided to have `ContextLoaderListener` initialize all configurations of Spring Security.

The springSecurityFilterChain filter

The next step is to configure `springSecurityFilterChain` to intercept all requests by creating an implementation of `AbstractSecurityWebApplicationInitializer`. It is critical for `springSecurityFilterChain` to be declared first, to ensure the request is secured before any other logic is invoked.

The following diagram shows the role of SecurityFilterChain:

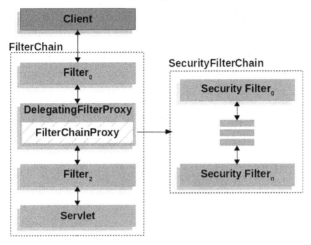

Figure 2.2 – Role of SecurityFilterChain

To ensure springSecurityFilterChain gets loaded first, we can use @Order(1), as shown in the following configuration:

```
//src/main/java/com/packtpub/springsecurity/web/configuration/
SecurityWebAppInitializer
@Order(1)
public class SecurityWebAppInitializer
        extends AbstractSecurityWebApplicationInitializer {
    public SecurityWebAppInitializer() {
        super();
    }


}
```

The SecurityWebAppInitializer class will automatically register the springSecurityFilterChain filter for every URL in your application and will add ContextLoaderListener, which loads SecurityConfig.

The DelegatingFilterProxy class

The o.s.web.filter.DelegatingFilterProxy class is a Servlet Filter provided by Spring Web that will delegate all work to a Spring bean from the ApplicationContext root, which must implement jakarta.servlet.Filter. Since, by default, the bean is looked up by name, using the <filter-name> value, we must ensure we use springSecurityFilterChain as the value of <filter-name>.

Here is a diagram of how DelegatingFilterProxy fits into the Filter instances and the FilterChain.

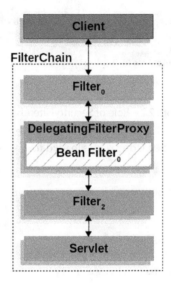

Figure 2.3 – Role of DelegatingFilterProxy

The pseudocode for how o.s.web.filter.DelegatingFilterProxy works for our web.xml file can be found in the following code snippet:

```
// DelegatingFilterProxy Pseudo Code
public void doFilter(ServletRequest request, ServletResponse response,
FilterChain chain) {
    Filter delegate = getFilterBean(someBeanName);
    delegate.doFilter(request, response);
}
```

This code showcases the interaction of the Spring Security `DelegatingFilterProxy` within a **Servlet**-based application.

The FilterChainProxy class

When working in conjunction with Spring Security, `o.s.web.filter.DelegatingFilterProxy` will delegate to the `o.s.s.web.FilterChainProxy` interface of Spring Security. The `FilterChainProxy` class allows Spring Security to conditionally apply any number of `Servlet Filters` to the `Servlet Request`. We will learn more about each of the Spring Security filters, and their roles in ensuring that our application is properly secured, throughout the rest of the book.

The following diagram shows the role of `FilterChainProxy`:

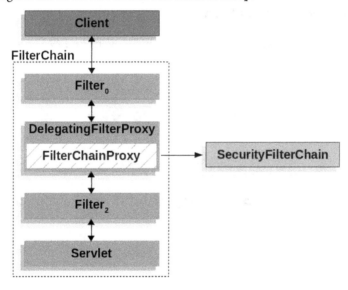

Figure 2.4 – Role of DelegatingFilterProxy

The pseudocode for how `FilterChainProxy` works is as follows:

```
public class FilterChainProxy implements Filter { void
doFilter(request, response, filterChain) {
// lookup all the Filters for this request
    List<Filter> delegates =   lookupDelegates(request,response)
// invoke each filter unless the delegate decided to stop
    for delegate in delegates { if continue processing
        delegate.doFilter(request,response,filterChain)
    }
```

```
// if all the filters decide it is ok allow the
// rest of the application to run if continue processing
    filterChain.doFilter(request,response) }
}
```

> **Important note**
>
> Due to the fact that both `DelegatingFilterProxy` and `FilterChainProxy` are the front doors to Spring Security when used in a web application, you would add a debug point when trying to figure out what is happening.

Running a secured application

If you have not already done so, restart the application and visit `http://localhost:8080/`. You will be presented with the following screen:

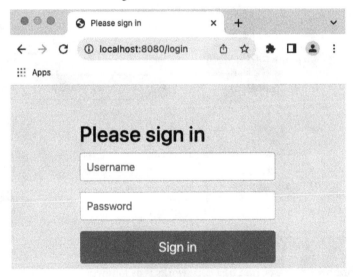

Figure 2.5 – Login page

Great job! We've implemented a basic layer of security in our application using Spring Security. At this point, you should be able to log in using `user1@example.com` as **Username** and `user1` as **Password**. You'll see the calendar welcome page, which describes at a high level what to expect from the application in terms of security.

> **Important note**
> Your code should now look like `chapter02.01-calendar`.

Common problems

Many users have trouble with the initial implementation of Spring Security in their applications. A few common issues and suggestions are listed next. We want to ensure that you can run the example application and follow along:

- Make sure you can build and deploy the application before putting Spring Security in place.

- Review some introductory samples and documentation on your servlet container if needed.

- It's usually easiest to use an IDE, such as IntelliJ IDEA, to run your servlet container. Not only is deployment typically seamless, but the console log is also readily available to review for errors. You can also set breakpoints at strategic locations to be triggered by exceptions to better diagnose errors.

- Make sure the versions of Spring and Spring Security that you're using match and that there aren't any unexpected Spring JAR files remaining as part of your application. As previously mentioned, when using Gradle, it can be a good idea to declare the Spring dependencies in the dependency management section.

In this section, we've established a Hello Spring Security application. We initiated the process by importing a sample application, subsequently updating dependencies, configuring Spring Security, managing web configurations, and finally, tackling common problems that may arise.

Moving forward, the following section will delve into the customization of the user experience post-login.

A little bit of polish

Stop at this point and think about what we've just built. You may have noticed some obvious issues that will require some additional work and knowledge of the Spring Security product before our application is production ready. Try to make a list of the changes that you think are required before this security implementation is ready to roll out on the public-facing website.

Applying the Hello World Spring Security implementation was blindingly fast and has provided us with a login page, username, and password-based authentication, as well as the automatic interception of URLs in our calendar application. However, there are gaps between what the automatic configuration setup provides and what our end goal is, which are listed as follows:

- While the login page is helpful, it's completely generic and doesn't look like the rest of our JBCP calendar application. We should add a login form that's integrated with our application's look and feel.

- There is no obvious way for a user to log out. We've locked down all pages in the application, including the Welcome page, which a potential user may want to browse anonymously. We'll need to redefine the roles required to accommodate anonymous, authenticated, and administrative users.

- We do not display any contextual information to indicate to the user that they are authenticated. It would be nice to display a greeting similar to "welcome `user1@example.com`."

- We've had to hardcode the username, password, and role information of the user in the `SecurityConfig` configuration file. Recall this section of the `userDetailsService()` method we added:

```
User.withDefaultPasswordEncoder()
        .username("user1@example.com")
        .password("user1")
        .roles("USER")
        .build();
```

- You can see that the username and password are right there in the file. It's unlikely that we'd want to add a new declaration to the file for every user of the system! To address this, we'll need to update the configuration with another type of authentication.

We'll explore different authentication options throughout the first half of the book.

Customizing login

We've seen how Spring Security makes it very easy to get started. Now, let's see how we can customize the login experience. In the following code snippet, we demonstrate the usage of some of the more common ways to customize login, but we encourage you to refer to the reference documentation of Spring Security, which includes the *Appendix, Additional Reference Material*, with all of the supported attributes.

Let's take a look at the following steps to customize login:

1. First, update your `SecurityConfig.java` file as follows:

```
//src/main/java/com/packtpub/springsecurity/configuration/
SecurityConfig.java
...
http
        .authorizeHttpRequests( authz -> authz
                .requestMatchers("/**")
                .hasRole("USER")
        )
        .formLogin(form -> form
                .loginPage("/login/form")
                .loginProcessingUrl("/login")
                .failureUrl("/login/form?error")
                .usernameParameter("username")
                .passwordParameter("password")
....
```

Let's take a look at the following methods depicted in the preceding code snippet:

- The `loginPage()` method specifies where Spring Security will redirect the browser if a protected page is accessed and the user is not authenticated. If a login page is not specified, Spring Security will redirect the user to `/spring_security_login`. Then, `o.s.s.web.filter.FilterChainProxy` will choose `o.s.s.web.authentication.ui.DefaultLoginPageGeneratingFilter`, which renders the default login page as one of the delegates, since `DefaultLoginPageGeneratingFilter` is configured to process `/spring_security_login` by default. Since we have chosen to override the default URL, we are in charge of rendering the login page when the `/login/form` URL is requested.

- The `loginProcessingUrl()` method defaults to `/j_spring_security_check` and specifies the URL that the login form (which should include the username and password) should be submitted to, using an HTTP post. When Spring Security processes this request, it will attempt to authenticate the user.

- The `failureUrl()` method specifies the page that Spring Security will redirect to if the username and password submitted to `loginProcessingUrl()` are invalid.

- The `usernameParameter()` and `passwordParameter()` methods default to `j_username` and `j_password` respectively, and specify the HTTP parameters that Spring Security will use to authenticate the user when processing the `loginProcessingUrl()` method.

> **Important note**
>
> It may be obvious, but if we only wanted to add a custom login page, we would only need to specify the `loginPage()` method. We would then create our login form using the default values for the remaining attributes. However, it is often good practice to override the values of anything visible to users, to prevent exposing that we are using Spring Security. Revealing what frameworks we are using is a type of information leakage, making it easier for attackers to determine potential holes in our security.

2. The next step is to create a login page. We can use any technology we want to render the login page, as long as the login form produces the HTTP request that we specified with our Spring Security configuration when submitted. By ensuring the HTTP request conforms to our configuration, Spring Security can authenticate the request for us. Create the `login.html` file, as shown in the following code snippet:

```
//src/main/webapp/WEB-INF/tempates/login.html
<div class="alert alert-danger" th:if="${param.error != null}">
    <strong>Failed to login.</strong>

    <span th:if="${session[SPRING_SECURITY_LAST_EXCEPTION] !=
```

```
null}">
        <span th:text="${session[SPRING_SECURITY_LAST_
EXCEPTION] .message}">Invalid credentials</span>
    </span>
</div>
<div class="alert alert-success" th:if="${param.logout !=
null}">
    You have been logged out.
</div>

<fieldset>
    <legend>Login Form</legend>
    <div class="mb-3">
        <label class="form-label" for="username">Username<//
label>
        <input autofocus="autofocus" class="form-control"
id="username"
                name="username"
                type="text"/>
    </div>
    <div class="mb-3">
        <label class="form-label" for="password">Password<//
label>
        <input class="form-control" id="password"
name="password"
                type="password"/>
    </div>

    <div class="mb-3">
        <input class="btn btn-primary" id="submit" name="submit"
type="submit"
                value="Login"/>
    </div>
```

> **Important note**
>
> Remember that if you are having problems typing anything in the book, you can refer to the
> solution at the next checkpoint (chapter02.02- calendar).

The following number of items are worth highlighting in the preceding login.html file:

- The form action should be /login, to match the value provided for the loginProcessingUrl()
  method we specified. For security reasons, Spring Security only attempts to authenticate when
  using POST by default.

- We can use `param.error` to see whether there was a problem logging in, since the value of our `failureUrl()` method, `/login/form?error`, contains the HTTP parameter error.

- The session attribute, `SPRING_SECURITY_LAST_EXCEPTION`, contains the last `o.s.s.core.AuthenticationException` exception, which can be used to display the reason for a failed login. The error messages can be customized by leveraging Spring's internationalization support.

- The input names for the `username` and `password` inputs are chosen to correspond to the values we specified for the `usernameParameter()` and `passwordParameter()` methods in our `SecurityConfig.java` configuration.

3. The last step is to make Spring MVC aware of our new URL. This can be done by adding the following method to `WebMvcConfig`:

```
//src/main/java/com/packtpub/springsecurity/web/configuration/
WebMvcConfig.java
@Configuration
@EnableWebMvc
public class WebMvcConfig implements WebMvcConfigurer {
...
    @Override
    public void addViewControllers(final ViewControllerRegistry
registry) {
        registry.addViewController("/login/form")
            .setViewName("login");
    }
...
}
```

This code adds the custom `login.html` login page and the related controller request mapping URL: `/login/form`.

Configuring logout

The `HttpSecurity` configuration of Spring Security automatically adds support for logging the user out. All that is needed is to create a link that points to `/j_spring_security_logout`. However, we will demonstrate how to customize the URL used to log the user out by performing the following steps:

1. Update the Spring Security configuration as follows:

```
//src/main/java/com/packtpub/springsecurity/configuration/
SecurityConfig.java
http
        .authorizeHttpRequests( authz -> authz
            .requestMatchers("/**")
```

```
                        .hasRole("USER")
            )
    . . .

            ).logout(form -> form
                    .logoutUrl("/logout")
                    .logoutSuccessUrl("/login?logout")
            )
    . . . .
```

2. You have to provide a link for the user to click on that will log them out. We will update the header.html file so that the Logout link appears on every page:

```
//src/main/webapp/WEB-INF/templates/fragments/header.html

<div id="navbar" ...>
        . . .
    <ul class="nav navbar-nav pull-right">
        <li><a id="navLogoutLink" th:href="@{/logout}"> Logout</
a></li>
    </ul>
        . . .
</div>
```

3. The last step is to update the login.html file to display a message indicating logout was successful when the logout parameter is present:

```
//src/main/webapp/WEB-INF/templates/login.html

<div th:if="${param.logout != null}" class="alert alert-
success"> You have been logged out.</div>
<label for="username">Username</label>
        . . .
```

> **Important note**
>
> Your code should now look like chapter02.02-calendar.

The page isn't redirecting properly

If you have not already, restart the application and visit http://localhost:8080; you will see an error, as shown in the following screenshot:

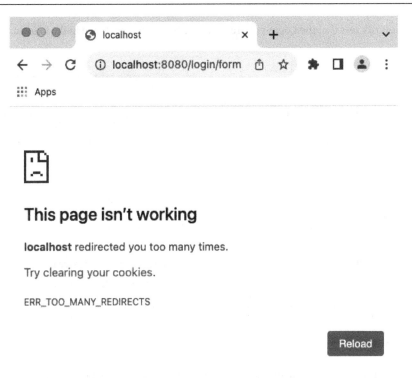

Figure 2.6 – Page not redirecting properly

What went wrong? The problem is, since Spring Security is no longer rendering the login page, we must allow everyone (not just the USER role) to access the **Login** page. Without granting access to the **Login** page, the following happens:

1. We request the **Welcome** page in the browser.
2. Spring Security sees that the **Welcome** page requires the USER role and that we are not authenticated, so it redirects the browser to the **Login** page.
3. The browser requests the **Login** page.
4. Spring Security sees that the **Login** page requires the USER role and that we are still not authenticated, so it redirects the browser to the **Login** page again.
5. The browser requests the **Login** page again.

6.   Spring Security sees that the **Login** page requires the USER role, as shown in the following diagram:

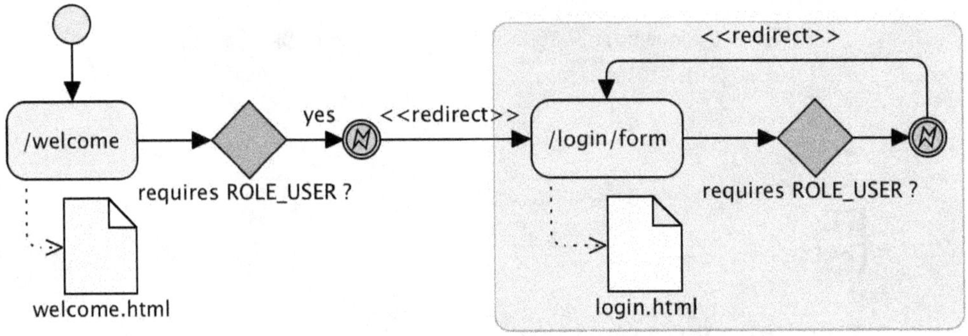

Figure 2.7 – Login process with Spring Security

The process could just keep repeating indefinitely. Fortunately for us, Firefox realizes that there are too many redirects occurring, stops performing the redirect, and displays a very informative error message. In the next section, we will learn how to fix this error by configuring URLs differently, depending on the access that they require.

Basic role-based authorization

We can expand on the Spring Security configuration from Hello Spring Security to vary the access controls by URL. In this section, you will find a configuration that allows more granular control over how resources can be accessed. In the configuration, Spring Security does the following tasks:

It completely ignores any request that starts with /resources/. This is beneficial since our images, CSS, and JavaScript do not need to use Spring Security.

It allows anonymous users to access the **Welcome**, **Login**, and **Logout** pages. It only allows administrators access to the **All Events** page.

It adds an administrator that can access the **All Events** page.

Take a look at the following code snippet:

```
//src/main/java/com/packtpub/springsecurity/configuration/
SecurityConfig.java
...
http
        .authorizeHttpRequests( authz -> authz
            .requestMatchers("/resources/**").permitAll()
            .requestMatchers("/webjars/**").permitAll()
            .requestMatchers("/").hasAnyRole("ANONYMOUS", "USER")
            .requestMatchers("/login/*").hasAnyRole("ANONYMOUS",
"USER")
```

```
                    .requestMatchers("/logout/*").hasAnyRole("ANONYMOUS",
"USER")

                    .requestMatchers("/admin/*").hasRole("ADMIN")
                    .requestMatchers("/events/").hasRole("ADMIN")
                    .requestMatchers("/**").hasRole("USER")
...
@Bean
public InMemoryUserDetailsManager userDetailsService() {
    UserDetails user = User.withDefaultPasswordEncoder()
            .username("user")
            .password("user")
            .roles("USER")
            .build();

    UserDetails admin = User.withDefaultPasswordEncoder()
            .username("admin")
            .password("admin")
            .roles("ADMIN")
            .build();

    UserDetails user1 = User.withDefaultPasswordEncoder()
            .username("user1@example.com")
            .password("user1")
            .roles("USER")
            .build();

    UserDetails admin1 = User.withDefaultPasswordEncoder()
            .username("admin1@example.com")
            .password("admin1")
            .roles("USER", "ADMIN")
            .build();

    return new InMemoryUserDetailsManager(user,admin, user1, admin1);
}
```

> **Important note**
>
> Notice that we do not include the application's context root, in the Spring Security configuration, because Spring Security takes care of the context root transparently for us. In this way, we do not need to update our configuration if we decide to deploy it to a different context root.

In Spring Security 6, you can specify multiple `requestMatchers` entries using a builder pattern that allows you to have greater control over how security is applied to different portions of your application. The first `RequestMatcher` object states that Spring Security should ignore any URL that starts with /resources/, and the second `RequestMatcher` object states that any other request will be processed by it. There are a few important things to note about using multiple `requestMatchers` methods, as follows:

- If no path attribute is specified, it is the equivalent of using a path of /**, which matches all requests.

- Each `requestMatchers()` method is considered in order, and only the first match is applied. So, the order in which they appear in your configuration file is important. The implication is that only the last `requestMatchers()` method can use a path that matches every request. If you do not follow this rule, Spring Security will produce an error. The following is invalid because the first matcher matches every request and will never get to the second mapping:

```
http.authorizeHttpRequests((authz) -> authz
        .requestMatchers("/**").hasRole("USER")
        .requestMatchers(("/admin/*").hasRole("ADMIN"))
```

- The default pattern is backed by `o.s.s.web.util.AntPathRequestMatcher`, which will compare the specified pattern to an **Ant Pattern** to determine whether it matches the `servletPath` and `pathInfo` methods of `HttpServletRequest`.

- The `RequestMatcher` interface is used to determine whether a request matches a given rule. We use `securityMatchers` to determine whether a given `HttpSecurity` should be applied to a given request. In the same way, we can use `requestMatchers` to determine the authorization rules that we should apply to a given request.

- The path attribute on the `requestMatchers()` method further refines the filtering of the request and allows access control to be applied. You can see that the updated configuration allows different types of access, depending on the URL pattern. The ANONYMOUS role is of particular interest since we have not defined it anywhere in `SecurityConfig.java`. This is the default authority assigned to a user that is not logged in. The following line, from the updates to our `SecurityConfig.java` file, is what allows anonymous (unauthenticated) users and users with the USER role authority to access the **Login** page. We will cover access control options in more detail in the second half of the book:

```
.requestMatchers("/login/*").hasAnyRole("ANONYMOUS", "USER")
```

When defining the `requestMatchers()` methods, there are several things to keep in mind, including the following:

- Just as each **http** method is considered from top to bottom, so are the `requestMatchers()` methods. This means it is important to specify the most specific elements first. The following example illustrates a configuration that does not specify the more specific pattern first, which will result in warnings from Spring Security at startup:

```
http.authorizeHttpRequests()

...

        // matches every request, so it will not continue
        .requestMatchers("/**").hasRole("USER")
        // below will never match
        .requestMatchers("/login/form").hasAnyRole("ANONYMOUS",
"USER")
```

- It is important to note that if `http.authorizeHttpRequests()` can have `anyRequest()`, there should be no child `requestMatchers()` method defined. This is because `anyRequest()` will match all requests that match this `http.authorizeHttpRequests()` tag. Defining a `requestMatchers()` child method with `anyRequest()` contradicts the `requestMatchers()` declaration. An example is as follows:

```
http.authorizeHttpRequests((authz) -> authz.anyRequest().
permitAll())
// This matcher will never be executed
// and not produce an error.
        .requestMatchers("/admin/*").hasRole("ADMIN"))
```

- The path attribute of the `requestMatchers()` element is independent and is not aware of the `anyRequest()` method.

If you have not done so already, restart the application and visit `http://localhost:8080`. Experiment with the application to see all the updates you have made, as follows:

1. Select a link that requires authentication and observes the new login page.

2. Try typing an invalid `username/password` and view the error message.

3. Try logging in as an admin (`admin1@example.com/admin1`), and view all of the events. Note that we are able to view all the events.

4. Try logging out and view the logout success message.

5. Try logging in as a regular user (`user1@example.com/user1`), and view all of the events. Note that we get an **Access Denied** page.

> **Important note**
>
> Your code should now look like `chapter02.03-calendar`.

Expression-based authorization

You may have noticed that granting access to everyone was not nearly as concise as we may have liked. Fortunately, Spring Security can leverage **Spring Expression Language (SpEL)** to determine whether a user has authorization. In the following code snippet, you can see the updates when using **SpEL** with Spring Security.

For Java configuration, `WebExpressionAuthorizationManager` is available to help use legacy SpEL:

```
//src/main/java/com/packtpub/springsecurity/configuration/
SecurityConfig.java


http
        .authorizeHttpRequests( authz -> authz
            .requestMatchers("/").access(new
WebExpressionAuthorizationManager("hasAnyRole('ANONYMOUS', 'USER')"))
                .requestMatchers("/resources/**").permitAll()
                .requestMatchers("/webjars/**").permitAll()
                .requestMatchers("/login/*").access(new
WebExpressionAuthorizationManager("hasAnyRole('ANONYMOUS', 'USER')"))
                .requestMatchers("/logout/*").access(new
WebExpressionAuthorizationManager("hasAnyRole('ANONYMOUS', 'USER')"))
                .requestMatchers("/errors/**").permitAll()
                .requestMatchers("/admin/*").access(new
WebExpressionAuthorizationManager("hasRole('ADMIN')"))
                .requestMatchers("/events/").access(new
WebExpressionAuthorizationManager("hasRole('ADMIN')"))
                .requestMatchers("/**").access(new
WebExpressionAuthorizationManager("hasRole('USER')"))
        )
```

> **Important note**
>
> You may notice that the `/events/` security constraint is brittle. For example, the `/events` URL is not protected by Spring Security to restrict the `ADMIN` role. This demonstrates the need to ensure that we provide multiple layers of security. We will exploit this sort of weakness in *Chapter 11, Fine-Grained Access Control*.

Changing the `access` attribute from `hasAnyRole('ANONYMOUS', 'USER')` to `permitAll()` might not seem like much, but this only scratches the surface of the power of Spring Security's expressions. We will go into much greater detail about access control and Spring expressions in the second half of the book. Go ahead and verify that the updates work by running the application.

> **Important note**
>
> It is recommended that you use type-safe authorization managers instead of SpEL. Your code should now look like `chapter02.04-calendar`.

Conditionally displaying authentication information

Currently, our application does not indicate whether we are logged in or not. It appears as though we are always logged in since the **Logout** link is always displayed. In this section, we will demonstrate how to display the authenticated user's username and conditionally display portions of the page using **Thymeleaf**'s Spring Security tag library. We do so by performing the following steps:

1. Update your dependencies to include the `thymeleaf-extras- springsecurity6` JAR file. Since we are using Gradle, we will add a new dependency declaration in our `build.gradle` file, as follows:

    ```
    //build.gradle
    dependencies{

    ...

        implementation 'org.thymeleaf.extras:thymeleaf-extras-
    springsecurity6:3.1.2.RELEASE'
    }
    ```

2. Next, we need to add `SpringSecurityDialect` to the Thymeleaf engine as follows:

    ```
    //src/com/packtpub/springsecurity/web/configuration/
    ThymeleafConfig.java
    @Bean
    public SpringTemplateEngine templateEngine(final
    ITemplateResolver templateResolver) {
        SpringTemplateEngine engine = new SpringTemplateEngine();
        engine.setTemplateResolver(templateResolver);
        engine.setAdditionalDialects(new HashSet<>() {{
            add(new LayoutDialect());
            add(new SpringSecurityDialect());
        }});
        return engine;
    }
    ```

3. The `sec:authorize` attribute determines whether the user is authenticated with the `isAuthenticated()` value, displays the HTML node if the user is authenticated, and hides the node in the event that the user is not authenticated. The `access` attribute should be rather familiar from the `requestMatchers().access()` element. In fact, both components leverage the same SpEL support.

    There are attributes in the **Thymeleaf** tag libraries that do not use expressions. However, using **SpEL** is typically the preferred method since it is more powerful.

    The `sec:authentication` attribute will look up the current `o.s.s.core.Authentication` object. The `property` attribute will find the principal attribute of the `o.s.s.core.Authentication` object, which in this case is `o.s.s.core.userdetails.UserDetails`. It then obtains the `UserDetails` username property and renders it to the page. Don't worry if the details of this are confusing. We are going to go over this in more detail in *Chapter 3, Custom Authentication*.

    Update the `header.html` file to leverage the Spring Security tag library. You can find the updates as follows:

```
//src/main/webapp/WEB-INF/templates/fragments/header.html
<html xmlns:th="http://www.thymeleaf.org" xmlns:sec="http://www.
thymeleaf.org/extras/spring-security">
<body>
...
                <div id="navbar" class="collapse navbar-
collapse">
...
                    <ul class="nav navbar-nav pull-right"
sec:authorize="isAuthenticated()">
                        <li>
                            <p class="navbar-text">Welcome <div
class="navbar-text" th:text="${#authentication.name}">User</
div></p>
                        </li>
                        <li>
                            <a id="navLogoutLink" class="btn
btn-default" role="button"  th:href="@{/logout}">Logout</a>
                        </li>
                        <li> | </li>
                    </ul>
                    <ul class="nav navbar-nav pull-right"
sec:authorize=" ! isAuthenticated()">
                        <li><a id="navLoginLink" class="btn
btn-default" role="button" th:href="@{/login/form}">Login</a></
li>
                        <li> | </li>
                    </ul>
```

```
                </div>
            </div>
        </nav>
    </div>
...
```

If you haven't done so already, restart the application to see the updates we have made. At this point, you may realize that we are still displaying links we do not have access to. For example, user1@ example.com should not see a link to the **All Events** page. Rest assured, we'll fix this when we cover the tags in greater detail in *Chapter 11, Fine-Grained Access Control*.

> **Important note**
>
> Your code should now look like this: chapter02.05-calendar.

Customizing behavior after login

We have already discussed how to customize a user's experience during login, but sometimes it is necessary to customize the behavior after login. In this section, we will discuss how Spring Security behaves after login and will provide a simple mechanism to customize this behavior.

In the default configuration, Spring Security has two different flows after successful authentication. The first scenario occurs if a user never visits a resource that requires authentication. In this instance, after a successful login attempt, the user will be sent to the defaultSuccessUrl() method chained to the formLogin() method. If left undefined, defaultSuccessUrl() will be the context root of the application.

If a user requests a protected page before being authenticated, Spring Security will remember the last protected page that was accessed before authenticating, using o.s.s.web.savedrequest.RequestCache. Upon successful authentication, Spring Security will send the user to the last protected page that was accessed before authentication. For example, if an unauthenticated user requests the My Events page, they will be sent to the Login page.

After successful authentication, they will be sent to the previously requested **My Events** page.

A common requirement is to customize Spring Security to send the user to a different defaultSuccessUrl() method, depending on the user's role. Let's take a look at how this can be accomplished by performing the following steps:

1. The first step is to configure the defaultSuccessUrl() method chained after the formLogin() method. Go ahead and update SecurityConfig.java:

    ```
    //src/main/java/com/packtpub/springsecurity/configuration/
    SecurityConfig.java
    .formLogin(form -> form
    ```

```
.loginPage("/login/form")
.loginProcessingUrl("/login")
.failureUrl("/login/form?error")
.usernameParameter("username")
.passwordParameter("password")
.defaultSuccessUrl("/default")
.permitAll()
```

2. The next step is to create a controller that processes /default. In the following code, you will find a sample Spring MVC controller, DefaultController, which demonstrates how to redirect administrators to the **All Events** page and other users to the **Welcome** page. Create a new file in the following location:

```java
//src/main/java/com/packtpub/springsecurity/web/controllers/
DefaultController.java
@Controller
public class DefaultController {

    @RequestMapping("/default")
    public String defaultAfterLogin(HttpServletRequest request)
{
        if (request.isUserInRole("ADMIN")) {
            return "redirect:/events/";
        }
        return "redirect:/";
    }
}
```

> **Important note**
> In IntelliJ IDEA, you can press *Alt + Enter* to automatically add the missing imports.

There are a few things to point out about DefaultController and how it works. The first is that Spring Security makes the HttpServletRequest parameter aware of the currently logged-in user. In this instance, we can inspect which role the user belongs to without relying on any of Spring Security's APIs.

This is good because if Spring Security's APIs change or we decide we want to switch our security implementation, we have less code that needs to be updated. It should also be noted that while we implement this controller with a Spring MVC controller, our defaultSuccessUrl() method can be handled by any controller implementation (for example, Struts, a standard servlet, and so on) if we desire.

3. If you wish to always go to the `defaultSuccessUrl()` method, you can leverage the second parameter to the `defaultSuccessUrl()` method, which is a `Boolean` for always use. We will not do this in our configuration, but you can see an example of it as follows:

```
.defaultSuccessUrl("/default", true)
```

4. You are now ready to give it a try. Restart the application and go directly to the **My Events** page, then log in; you will see that you are on the **My Events** page.

5. Next, log out and try logging in as `user1@example.com`.

6. You should be on the **Welcome** page. Log out and log in as `admin1@example.com`, and you will be sent to the **All Events** page.

> **Important note**
>
> Your code should now look like `chapter02.06-calendar`.

Summary

In this chapter, we have applied a very basic Spring Security configuration. The main objective was to outline the steps and key concepts involved in implementing Spring Security in Spring 6 applications. To achieve this goal, we started by importing the sample application, then updating dependencies, configuring Spring Security, and managing web configurations, and in the end, we addressed common issues.

Additionally, we explained how to customize the user's login and logout experience and demonstrated how to display basic information based on roles and SpEL. We finished this chapter by customizing behavior after login.

In the next chapter, we will discuss how authentication in Spring Security works and how we can customize it to our needs.

3

Custom Authentication

In *Chapter 2, Getting Started with Spring Security*, we demonstrated how to use an in-memory datastore to authenticate the user. In this chapter, we'll explore how to solve some common, real-world problems by extending Spring Security's authentication support to use our existing set of APIs. Through this exploration, we'll get an understanding of each of the building blocks that Spring Security uses in order to authenticate users.

During this chapter, we will cover the following topics:

- Leveraging Spring Security's annotations and Java-based configuration
- Discovering how to obtain the details of the currently logged-in user
- Adding the ability to log in after creating a new account
- Learning the simplest method for indicating to Spring Security that a user is authenticated
- Creating custom `UserDetailsService` and `AuthenticationProvider` implementations that properly decouple the rest of the application from Spring Security
- Adding domain-based authentication to demonstrate how to authenticate with more than just a username and password

This chapter's code in action link is here: `https://packt.link/5tPFD`.

Authentication architecture in Spring Security

The realm of application security essentially involves addressing two largely independent issues: authentication (identifying *who you are*) and authorization (determining *what you are allowed to do*).

Occasionally, individuals may use the term *access control* interchangeably with *authorization*, adding a layer of potential confusion.

However, framing it as *access control* can offer clarity, considering the multifaceted use of the term *authorization* elsewhere.

Spring Security adopts an architecture deliberately crafted to segregate authentication from authorization, providing distinct strategies and extension points for each. We will uncover in this section the main architectural components of Spring Security used for authentication.

The SecurityContextHolder class

At the heart of Spring Security's authentication model is the `SecurityContextHolder`. It contains the `SecurityContext`.

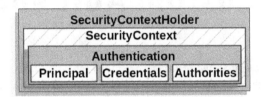

Figure 3.1 – Spring Security's SecurityContextHolder

The SecurityContext interface

`SecurityContextHolder` is where Spring Security stores the details of who is authenticated. Spring Security does not care how `SecurityContextHolder` is populated. If it contains a value, it is used as the currently authenticated user.

The Authentication interface

The **Authentication** interface in Spring Security serves dual purposes:

- It acts as input for `AuthenticationManager`, supplying the user's provided credentials for authentication. In this context, the method `isAuthenticated()` returns false.

- It serves as a representation of the presently authenticated user, retrievable from the `SecurityContext`.

Key components within the **Authentication** interface include the following:

- **Principal**: This identifies the user, often represented as an instance of `UserDetails`, especially in username/password authentication.

- **Credentials**: This typically encompasses a password. In many instances, this information is cleared post-authentication to prevent inadvertent leakage.

- **Authorities**: This comprises `GrantedAuthority` instances denoting high-level permissions granted to the user. Examples include roles and scopes.

The AuthenticationManager interface

AuthenticationManager serves as the API specifying how authentication is conducted by Spring Security's Filters. The resulting authentication is subsequently established on SecurityContextHolder by the invoking controller (i.e., Spring Security's Filters instances).

If you're not integrating with Spring Security's Filters instances, you have the option to directly set SecurityContextHolder without the need for AuthenticationManager.

Although the AuthenticationManager's implementation can vary, the prevalent choice is often ProviderManager.

The ProviderManager class

ProviderManager stands out as the frequently employed realization of AuthenticationManager. It delegates responsibilities to a list of AuthenticationProvider instances. Each AuthenticationProvider possesses the capability to express whether authentication should succeed, fail, or delegate the decision-making to a subsequent AuthenticationProvider. In the event that none of the configured AuthenticationProvider instances can authenticate, the authentication process results in a ProviderNotFoundException. This particular AuthenticationException signifies that the ProviderManager lacked configuration to support the specific **authentication** type provided to it.

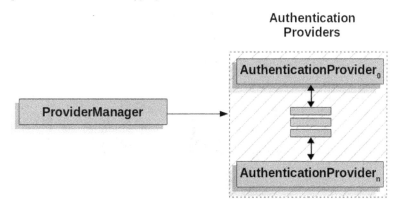

Figure 3.2 – Spring Security SecurityContextHolder

In practical terms, every AuthenticationProvider is equipped to carry out a distinct authentication method. For instance, one AuthenticationProvider may validate a username/password, while another is capable of authenticating a SAML Assertion. This setup empowers each AuthenticationProvider to handle a specialized form of authentication, accommodating various authentication types and presenting only a singular AuthenticationManager bean.

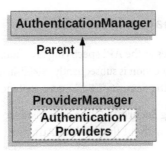

Figure 3.3 – Spring Security SecurityContextHolder

Additionally, `ProviderManager` enables the configuration of an optional parent `AuthenticationManager`. This parent `AuthenticationManager` is consulted when no `AuthenticationProvider` is able to execute authentication. The parent can take on any form of `AuthenticationManager`, with `ProviderManager` frequently being the chosen type.

Multiple instances of `ProviderManager` can have a shared parent `AuthenticationManager`. This occurrence is quite typical in situations where multiple `SecurityFilterChain` instances share a common authentication process (represented by the shared parent `AuthenticationManager`). However, these instances may also employ different authentication mechanisms, each managed by distinct `ProviderManager` instances.

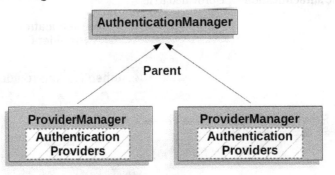

Figure 3.4 – Spring Security's SecurityContextHolder

By default, `ProviderManager` attempts to remove any sensitive credential information from the `Authentication` object returned upon a successful authentication request. This precautionary measure ensures that sensitive details, such as passwords, are not stored in the `HttpSession` for longer than necessary.

The AuthenticationProvider interface

It is possible to inject multiple instances of `AuthenticationProviders` into `ProviderManager`. Each `AuthenticationProvider` is responsible for a specific form of authentication. For instance, `DaoAuthenticationProvider` is designed for username/password-based authentication, while `JwtAuthenticationProvider` specializes in authenticating JSON Web Tokens.

Exploring the JBCP calendar architecture

We will start this chapter by analyzing the domain model within the **JBPC Calendar** architecture.

In *Chapter 1*, *Anatomy of an Unsafe Application*, and *Chapter 2*, *Getting Started with Spring Security*, we used the Spring **Bill Of Materials** (**BOM**) to assist in dependency management, but the rest of the code in the projects used the core Spring Framework and required manual configuration. Starting with this chapter, we will be using Spring Boot for the rest of the applications, to simplify the application configuration process. The Spring Security configuration we will be creating will be the same for both a Spring Boot and a non-Boot application. We will cover more details on Spring IO and Spring Boot in the *Appendix*, *Additional Reference Material*.

In the upcoming sections, we will delve into the domain model of the JBCP calendar application. We aim to gain insights into the process of incorporating Spring Security with personalized user configurations and APIs.

The CalendarUser object

Our calendar application uses a domain object named `CalendarUser`, which contains information about our users, as follows:

```
//src/main/java/com/packtpub/springsecurity/domain/CalendarUser.java
public class CalendarUser implements Serializable {
    private Integer id;

    private String firstName;

    private String lastName;

    private String email;

    private String password;
... accessor methods omitted ..
}
```

The Event object

Our application has an `Event` object that contains information about each event, as follows:

```
//src/main/java/com/packtpub/springsecurity/domain/Event.java
public record Event(
        Integer id,
        @NotEmpty(message = "Summary is required") String summary,
        @NotEmpty(message = "Description is required") String
description,
        @NotNull(message = "When is required") Calendar dateWhen,
        @NotNull(message = "Owner is required") CalendarUser owner,
        CalendarUser attendee
) {}
```

The CalendarService interface

Our application contains a `CalendarService` interface that can be used to access and store our domain objects. The code for `CalendarService` is as follows:

```
//src/main/java/com/packtpub/springsecurity/service/CalendarService.
java
public interface CalendarService {
    CalendarUser getUser(int id);

    CalendarUser findUserByEmail(String email);

    List<CalendarUser> findUsersByEmail(String partialEmail);

    int createUser(CalendarUser user);

    Event getEvent(int eventId);

    int createEvent(Event event);

    List<Event> findForUser(int userId);

    List<Event> getEvents();
}
```

We won't go over the methods used in `CalendarService`, but they should be fairly straightforward. If you would like details about what each method does, please consult the Javadoc in the sample code.

The UserContext interface

Like most applications, our application requires us to interact with the currently logged-in user. We have created a very simple interface called UserContext to manage the currently logged-in user, as follows:

```
//src/main/java/com/packtpub/springsecurity/service/UserContext.java
public interface UserContext {
    CalendarUser getCurrentUser();

    void setCurrentUser(CalendarUser user);
}
```

This means that our application can call UserContext.getCurrentUser() to obtain the details of the currently logged-in user. It can also call UserContext.setCurrentUser(CalendarUser) to specify which user is logged in. Later in this chapter, we will explore how we can write an implementation of this interface that uses Spring Security to access our current user and obtain their details using SecurityContextHolder.

Spring Security provides quite a few different methods for authenticating a user. However, the net result is that Spring Security will populate o.s.s.core.context.SecurityContext with o.s.s.core.Authentication. The Authentication object represents all the information we gathered at the time of authentication (username, password, roles, and so on). The SecurityContext interface is then set on the o.s.s.core.context.SecurityContextHolder interface. This means that Spring Security and developers can use SecurityContextHolder to obtain information about the currently logged-in user. An example of obtaining the current username is illustrated as follows:

```
String username = SecurityContextHolder.getContext()
        .getAuthentication()
        .getName();
```

> **Important note**
>
> It should be noted that null checks should always be done on the Authentication object, as this could be null if the user is not logged in.

The SpringSecurityUserContext interface

The current UserContext implementation, UserContextStub, is a stub that always returns the same user. This means that the **My Events** page will always display the same user no matter who is logged in. Let's update our application to utilize the current Spring Security user's username, to determine which events to display on the **My Events** page.

> **Important note**
>
> You should be starting with the sample code in `chapter03.00- calendar`.

Take a look at the following steps:

1. The first step is to comment out the `@Component` attribute on `UserContextStub`, so that our application no longer uses our scanned results.

> **Important note**
>
> The `@Component` annotation is used in conjunction with the `@Configuration` annotation found in `com/packtpub/springsecurity/web/configuration/WebMvcConfig.java`, to automatically create a Spring bean rather than creating an explicit XML or Java configuration for each bean. You can learn more about the classpath of Spring scanning at `https://docs.spring.io/spring-framework/reference/core/beans/classpath-scanning.html`.

    Take a look at the following code snippet:

    ```
    ...
    @Component
    public class UserContextStub implements UserContext {
    ...
    ```

2. The next step is to utilize `SecurityContext` to obtain the currently logged-in user. We have included `SpringSecurityUserContext` within this chapter's code, which is wired up with the necessary dependencies but contains no actual functionality.

3. Open the `SpringSecurityUserContext.java` file and add the `@Component` annotation. Next, replace the `getCurrentUser` implementation, as illustrated in the following code snippet:

    ```
    //src/main/java/com/packtpub/springsecurity/service/
    SpringSecurityUserContext.java
    @Component
    public class SpringSecurityUserContext implements UserContext {

        private final CalendarService calendarService;

        private final UserDetailsService userDetailsService;

        public SpringSecurityUserContext(final CalendarService
    calendarService,
                final UserDetailsService userDetailsService) {
            if (calendarService == null) {
    ```

```java
            throw new IllegalArgumentException("calendarService
cannot be null");
        }
        if (userDetailsService == null) {
            throw new IllegalArgumentException("userDetailsService
cannot be null");
        }
        this.calendarService = calendarService;
        this.userDetailsService = userDetailsService;
    }

    @Override
    public CalendarUser getCurrentUser() {

        SecurityContext context = SecurityContextHolder.
getContext();
        Authentication authentication = context.
getAuthentication();

        if (authentication == null) {
            return null;
        }

        User user = (User) authentication.getPrincipal();
        String email = user.getUsername();

        if (email == null) {
            return null;
        }

        CalendarUser result = calendarService.
findUserByEmail(email);
        if (result == null) {
            throw new IllegalStateException(
                "Spring Security is not in synch with
CalendarUsers. Could not find user with email " + email);
        }
        return result;
    }

    @Override
    public void setCurrentUser(CalendarUser user) {
        throw new UnsupportedOperationException();
    }
}
```

Our code obtains the username from the current Spring Security `Authentication` object and utilizes that to look up the current `CalendarUser` object by email address. Since our Spring Security username is an email address, we can use the email address to link `CalendarUser` with the Spring Security user. Note that if we were to link accounts, we would normally want to do this with a key that we generated rather than something that may change (that is, an email address). We follow the good practice of returning only our domain object to the application. This ensures that our application is only aware of our `CalendarUser` object and thus is not coupled to Spring Security.

This code may seem eerily similar to when we used the `sec:authorize= "isAuthenticated()"` tag attribute in *Chapter 2, Getting Started with Spring Security*, to display the current user's username. In fact, the Spring Security tag library uses `SecurityContextHolder` in the same manner as we have done here. We could use our `UserContext` interface to place the current user on `HttpServletRequest` and thus remove our dependency on the Spring Security tag library.

4. Start up the application, visit `http://localhost:8080/`, and log in with `admin1@ example.com` as the username and `admin1` as the password.

5. Visit the **My Events** page and you will see that only the events for that current user, who is the owner or the attendee, are displayed.

6. Try creating a new event; you will observe that the owner of the event is now associated with the logged-in user.

7. Log out of the application and repeat these steps with `user1@example.com` as the username and `user1` as the password.

> **Important note**
> Your code should now look like `chapter03.01-calendar`.

In this section, we have covered the JBCP calendar architecture. In the next session, we will see how to manage new users using `SecurityContextHolder`.

Logging in new users using SecurityContextHolder

A common requirement is to allow users to create a new account and then automatically log them into the application. In this section, we'll describe the simplest method for indicating that a user is authenticated, by utilizing `SecurityContextHolder`.

Managing users in Spring Security

The application provided in *Chapter 1, Anatomy of an Unsafe Application*, provides a mechanism for creating a new `CalendarUser` object, so it should be fairly easy to create our `CalendarUser` object after a user signs up. However, Spring Security has no knowledge of `CalendarUser`. This means that we will need to add a new user in Spring Security, too. Don't worry, we will remove the need for the dual maintenance of users later in this chapter.

Spring Security provides an `o.s.s.provisioning.UserDetailsManager` interface for managing users. Remember our in-memory Spring Security configuration?

```
auth.inMemoryAuthentication(). withUser("user").password("user").
roles("USER");
```

The `SecurityConfig.userDetailsService()` method creates an in-memory implementation of `UserDetailsManager`, named `o.s.s.provisioning.InMemoryUserDetailsManager`, which can be used to create a new Spring Security user.

Let's see how we can manage users in Spring Security by performing the following steps:

1. To expose `UserDetailsManager` using a Java-based configuration, we need to create `InMemoryUserDetailsManager`:

   ```
   //src/main/java/com/packtpub/springsecurity/configuration/
   SecurityConfig.java

   @Bean
   public InMemoryUserDetailsManager userDetailsService() {
           UserDetails user1 = User.withDefaultPasswordEncoder()
           .username("user1@example.com")
           .password("user1")
           .roles("USER")
           .build();

           UserDetails admin1 = User.withDefaultPasswordEncoder()
           .username("admin1@example.com")
           .password("admin1")
           .roles("USER", "ADMIN")
           .build();
           return new InMemoryUserDetailsManager(user1, admin1);
   }
   ```

2. Once we have an exposed `UserDetailsManager` interface in our Spring configuration, all we need to do is update our existing `CalendarService` implementation, `DefaultCalendarService`, to add a user in Spring Security. Make the following updates to the `DefaultCalendarService.java` file:

```
//src/main/java/com/packtpub/springsecurity/service/
DefaultCalendarService.java
public int createUser(final CalendarUser user) {
        List<GrantedAuthority> authorities = AuthorityUtils.
createAuthorityList("ROLE_USER");
        UserDetails userDetails = new User(user.getEmail(), user.
getPassword(), authorities);
        userDetailsManager.createUser(userDetails);
        return userDao.createUser(user);
        }
```

3. To leverage `UserDetailsManager`, we first convert `CalendarUser` into the `UserDetails` object of Spring Security.

4. Later, we use `UserDetailsManager` to save the `UserDetails` object. The conversion is necessary because Spring Security has no understanding of how to save our custom `CalendarUser` object, so we must map `CalendarUser` to an object Spring Security understands. You will notice that the `GrantedAuthority` object corresponds to the `authorities` attribute of our `SecurityConfig` file. We hardcode this for simplicity and because there is no concept of roles in our existing system.

Logging in a new user to an application

Now that we can add new users to the system, we need to indicate that the user is authenticated. Update `SpringSecurityUserContext` to set the current user on the `SecurityContextHolder` object of Spring Security, as follows:

```
@Override
public void setCurrentUser(CalendarUser user) {
    if (user == null) {
        throw new IllegalArgumentException("user cannot be null");
    }
    UserDetails userDetails = userDetailsService.
loadUserByUsername(user.getEmail());
    UsernamePasswordAuthenticationToken authentication = new
UsernamePasswordAuthenticationToken(userDetails,
        user.getPassword(), userDetails.getAuthorities());
    SecurityContextHolder.getContext().
setAuthentication(authentication);
}
```

The first step we perform is to convert our `CalendarUser` object into the `UserDetails` object of Spring Security. This is necessary because, just as Spring Security didn't know how to save our custom `CalendarUser` object, Spring Security also does not understand how to make security decisions with our custom `CalendarUser` object. We use Spring Security's `o.s.s.core.userdetails.UserDetailsService` interface to obtain the same `UserDetails` object we saved with `UserDetailsManager`. The `UserDetailsService` interface provides a subset of the functionality provided by Spring Security's `UserDetailsManager` object that we have already seen.

Next, we create a `UsernamePasswordAuthenticationToken` object and place `UserDetails`, the password, and `GrantedAuthority` in it. Lastly, we set the authentication on `SecurityContextHolder`. In a web application, Spring Security will automatically associate the `SecurityContext` object in `SecurityContextHolder` to our HTTP session for us.

> **Important note**
>
> It is important that Spring Security must not be instructed to ignore a URL (that is, using the `permitAll()` method), as discussed in *Chapter 2*, *Getting Started with Spring Security*, in which `SecurityContextHolder` is accessed or set. This is because Spring Security will ignore the request and thus not persist `SecurityContext` for subsequent requests. The session management support is composed of a few components that work together to provide the functionality, so we used `securityContext.requireExplicitSave(false)` to persist the session.

The advantage of this approach is that there is no need to hit the datastore again. In our case, the datastore is an in-memory datastore, but this could be backed by a database, which could have some security implications. The disadvantage of this approach is that we do not get to reuse the code much. Since this method is invoked infrequently, we opt for reusing the code. In general, it is best to evaluate each situation separately to determine which approach makes the most sense.

Updating SignupController

The application has a `SignupController` object, which is what processes the HTTP request to create a new `CalendarUser` object. The last step is to update `SignupController` to create our user and then indicate that they are logged in. Make the following updates to `SignupController`:

```
//src/main/java/com/packtpub/springsecurity/web/controllers/
SignupController.java

@PostMapping("/signup/new")
public String signup(final @Valid SignupForm signupForm,
final BindingResult result,
    RedirectAttributes redirectAttributes) {
    if (result.hasErrors()) {
    return "signup/form";
```

```
        }

        String email = signupForm.getEmail();
        if (calendarService.findUserByEmail(email) != null) {
        result.rejectValue("email", "errors.signup.email", "Email
address is already in use. FOO");
        redirectAttributes.addFlashAttribute("error", "Email address is
already in use. FOO");
        return "signup/form";
        }

        CalendarUser user = new CalendarUser(null, signupForm.
getFirstName(), signupForm.getLastName(), email, signupForm.
getPassword());
        int id = calendarService.createUser(user);
        user.setId(id);
        userContext.setCurrentUser(user);

        redirectAttributes.addFlashAttribute("message", "You have
successfully signed up and logged in.");
        return "redirect:/";
        }
```

If you have not done so already, restart the application, visit http://localhost:8080/, create a new user, and you can see that the new user is automatically logged in.

> **Important note**
> Your code should now look like chapter03.02-calendar.

In this section, we have covered the new user signup workflow. In the next section, we will create a custom UserDetailsService object.

Creating a custom UserDetailsService object

While we can link our domain model (CalendarUser) with Spring Security's domain model (UserDetails), we have to maintain multiple representations of the user. To resolve this dual maintenance, we can implement a custom UserDetailsService object to translate our existing CalendarUser domain model into an implementation of Spring Security's UserDetails interface. By translating our CalendarUser object into UserDetails, Spring Security can make security decisions using our custom domain model. This means that we will no longer need to manage two different representations of a user.

The CalendarUserDetailsService class

Up to this point, we have needed two different representations of users: one for Spring Security to make security decisions, and one for our application to associate our domain objects to. Create a new class named `CalendarUserDetailsService` that will make Spring Security aware of our `CalendarUser` object. This will ensure that Spring Security can make decisions based on our domain model. Create a new file named `CalendarUserDetailsService.java`, as follows:

```java
//src/main/java/com/packtpub/springsecurity/service/
CalendarUserDetailsService.java

@Component
public class CalendarUserDetailsService implements UserDetailsService
{

    private static final Logger logger = LoggerFactory
        .getLogger(CalendarUserDetailsService.class);

    private final CalendarUserDao calendarUserDao;

    public CalendarUserDetailsService(final CalendarUserDao
calendarUserDao) {
        if (calendarUserDao == null) {
            throw new IllegalArgumentException("calendarUserDao cannot
be null");
        }
        this.calendarUserDao = calendarUserDao;
    }


    @Override
    public UserDetails loadUserByUsername(String username) throws
UsernameNotFoundException {
        CalendarUser user = calendarUserDao.findUserByEmail(username);
        if (user == null) {
            throw new UsernameNotFoundException("Invalid username/
password.");
        }
        Collection<? extends GrantedAuthority> authorities =
CalendarUserAuthorityUtils.createAuthorities(user);
        return new User(user.getEmail(), user.getPassword(),
authorities);
    }
}
```

Here, we utilize `CalendarUserDao` to obtain `CalendarUser` by using the email address. We take care not to return a null value; instead, a `UsernameNotFoundException` exception should be thrown, as returning `null` breaks the `UserDetailsService` interface.

We then convert `CalendarUser` into `UserDetails`, implemented by the user, as we did in the previous sections.

We now utilize a utility class named `CalendarUserAuthorityUtils` that we provided in the sample code. This will create `GrantedAuthority` based on the email address so that we can support users and administrators. If the email starts with `admin`, the user is treated as `ROLE_ADMIN` and `ROLE_USER`. Otherwise, the user is treated as `ROLE_USER`. Of course, we would not do this in a real application, but it's this simplicity that allows us to focus on this lesson.

Configuring UserDetailsService

Now that we have a new `UserDetailsService` object, let's update the Spring Security configuration to utilize it. Our `CalendarUserDetailsService` class is added to our Spring configuration automatically since we leverage `classpath` scanning and the `@Component` annotation. This means we only need to update Spring Security to refer to the `CalendarUserDetailsService` class we just created. `userDetailsService()` methods, Spring Security's in-memory implementation of `UserDetailsService` since we are now providing our own `UserDetailsService` implementation.

Update the `SecurityConfig.java` file, as follows to declare a `DelegatingPasswordEncoder` with default mappings. Additional mappings may be added and the encoding will be updated to conform with best practices. However, due to the nature of `DelegatingPasswordEncoder`, the updates should not impact users:

```
@Configuration
@EnableWebSecurity
public class SecurityConfig {

    @Bean
    public SecurityFilterChain filterChain(HttpSecurity http) throws
Exception {
...
    }
    @Bean
    public PasswordEncoder encoder() {
        return PasswordEncoderFactories.
createDelegatingPasswordEncoder();
    }

}
```

Removing references to UserDetailsManager

We need to remove the code we added in `DefaultCalendarService` that used `UserDetailsManager` to synchronize the Spring Security `o.s.s.core.userdetails.User` interface and `CalendarUser`. First, the code is not necessary since Spring Security now refers to `CalendarUserDetailsService`. Second, since we removed the `inMemoryAuthentication()` method, there is no `UserDetailsManager` object defined in our Spring configuration. Go ahead and remove all references to `UserDetailsManager` found in `DefaultCalendarService`. The updates will look similar to the following sample snippets:

Start up the application and see that Spring Security's in-memory `UserDetailsManager` object is no longer necessary (we removed it from our `SecurityConfig.java` file).

> **Important note**
> Your code should now look like `chapter03.03-calendar`.

The CalendarUserDetails object

We have successfully eliminated the need to manage both Spring Security users and our `CalendarUser` objects. However, it is still cumbersome for us to continually need to translate between the two objects. Instead, we will create a `CalendarUserDetails` object, which can be referred to as both `UserDetails` and `CalendarUser`. Update `CalendarUserDetailsService` to use `CalendarUserDetails`, as follows:

```
@Component
public class CalendarUserDetailsService implements UserDetailsService
{

    private final CalendarUserDao calendarUserDao;

    public CalendarUserDetailsService(CalendarUserDao calendarUserDao)
{
        if (calendarUserDao == null) {
            throw new IllegalArgumentException("calendarUserDao cannot
be null");
        }
        this.calendarUserDao = calendarUserDao;
    }


    @Override
    public UserDetails loadUserByUsername(String username) throws
UsernameNotFoundException {
        . . .
```

```
        }
        return new CalendarUserDetails(user);
    }


    private final class CalendarUserDetails extends CalendarUser
implements UserDetails {

        CalendarUserDetails(CalendarUser user) {
            super(user.getId(), user.getFirstName(), user.getLastName(),
user.getEmail(), user.getPassword());
        }

        @Override
        public Collection<? extends GrantedAuthority> getAuthorities()
{
            return CalendarUserAuthorityUtils.createAuthorities(this);
        }

        @Override
        public String getUsername() {
            return getEmail();}

        @Override
        public boolean isAccountNonExpired() {
            return true;}

        @Override
        public boolean isAccountNonLocked() {
            return true;}

        @Override
        public boolean isCredentialsNonExpired() {
            return true; }

        @Override
        public boolean isEnabled() {
            return true;    }
    }
}
```

In the next section, we will see that our application can now refer to the principal authentication on the current `CalendarUser` object. However, Spring Security can continue to treat `CalendarUserDetails` as a `UserDetails` object.

The SpringSecurityUserContext simplifications

We have updated `CalendarUserDetailsService` to return a `UserDetails` object that extends `CalendarUser` and implements `UserDetails`. This means that, rather than having to translate between the two objects, we can simply refer to a `CalendarUser` object. Update `SpringSecurityUserContext` as follows:

```
@Component
public class SpringSecurityUserContext implements UserContext {

    @Override
    public CalendarUser getCurrentUser() {
        SecurityContext context = SecurityContextHolder.getContext();
        Authentication authentication = context.getAuthentication();
        if (authentication == null) {
            return null;
        }
        return (CalendarUser) authentication.getPrincipal();
    }


    @Override
    public void setCurrentUser(CalendarUser user) {
        if (user == null) {
            throw new IllegalArgumentException("user cannot be null");
        }
        Collection<? extends GrantedAuthority> authorities =
CalendarUserAuthorityUtils.createAuthorities(user);
        UsernamePasswordAuthenticationToken authentication = new
UsernamePasswordAuthenticationToken(user,
                user.getPassword(), authorities);
        SecurityContextHolder.getContext().
setAuthentication(authentication);
    }
}
```

The updates no longer require the use of `CalendarUserDao` or Spring Security's `UserDetailsService` interface. Remember our `loadUserByUsername` method from the previous section? The result of this method call becomes the principal of the authentication. Since our updated `loadUserByUsername` method returns an object that extends `CalendarUser`, we can safely cast the principal of the `Authentication` object to `CalendarUser`. We can pass a `CalendarUser` object as the principal into the constructor for `UsernamePasswordAuthenticationToken` when invoking the `setCurrentUser` method. This allows us to still cast the principal to a `CalendarUser` object when invoking the `getCurrentUser` method.

Displaying custom user attributes

Now that `CalendarUser` is populated into Spring Security's authentication, we can update our UI to display the name of the current user rather than the email address. Update the `header.html` file with the following code:

```
//src/main/resources/templates/fragments/header.html
<li class="nav-item">
    <a class="nav-link" href="#">Welcome <span class="navbar-text"

th:text="${#authentication.getPrincipal().getName()}"> </span></a>
</li>
```

Internally, the `"${#authentication.getPrincipal().getName()}"` tag attribute executes the following code. Observe that the highlighted values correlate to the `property` attribute of the `authentication` tag we specified in the `header.html` file:

```
SecurityContext context = SecurityContextHolder.getContext();
Authentication authentication = context.getAuthentication();
CalendarUser user = (CalendarUser) authentication.getPrincipal();
String firstAndLastName = user.getName();
```

Restart the application, visit `http://localhost:8080/`, and log in to view the updates. Instead of seeing the current user's email, you should now see their first and last names.

> **Important note**
> Your code should now look like `chapter03.04-calendar`.

After configuring the `CalendarUserDetailsService` and `UserDetailsService` in the section and simplifying `SpringSecurityUserContext` to display custom user attributes, in the next section, we will explore how to create a custom `AuthenticationProvider`.

Creating a custom AuthenticationProvider object

Spring Security delegates to an `AuthenticationProvider` object to determine whether a user is authenticated or not. This means we can write custom `AuthenticationProvider` implementations to inform Spring Security how to authenticate in different ways. The good news is that Spring Security provides quite a few `AuthenticationProvider` objects, so more often than not you will not need to create one. In fact, up until this point, we have been utilizing Spring Security's `o.s.s.authentication.dao.DaoAuthenticationProvider` object, which compares the username and password returned by `UserDetailsService`.

Creating CalendarUserAuthenticationProvider

Throughout the rest of this section, we are going to create a custom `AuthenticationProvider` object named `CalendarUserAuthenticationProvider` that will replace `CalendarUserDetailsService`. Then, we will use `CalendarUserAuthenticationProvider` to consider an additional parameter to support authenticating users from multiple domains.

> **Important note**
>
> We must use an `AuthenticationProvider` object rather than `UserDetailsService` because the `UserDetails` interface has no concept of a domain parameter.

Create a new class named `CalendarUserAuthenticationProvider`, as follows:

```
//src/main/java/com/packtpub/springsecurity/authentication/
CalendarUserAuthenticationProvider.java
@Component
public class CalendarUserAuthenticationProvider implements
AuthenticationProvider {
    private final CalendarService calendarService;


    public CalendarUserAuthenticationProvider(final CalendarService
calendarService) {
        if (calendarService == null) {
            throw new IllegalArgumentException("calendarService cannot
be null");
        }
        this.calendarService = calendarService;
    }


    @Override
    public Authentication authenticate(final Authentication
authentication) throws AuthenticationException {
        UsernamePasswordAuthenticationToken token =
(UsernamePasswordAuthenticationToken) authentication;
        String email = token.getName();
        CalendarUser user = email == null ? null : calendarService.
findUserByEmail(email);
        if (user == null) {
            throw new UsernameNotFoundException("Invalid username/
password");
        }
        String password = user.getPassword();
        if (!password.equals(token.getCredentials())) {
            throw new BadCredentialsException("Invalid username/
```

```
password");
        }
        Collection<? extends GrantedAuthority> authorities =
CalendarUserAuthorityUtils.createAuthorities(user);
        return new UsernamePasswordAuthenticationToken(user, password,
authorities);
    }

    @Override
    public boolean supports(final Class<?> authentication) {
        return UsernamePasswordAuthenticationToken.class.
equals(authentication);
    }
}
```

> **Important note**
>
> Remember that you can use your **Integrated Development Environment** (IDE) to add the missing imports. Alternatively, you can copy the implementation from chapter03.05-calendar.

Before Spring Security can invoke the authenticate method, the supports method must return true for the Authentication class that will be passed in. In this case, AuthenticationProvider can authenticate a username and password. We do not accept subclasses of UsernamePasswordAuthenticationToken since there may be additional fields that we do not know how to validate.

The authenticate method accepts an Authentication object as an argument that represents an authentication request. In practical terms, it is the input from the user that we need to attempt to validate. If authentication fails, the method should throw an o.s.s.core.AuthenticationException exception. If authentication succeeds, it should return an Authentication object that contains the proper GrantedAuthority objects for the user. The returned Authentication object will be set on SecurityContextHolder. If authentication cannot be determined, the method should return null.

The first step in authenticating the request is to extract the information from the Authentication object that we need to authenticate the user. In our case, we extract the username and lookup CalendarUser by email address, just as CalendarUserDetailsService did. If the provided username and password match CalendarUser, we will return a UsernamePasswordAuthenticationToken object with proper GrantedAuthority. Otherwise, we will throw an AuthenticationException exception.

Remember how the login page leveraged SPRING_SECURITY_LAST_EXCEPTION to explain why the login failed? The message for the AuthenticationException exception thrown in AuthenticationProvider is the last AuthenticationException exception and will be displayed on our login page in the event of a failed login.

Configuring the CalendarUserAuthenticationProvider object

Let's perform the following steps to configure `CalendarUserAuthenticationProvider`:

1. Update the `SecurityConfig.java` file to refer to our newly created `CalendarUserAuthenticationProvider` object, and remove the reference to `CalendarUserDetailsService`, as shown in the following code snippet:

```
//src/main/java/com/packtpub/springsecurity/configuration/
SecurityConfig.java
@EnableWebSecurity
public class SecurityConfig {

    private final CalendarUserAuthenticationProvider cuap;

    public SecurityConfig(CalendarUserAuthenticationProvider
cuap) {
        this.cuap = cuap;
    }


    @Bean
    public AuthenticationManager authManager(HttpSecurity http)
throws Exception {
        AuthenticationManagerBuilder authenticationManagerBuilder
=
                http.getSharedObject(AuthenticationManagerBuilder.
class);
        authenticationManagerBuilder.
authenticationProvider(cuap);
        return authenticationManagerBuilder.build();
    }
    ...
}
```

2. Update the `SecurityConfig.java` file, as follows, by removing the `PasswordEncoder` bean:

```
@Configuration
@EnableWebSecurity
public class SecurityConfig {

    @Bean
    public SecurityFilterChain filterChain(HttpSecurity http)
throws Exception {
        ...
    }
```

```
    // We removed the PasswordEncoder

    }
```

3. Restart the application and ensure everything is still working. As a user, we do not notice anything different. However, as a developer, we know that `CalendarUserDetails` is no longer required; we are still able to display the current user's first and last names, and Spring Security is still able to leverage `CalendarUser` for authentication.

> **Important note**
>
> Your code should now look like `chapter03.05-calendar`.

Authenticating with different parameters

One of the strengths of `AuthenticationProvider` is that it can authenticate with any parameters you wish. For example, maybe your application uses a random identifier for authentication, or perhaps it is a multitenant application and requires a username, password, and domain. In the following section, we will update `CalendarUserAuthenticationProvider` to support multiple domains.

> **Important note**
>
> A domain is a way to scope our users. For example, if we deploy our application once but have multiple clients using the same deployment, each client may want a user with the username `admin`. By adding a domain to our user object, we can ensure that each user is distinct and still supports this requirement.

The DomainUsernamePasswordAuthenticationToken class

When a user authenticates, Spring Security submits an `Authentication` object to `AuthenticationProvider` with the information provided by the user. The current `UsernamePasswordAuthentication` object only contains a username and password field. Create a `DomainUsernamePasswordAuthenticationToken` object that contains a `domain` field, as shown in the following code snippet:

```
//src/main/java/com/packtpub/springsecurity/authentication/
DomainUsernamePasswordAuthenticationToken.java
public final class DomainUsernamePasswordAuthenticationToken extends
        UsernamePasswordAuthenticationToken {
    private final String domain;

    // used for attempting authentication
    public DomainUsernamePasswordAuthenticationToken(String
```

```
            principal, String credentials, String domain) {
        super(principal, credentials);
        this.domain = domain;
    }

    // used for returning to Spring Security after being
    //authenticated
    public DomainUsernamePasswordAuthenticationToken(CalendarUser
            principal, String credentials, String domain,
            Collection<? extends GrantedAuthority> authorities) {
        super(principal, credentials, authorities);
        this.domain = domain;
    }

    public String getDomain() {
        return domain;
    }
}
}
```

Updating CalendarUserAuthenticationProvider

Let's take a look at the following steps for updating the CalendarUserAuthenticationProvider.
java file:

1. Now, we need to update CalendarUserAuthenticationProvider to utilize the
    domain field, as follows:

    ```
    @Component
    public class CalendarUserAuthenticationProvider implements
    AuthenticationProvider {

        private static final Logger logger = LoggerFactory
                .getLogger(CalendarUserAuthenticationProvider.class);

        private final CalendarService calendarService;

        @Autowired
        public CalendarUserAuthenticationProvider(CalendarService
    calendarService) {
            if (calendarService == null) {
                throw new IllegalArgumentException("calendarService
    cannot be null");
            }
            this.calendarService = calendarService;
        }
    ```

```
    @Override
    public Authentication authenticate(Authentication
authentication) throws AuthenticationException {
        DomainUsernamePasswordAuthenticationToken token =
(DomainUsernamePasswordAuthenticationToken) authentication;
        String userName = token.getName();
        String domain = token.getDomain();
        String email = userName + "@" + domain;

        CalendarUser user = calendarService.
findUserByEmail(email);
        logger.info("calendarUser: {}", user);

        if (user == null) {
            throw new UsernameNotFoundException("Invalid username/
password");
        }
        String password = user.getPassword();
        if (!password.equals(token.getCredentials())) {
            throw new BadCredentialsException("Invalid username/
password");
        }
        Collection<? extends GrantedAuthority> authorities =
CalendarUserAuthorityUtils.createAuthorities(user);
        logger.info("authorities: {}", authorities);
        return new
DomainUsernamePasswordAuthenticationToken(user, password,
domain, authorities);
    }

    @Override
    public boolean supports(Class<?> authentication) {
        return DomainUsernamePasswordAuthenticationToken.class.
equals(authentication);
    }
}
```

2.   We first update the support method so that Spring Security will pass `DomainUsername`
     `PasswordAuthenticationToken` into our `authenticate` method.

3.   We then use the domain information to create our email address and authenticate, as we had
     previously done. Admittedly, this example is contrived. However, the example can illustrate
     how to authenticate with an additional parameter.

4. The `CalendarUserAuthenticationProvider` interface can now use the new domain field. However, there is no way for a user to specify the domain. For this, we must update our `login.html` file.

Adding domain to the login page

Open up the `login.html` file and add a new input named `domain`, as follows:

```
//src/main/resources/templates/login.html
<div class="mb-3">
<label class="form-label" for="username">Username</label>
<input autofocus="autofocus" class="form-control" id="username"
        name="username"
        type="text"/>
</div>
<div class="mb-3">
<label class="form-label" for="password">Password</label>
<input class="form-control" id="password" name="password"
        type="password"/>
</div>

<div class="mb-3">
<label class="form-label" for="domain">Domain</label>
<input class="form-control" id="domain" name="domain" type="text"/>
</div>
```

Now, a domain will be submitted when users attempt to log in. However, Spring Security is unaware of how to use that domain to create a `DomainUsernamePasswordAuthenticationToken` object and pass it into `AuthenticationProvider`. To fix this, we will need to create `DomainUsernamePasswordAuthenticationFilter`.

The DomainUsernamePasswordAuthenticationFilter class

Spring Security provides a number of `Servlet Filters` that act as controllers for authenticating users. The filters are invoked as one of the delegates of the `FilterChainProxy` object that we discussed in *Chapter 2, Getting Started with Spring Security*. Previously, the `formLogin()` method instructed Spring Security to use `o.s.s.web.authentication.UsernamePasswordAuthenticationFilter` to act as a login controller. The filter's job is to perform the following tasks:

*   Obtain a username and password from the HTTP request.

*   Create a `UsernamePasswordAuthenticationToken` object with the information obtained from the HTTP request.

*   Request that Spring Security validates `UsernamePasswordAuthenticationToken`.

- If the token is validated, it will set the authentication returned to it on `SecurityContext Holder`, just as we did when a new user signed up for an account. We will need to extend `UsernamePasswordAuthenticationFilter` to leverage our newly created `DoainUsernamePasswordAuthenticationToken` object.

- Create a `DomainUsernamePasswordAuthenticationFilter` object, as follows:

```
//src/main/java/com/packtpub/springsecurity/web/authentication/
DomainUsernamePasswordAuthenticationFilter.java
public final class DomainUsernamePasswordAuthenticationFilter
extends UsernamePasswordAuthenticationFilter {

    public DomainUsernamePasswordAuthenticationFilter(final
AuthenticationManager authenticationManager) {
        super.setAuthenticationManager(authenticationManager);
    }

    public Authentication attemptAuthentication
            (HttpServletRequest request, HttpServletResponse
response) throws
            AuthenticationException {
        if (!request.getMethod().equals("POST")) {
            throw new AuthenticationServiceException
                    ("Authentication method not supported: "
                        + request.getMethod());
        }
        String username = obtainUsername(request);
        String password = obtainPassword(request);
        String domain = request.getParameter("domain");
        DomainUsernamePasswordAuthenticationToken authRequest
            = new
DomainUsernamePasswordAuthenticationToken(username,
            password, domain);
        setDetails(request, authRequest);
        return this.getAuthenticationManager()
            .authenticate(authRequest);
    }
}
```

The new `DomainUsernamePasswordAuthenticationFilter` object will perform the following tasks:

- Obtain a username, password, and domain from the `HttpServletRequest` method.

- Create our `DomainUsernamePasswordAuthenticationToken` object with information obtained from the HTTP request.

- Request that Spring Security validates DomainUsernamePasswordAuthenticationToken. The work is delegated to CalendarUserAuthenticationProvider.

- If the token is validated, its superclass will set the authentication returned by CalendarUserAuthenticationProvider on SecurityContextHolder, just as we did to authenticate a user after they created a new account.

Updating our configuration

Now that we have created all the code required for an additional parameter, we need to configure Spring Security to be aware of it. The following code snippet includes the required updates to our SecurityConfig.java file to support our additional parameter:

```java
//src/main/java/com/packtpub/springsecurity/configuration/
SecurityConfig.java
@Configuration
@EnableWebSecurity
public class SecurityConfig {

    private final CalendarUserAuthenticationProvider cuap;

    public SecurityConfig(CalendarUserAuthenticationProvider cuap) {
        this.cuap = cuap;
    }



    @Bean
    public SecurityFilterChain filterChain(HttpSecurity http,
AuthenticationManager authManager) throws Exception {
        http.authorizeRequests((authz) -> authz
                    .requestMatchers(antMatcher("/webjars/**")).
permitAll()
                    .requestMatchers(antMatcher("/css/**")).permitAll()
                    .requestMatchers(antMatcher("/favicon.ico")).
permitAll()
                    // H2 console:
                    .requestMatchers(antMatcher("/admin/h2/**")).
permitAll()
                    .requestMatchers(antMatcher("/")).permitAll()
                    .requestMatchers(antMatcher("/login/*")).
permitAll()
                    .requestMatchers(antMatcher("/logout")).permitAll()
                    .requestMatchers(antMatcher("/signup/*")).
permitAll()
                    .requestMatchers(antMatcher("/errors/**")).
permitAll()
```

```
                    .requestMatchers(antMatcher("/admin/*")).
hasRole("ADMIN")
                    .requestMatchers(antMatcher("/events/")).
hasRole("ADMIN")
                    .requestMatchers(antMatcher("/**")).
hasRole("USER"))

            .exceptionHandling(exceptions -> exceptions
                .accessDeniedPage("/errors/403")
                .authenticationEntryPoint(new
LoginUrlAuthenticationEntryPoint("/login/form")))
            .logout(form -> form
                .logoutUrl("/logout")
                .logoutSuccessUrl("/login/form?logout")
                .permitAll())
            // CSRF is enabled by default, with Java Config
            .csrf(AbstractHttpConfigurer::disable)
            // Add custom DomainUsernamePasswordAuthenticationFilter
            .
addFilterAt(domainUsernamePasswordAuthenticationFilter(authManager),
UsernamePasswordAuthenticationFilter.class);

    http.securityContext((securityContext) -> securityContext.
requireExplicitSave(false));
    http.headers(headers -> headers.
frameOptions(FrameOptionsConfig::disable));
    return http.build();
}


@Bean
public DomainUsernamePasswordAuthenticationFilter
domainUsernamePasswordAuthenticationFilter(AuthenticationManager
authManager) {
    DomainUsernamePasswordAuthenticationFilter dupaf = new
        DomainUsernamePasswordAuthenticationFilter(authManager);
    dupaf.setFilterProcessesUrl("/login");
    dupaf.setUsernameParameter("username");
    dupaf.setPasswordParameter("password");
    dupaf.setAuthenticationSuccessHandler(new
SavedRequestAwareAuthenticationSuccessHandler() {{
        setDefaultTargetUrl("/default");
    }});
    dupaf.setAuthenticationFailureHandler(new
SimpleUrlAuthenticationFailureHandler() {{
        setDefaultFailureUrl("/login/form?error");
    }});
```

```
        dupaf.afterPropertiesSet();
        return dupaf;
    }


    @Bean
    public AuthenticationManager authManager(HttpSecurity http) throws
Exception {
        AuthenticationManagerBuilder authenticationManagerBuilder =
            http.getSharedObject(AuthenticationManagerBuilder.class);
        authenticationManagerBuilder.authenticationProvider(cuap);
        return authenticationManagerBuilder.build();
    }


}
```

> **Important note**
>
> The preceding code snippet configures standard beans in our Spring Security configuration. We have shown this to demonstrate that it can be done. However, throughout much of the rest of the book, we include standard bean configuration in its own file, as this makes the configuration less verbose. If you are having trouble, or prefer not to type all of this, you may copy it from chapter03.06-calendar.

The following are a few highlights from the configuration updates:

- We overrode defaultAuthenticationEntryPoint and added a reference to o.s.s.web.authentication.LoginUrlAuthenticationEntryPoint, which determines what happens when a request for a protected resource occurs and the user is not authenticated. In our case, we are redirected to a login page.

- We removed the formLogin() method and used a .addFilterAt() method to insert our custom filter into FilterChainProxy. The position indicates the order in which the delegates of FilterChain are considered and cannot overlap with another filter, but can replace the filter at the current position. We replaced UsernamePasswordAuthenticationFilter with our custom filter.

Take a look at the following diagram for your reference:

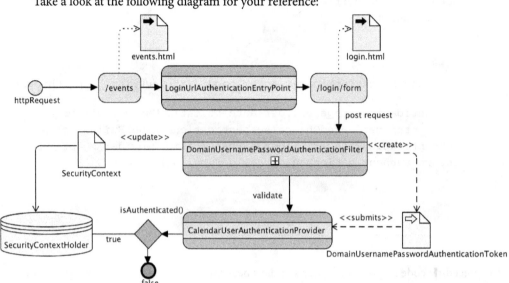

Figure 3.5 – Custom Authentication implementation

You can now restart the application and try the following steps, depicted in the preceding diagram, to understand how all the pieces fit together:

1. Visit http://localhost:8080/events.

2. Spring Security will intercept the secured URL and use the LoginUrlAuthentication EntryPoint object to process it.

3. The LoginUrlAuthenticationEntryPoint object will send the user to the login page. Enter admin1 as the username, example.com as the domain, and admin1 as the password.

4. The DomainUsernamePasswordAuthenticationFilter object will intercept the process of the login request. It will then obtain the username, domain, and password from the HTTP request and create a DomainUsernamePasswordAuthenticationToken object.

5. The DomainUsernamePasswordAuthenticationFilter object submits DomainUsernamePasswordAuthenticationToken to CalendarUser AuthenticationProvider.

6. The CalendarUserAuthenticationProvider interface validates DomainUsername PasswordAuthenticationToken and then returns an authenticated DomainUsername PasswordAuthenticationToken object (that is, isAuthenticated() returns true).

7. The `DomainUserPasswordAuthenticationFilter` object updates `SecurityContext` with `DomainUsernamePasswordAuthenticationToken` and places it on `SecurityContextHolder`.

> **Important note**
>
> Your code should now look like `chapter03.06-calendar`.

Now that we have covered the users sign-up workflow and how to create custom `UserDetailsService` and `AuthenticationProvider` objects, we will discuss in the next section which authentication method should be used.

Which authentication method should you use?

We have covered the three main methods of authenticating, so, which one is the best? Like all solutions, each comes with its pros and cons. You can find a summary of when to use a specific type of authentication by referring to the following list:

- `SecurityContextHolder`: Interacting directly with `SecurityContextHolder` is certainly the easiest way of authenticating a user. It works well when you are authenticating a newly created user or authenticating unconventionally. By using `SecurityContextHolder` directly, we do not have to interact with so many Spring Security layers. The downside is that we do not get some of the more advanced features that Spring Security provides automatically. For example, if we want to send the user to the previously requested page after logging in, we will have to manually integrate that into our controller.

- `UserDetailsService`: Creating a custom `UserDetailsService` object is an easy mechanism that allows Spring Security to make security decisions based on our custom domain model. It also provides a mechanism to hook into other Spring Security features. For example, Spring Security requires `UserDetailsService` to use the built-in remember-me support covered in *Chapter 7, Remember-me Services*. The `UserDetailsService` object does not work when authentication is not based on a username and password.

- `AuthenticationProvider`: This is the most flexible method for extending Spring Security. It allows a user to authenticate with any parameters that they wish. However, if we wish to leverage features such as Spring Security's remember-me, we will still need `UserDetailsService`.

Summary

This chapter has used real-world problems to introduce the basic building blocks used in Spring Security. It also demonstrates to us how we can make Spring Security authenticate against our custom domain objects by extending those basic building blocks. In short, we have learned that the `SecurityContextHolder` interface is the central location for determining the current user. Not only can it be used by developers to access the current user, but also to set the currently logged-in user.

We also explored how to create custom `UserDetailsService` and `AuthenticationProvider` objects and how to perform authentication with more than just a username and password.

In the next chapter, we will explore some of the built-in support for **Java Database Connectivity (JDBC)**-based authentication.

Part 2:
Authentication Techniques

In this part, we explore various authentication methods and services provided by Spring Security. First, we delve into authenticating users against a database using Spring Security's JDBC support. Additionally, we discuss securing passwords with Spring Security's cryptography module to enhance security.

Moving on, we will explore the integration of Spring Data with Spring Security, utilizing JPA to authenticate against relational databases and MongoDB for document databases.

Next, we will introduce the **Lightweight Directory Access Protocol (LDAP)** and its integration with Spring Security. We explore how LDAP can provide authentication, authorization, and user information services within a Spring Security-enabled application.

Then, we uncover the functionality of the remember-me feature in Spring Security and its configuration. Additionally, we address considerations when implementing the remember-me functionality, enabling applications to remember users even after session expiration and browser closure.

Lastly, we will explore alternative authentication methods that are available to accommodate various credential types. We transcend traditional form-based authentication to delve into the realm of authentication using trusted client-side certificates. Spring Security provides robust support for these diverse authentication requirements, offering a framework to implement and manage authentication using client-side certificates, thus enhancing security measures within applications.

This part has the following chapters:

- *Chapter 4, JDBC-based Authentication*
- *Chapter 5, Authentication with Spring Data*
- *Chapter 6, LDAP Directory Services*
- *Chapter 7, Remember-me Services*
- *Chapter 8, Client Certificate Authentication with TLS*

4

JDBC-based Authentication

In the previous chapter, we saw how we can extend Spring Security to utilize our `CalendarDao` interface and our existing domain model to authenticate users. In this chapter, we will see how we can use Spring Security's built-in JDBC support. To keep things simple, this chapter's sample code is based on our Spring Security setup from *Chapter 2, Getting Started with Spring Security*. In this chapter, we will cover the following topics:

- Using Spring Security's built-in JDBC-based authentication support
- Utilizing Spring Security's group-based authorization to make administering users easier
- Learning how to use Spring Security's UserDetailsManager interface
- Configuring Spring Security to utilize the existing CalendarUser schema to authenticate users
- Learning how we can secure passwords using Spring Security's new cryptography module
- Using Spring Security's default JDBC authentication

If your application has not yet implemented security, or if your security infrastructure is using a database, Spring Security provides out-of-the-box support that can simplify the solving of your security needs. Spring Security provides a default schema for users, authorities, and groups. If that does not meet your needs, it allows for the querying and managing of users to be customized. In the next section, we are going to go through the basic steps for setting up JDBC authentication with Spring Security.

This chapter's code in action link is here: `https://packt.link/of0XA`.

Installing the required dependencies

Our application has already defined all the necessary dependencies required for this chapter. However, if you are using Spring Security's JDBC support, you are likely going to want the following dependencies listed in your `build.gradle` file. It is important to highlight that the JDBC driver that you will use will depend on which database you are using. Consult your database vendor's documentation for details on which driver is needed for your database.

> **Important note**
>
> Remember that all the Spring versions need to match, and all Spring Security versions need to match (this includes transitive dependency versions). If you are having difficulty getting this to work in your own application, you may want to define the dependency management section in `build.gradle` to enforce this, as shown in *Chapter 2, Getting Started with Spring Security*. As previously mentioned, you will not need to worry about this when using the sample code, since we have already set up the necessary dependencies for you.

The following snippet defines the required dependencies needed for this chapter, including Spring Security and JDBC dependencies:

```
//build.gradle
dependencies {
    ...
    // spring-jdbc
    implementation 'org.springframework.boot:spring-boot-starter-data-jdbc'
    // H2 db
    implementation 'com.h2database:h2'
    // spring-security
    implementation 'org.springframework.boot:spring-boot-starter-security'
    ...
}
```

The main change here of the `build.gradle`, is to add the `spring-boot-starter-data-jdbc` dependency, to enable the Spring JDBC support.

Using the H2 database

The first portion of this exercise involves setting up an instance of the Java-based H2 which is an open-source, in-memory and embedded relational database written in Java. It is designed to be fast, lightweight, and easy to use. H2 database will be populated with the Spring Security default schema. We'll configure H2 to run in memory using Spring's `EmbeddedDatabase` configuration feature a significantly simpler method of configuration than setting up the database by hand. You can find additional information on the H2 website at `http://www.h2database.com/`.

Keep in mind that in our sample application, we'll primarily use H2 due to its ease of setup. Spring Security will work with any database that supports ANSI SQL out of the box. We encourage you to tweak the configuration and use the database of your preference if you're following along with the examples. As we didn't want this portion of the book to focus on the complexities of database setup, we chose convenience over realism for the purpose of the exercises.

In the following subsections, we will provide sample SQL scripts for our JBCP Calendar application. The scripts will be configured using H2 embedded database.

Finally, we will enable the spring-security support, we will have to add a custom implementation of UserDetailsManager.

Provided JDBC scripts

We've supplied all the SQL files that are used for creating the schema and data in an H2 database for this chapter in the src/main/resources/database/h2/ folder. Any files prefixed with **security** are to support Spring Security's default JDBC implementation. Any SQL files prefixed with calendar are custom SQL files for the JBCP calendar application. Hopefully, this will make running the samples a little easier. If you're following along with your own database instance, you may have to adjust the schema definition syntax to fit your particular database. Additional database schemas can be found in the Spring Security reference. You can find a link to the Spring Security Reference in the book's *Appendix*, *Additional Reference Material*.

Configuring the H2 embedded database

To configure the H2 embedded database, we need to create a DataSource and run SQL to create the Spring Security table structure. We will need to update the SQL that is loaded at startup to include Spring Security's basic schema definition, Spring Security user definitions, and the authority mappings for users. You can find the DataSource definition and the relevant updates in the following code snippet:

```
//src/main/java/com/packtpub/springsecurity/configuration/
DataSourceConfig. Java
@Bean
public DataSource dataSource() {
    return new EmbeddedDatabaseBuilder()
            .setName("dataSource")
            .setType(EmbeddedDatabaseType.H2)
            .addScript("/database/h2/calendar-schema.sql")
            .addScript("/database/h2/calendar-data.sql")
            .addScript("/database/h2/security-schema.sql")
            .addScript("/database/h2/security-users.sql")
            .addScript("/database/h2/security-user-authorities.sql")
            .build();
}
```

Remember that the EmbeddedDatabaseBuilder() method creates this database only in memory, so you won't see anything on the disk, and you won't be able to use standard tools to query it. However, you can use the H2 console that is embedded in the application to interact with the database. See the instructions on the **Welcome** page of our application to learn how to use it.

Configuring a JDBC UserDetailsManager implementation

We'll modify the `SecurityConfig.java` file to declare that we're using a JDBC `User DetailsManager` implementation, instead of the Spring Security in-memory `User DetailsService` implementation that we configured in *Chapter 2, Getting Started with Spring Security*. This is done with a simple change to the `UserDetailsManager` declaration, as follows:

```
//src/main/java/com/packtpub/springsecurity/configuration/
SecurityConfig.java
@Bean
public UserDetailsManager userDetailsService(DataSource dataSource) {
    return new JdbcUserDetailsManager(dataSource);
}
```

We replace the previous `configure(AuthenticationManagerBuilder)` method, along with all of the child elements, with the `userDetailsService()` method, as shown in the preceding code snippet.

In this section, we have been able to configure H2 database with custom `UserDetailsManager` implementation, to enable Spring Security support.

The default user schema of Spring Security

Let's take a look at each of the SQL files used to initialize the database. The first script we added contains the default Spring Security schema definition for users and their authorities. The following script has been adapted from Spring Security's Reference, which is listed in the *Appendix, Additional Reference Material* to have explicitly named constraints, to make troubleshooting easier:

```
//src/main/resources/database/h2/security-schema.sql
create table users
(
    username varchar(256) not null primary key,
    password varchar(256) not null,
    enabled  boolean      not null
);

create table authorities
(
    username  varchar(256) not null,
    authority varchar(256) not null,
    constraint fk_authorities_users foreign key (username) references
users (username)
);
create unique index ix_auth_username on authorities (username,
authority);
```

Defining users

The next script is in charge of defining the users in our application. The included SQL statement creates the same users that we have used throughout the entire book so far. The file also adds an additional user, disabled1@example.com, who will not be able to log in since we indicate the user as disabled:

```
//src/main/resources/database/h2/security-users.sql
insert into users (username, password, enabled)
values ('user1@example.com', '{noop}user1', 1);
insert into users (username, password, enabled)
values ('admin1@example.com', '{noop}admin1', 1);
insert into users (username, password, enabled)
values ('user2@example.com', '{noop}admin1', 1);
insert into users (username, password, enabled)
values ('disabled1@example.com', '{noop}disabled1', 0);
insert into users (username, password, enabled)
values ('admin', '{noop}admin', 1);
```

Defining user authorities

You may have noticed that there is no indication if a user is an administrator or a regular user. The next file specifies a direct mapping of the user to the corresponding authorities. If a user did not have an authority mapped to it, Spring Security would not allow that user to be logged in:

```
//src/main/resources/database/h2/security-user-authorities.sql
insert into authorities(username, authority)
values ('user1@example.com', 'ROLE_USER');
insert into authorities(username, authority)
values ('admin1@example.com', 'ROLE_ADMIN');
insert into authorities(username, authority)
values ('admin1@example.com', 'ROLE_USER');
insert into authorities(username, authority)
values ('user2@example.com', 'ROLE_USER');
insert into authorities(username, authority)
values ('disabled1@example.com', 'ROLE_USER');
```

After the SQL is added to the embedded database configuration, we should be able to start the application and log in. Try logging in with the new user using disabled1@example.com as the username and disabled1 as the password. Notice that Spring Security does not allow the user to log in and provides the error message Reason: User is disabled.

> **Important note**
>
> Your code should now look like this: calendar04.01-calendar.

In this section, we have used the default Spring Security user schema and authorities. In the next section, we will explore how we can define **Group Based Access Control (GBAC)**.

Exploring UserDetailsManager interface

We have already leveraged the `InMemoryUserDetailsManager` class in Spring Security in *Chapter 3*, *Custom Authentication*, to look up the current `CalendarUser` application in our `SpringSecurityUserContext` implementation of `UserContext`. This allowed us to determine which `CalendarUser` should be used when looking up the events for the **My Events** page. *Chapter 3*, *Custom Authentication*, also demonstrated how to update the `DefaultCalendarService.java` file to utilize `InMemoryUserDetailsManager`, to ensure that we created a new Spring Security user when we created `CalendarUser`. This chapter reuses exactly the same code. The only difference is that the `UserDetailsManager` implementation is backed by the `JdbcUserDetailsManager` class of Spring Security, which uses a database instead of an in-memory datastore.

What other features does `UserDetailsManager` provide out of the box?

Although these types of functions are relatively easy to write with additional JDBC statements, Spring Security actually provides out-of-the-box functionality to support many common **Create, Read, Update, and Delete (CRUD)** operations on users in JDBC databases. This can be convenient for simple systems, and a good base to build on for any custom requirements that a user may have:

Method	Description
`void createUser(UserDetails user)`	It creates a new user with the given `UserDetails` information, including any declared `GrantedAuthority` authorities.
`void updateUser(final UserDetails user)`	It updates a user with the given `UserDetails` information. It updates `GrantedAuthority` and removes the user from the user cache.
`void deleteUser(String username)`	It deletes the user with the given username and removes the user from the user cache.
`boolean userExists(String username)`	It indicates whether or not a user (active or inactive) exists with the given username.
`void changePassword(String oldPassword, String newPassword)`	It changes the password of the currently logged-in user. The user must then supply the correct password in order for the operation to succeed.

Table 4.1 – Custom database requirements settings

If UserDetailsManager does not provide all the methods that are necessary for your application, you can extend the interface to provide these custom requirements. For example, if you needed the ability to list all of the possible users in an administrative view, you could write your own interface with this method and provide an implementation that points to the same datastore as the UserDetailsManager implementation you are currently using.

Group-based access control

The JdbcUserDetailsManager class supports the ability to add a level of indirection between the users and the GrantedAuthority declarations by grouping GrantedAuthority into logical sets called groups.

Users are then assigned one or more groups, and their membership confers a set of the Granted Authority declarations:

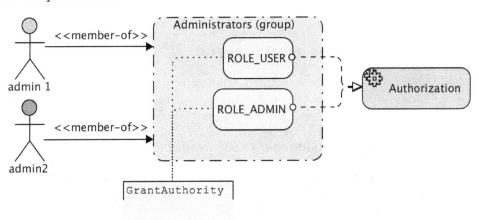

Figure 4.1 – Group-based access control sample

As you can see in the preceding diagram, this indirection allows the assignment of the same set of roles to multiple users, by simply assigning any new users to existing groups. This is different behavior that we've seen so far, where previously we assigned GrantedAuthority directly to individual users.

This bundling of common sets of authorities can be helpful in the following scenarios:

- You need to segregate users into communities, with some overlapping roles between groups.
- You want to globally change the authorization for a class of user. For example, if you have a supplier group, you might want to enable or disable their access to particular portions of the application.
- You have a large number of users, and you don't need user-level authority configuration.

Unless your application has a very small user base, there is a very high likelihood that you'll be using group-based access control. While group-based access control is slightly more complex than other strategies, the flexibility and simplicity of managing a user's access makes this complexity worthwhile. This indirect technique of aggregating user privileges by group is commonly referred to as GBAC.

GBAC is an approach common to almost every secured operating system or software package on the market. **Microsoft Active Directory (AD)** is one of the most visible implementations of large-scale GBAC, due to its design of slotting AD users into groups and assigning privileges to those groups. Management of privileges in large AD-based organizations is made exponentially simpler through the use of GBAC.

Try to think of the security models of the software you use—how are the users, groups, and privileges managed? What are the pros and cons of the way the security model is written?

Let's add a level of abstraction to the JBCP calendar application and apply the concept of group-based authorization to the site.

Configuring group-based access control

We'll add two groups to the application: regular users, which we'll call Users, and administrative users, which we'll call Administrators. Our existing accounts will be associated with the appropriate groups through an additional SQL script.

Configuring JdbcUserDetailsManager to use groups

By default, Spring Security does not use GBAC. Therefore, we must instruct Spring Security to enable the use of groups. Modify the SecurityConfig.java file to use GROUP_AUTHORITIES_BY_USERNAME_QUERY, as follows:

```
//src/main/java/com/packtpub/springsecurity/configuration/
SecurityConfig.ja va
private static String CUSTOM_GROUP_AUTHORITIES_BY_USERNAME_QUERY =
"select g.id, g.group_name, ga.authority " +
        "from groups g, group_members gm, " +
        "group_authorities ga where gm.username = ? " +
        "and g.id = ga.group_id and g.id = gm.group_id";


@Bean
public UserDetailsManager userDetailsService(DataSource dataSource) {
    JdbcUserDetailsManager jdbcUserDetailsManager = new
JdbcUserDetailsManager(dataSource);
    jdbcUserDetailsManager.setEnableGroups(true);
    jdbcUserDetailsManager.setGroupAuthoritiesByUsernameQuery(CUSTOM_
GROUP_AUTHORITIES_BY_USERNAME_QUERY);
```

```
        return jdbcUserDetailsManager;
}
```

Utilizing GBAC JDBC scripts

Next, we need to update the scripts that are being loaded at startup. We need to remove the `security-user-authorities.sql` mapping so that our users no longer obtain their authorities with direct mapping. We then need to add two additional SQL `scripts`. Update the `DataSource` bean configuration to load the SQL required for GBAC, as follows:

```java
//src/main/java/com/packtpub/springsecurity/configuration/
DataSourceConfig. java
@Bean
public DataSource dataSource() {
    return new EmbeddedDatabaseBuilder()
            .setName("dataSource")
            .setType(EmbeddedDatabaseType.H2)
            .addScript("/database/h2/calendar-schema.sql")
            .addScript("/database/h2/calendar-data.sql")
            .addScript("/database/h2/security-schema.sql")
            .addScript("/database/h2/security-users.sql")
            .addScript("/database/h2/security-groups-schema.sql")
            .addScript("/database/h2/security-groups-mappings.sql")
            .build();
}
```

The group-based schema

It may be obvious, but the first SQL file we added contains updates to the schema to support group-based authorization. You can find the contents of the file in the following code snippet:

```sql
//src/main/resources/database/h2/security-groups-schema.sql
create table groups
(
    id          bigint generated by default as identity (start with 0)
primary key,
    group_name varchar(256) not null
);

create table group_authorities
(
    group_id  bigint       not null,
    authority varchar(50) not null,
    constraint fk_group_authorities_group foreign key (group_id)
references groups (id)
```

```
);

create table group_members
(
    id        bigint generated by default as identity (start with 0)
primary key,
    username varchar(50) not null,
    group_id bigint       not null,
    constraint fk_group_members_group foreign key (group_id)
references groups (id)
);
```

Group authority mappings

Now we need to map our existing users to groups, and the groups to authorities. This is done in the
`security-groups-mappings.sql` file. Mapping based on groups can be convenient because
often, organizations already have a logical group of users for various reasons. By utilizing the existing
groupings of users, we can drastically simplify our configuration. This is how a layer of indirection
helps us. We have included the group definitions, group to authority mappings, and a few users in
the following group mapping:

```
//src/main/resources/database/h2/security-groups-mappings.sql
-----
-- Create the Groups
insert into groups(group_name)
values ('Users');
insert into groups(group_name)
values ('Administrators');


-----
-- Map the Groups to Roles
insert into group_authorities(group_id, authority)
select id, 'ROLE_USER'
from groups
where group_name = 'Users';
-- Administrators are both a ROLE_USER and ROLE_ADMIN
insert into group_authorities(group_id, authority)
select id, 'ROLE_USER'
from groups
where group_name = 'Administrators';
insert into group_authorities(group_id, authority)
select id, 'ROLE_ADMIN'
from groups
where group_name = 'Administrators';
```

```
-----
-- Map the users to Groups
insert into group_members(group_id, username)
select id, 'user1@example.com'
from groups
where group_name = 'Users';
insert into group_members(group_id, username)
select id, 'admin1@example.com'
from groups
where group_name = 'Administrators';
insert into group_members(group_id, username)
select id, 'user2@example.com'
from groups
where group_name = 'Users';
insert into group_members(group_id, username)
select id, 'disabled1@example.com'
from groups
where group_name = 'Users';

insert into group_members(group_id, username)
select id, 'admin'
from groups
where group_name = 'Administrators';
```

Go ahead and start the application, and it will behave just as before; however, the additional layer of abstraction between the users and roles simplifies the managing of large groups of users.

> **Important note**
> Your code should now look like this: `calendar04.02-calendar`.

After exploring in this section how we can define GBAC, we will define in the following section a custom database queries to retrieve users and authorities.

Support for a custom schema

It's common for new users of Spring Security to begin their experience by adapting the JDBC user, group, or role mapping to an existing schema. Even though a legacy database doesn't conform to the expected Spring Security schema, we can still configure `JdbcDaoImpl` to map to it.

We will now update Spring Security's JDBC support to use our existing `CalendarUser` database along with a new `calendar_authorities` table.

We can easily change the configuration of `JdbcUserDetailsManager` to utilize this schema and override Spring Security's expected table definitions and columns, which we're using for the JBCP calendar application.

In the following subsections, we will update the SQL user and authorities scripts to insert custom roles. At the end, we will configure `JdbcUserDetailsManager` to use this custom SQL queries.

Determining the correct JDBC SQL queries

The `JdbcUserDetailsManager` class has three SQL queries that have a well-defined parameter and a set of returned columns. We must determine the SQL that we'll assign to each of these queries, based on the intended functionality. Each SQL query used by `JdbcUserDetailsManager` takes the username presented at login as its one and only parameter:

Namespace query attribute name	Description	Expected SQL columns
`users-by-username-query`	Returns one or more users matching the username; only the first user is used.	`Username (string)` `Password (string)` `Enabled (Boolean)`
`authorities-by-username-query`	Returns one or more granted authorities directly provided to the user. Typically used when GBAC is disabled.	`Username (string)` `GrantedAuthority (string)`
`group-authorities-by-username-query`	Returns granted authorities and group details provided to the user through group membership. Used when GBAC is enabled.	`Group Primary Key (any)` `Group Name (any)` `GrantedAuthority (string)`

Table 4.2 – JDBC queries in spring-security

Be aware that in some cases, the return columns are not used by the default `JdbcUserDetailsManager` implementation, but they must be returned anyway.

Updating the SQL scripts that are loaded

We need to initialize the `DataSource` with our custom schema, rather than with Spring Security's default schema. Update the `DataSourceConfig.java` file, as follows:

```
//src/main/java/com/packtpub/springsecurity/configuration/
DataSourceConfig. java
@Bean
public DataSource dataSource() {
    return new EmbeddedDatabaseBuilder()
            .setName("dataSource")
            .setType(EmbeddedDatabaseType.H2)
            .addScript("/database/h2/calendar-schema.sql")
            .addScript("/database/h2/calendar-data.sql")
            .addScript("/database/h2/calendar-authorities.sql")
            .build();
}
```

Notice that we have removed all of the scripts that start with security and replaced them with `calendar-authorities.sql`.

The CalendarUser authority SQL

You can view the `CalendarUser` authority mappings in the following code snippet:

```
//src/main/resources/database/h2/calendar-authorities.sql
create table calendar_user_authorities
(
    id IDENTITY NOT NULL PRIMARY KEY,
    calendar_user bigint        not null,
    authority      varchar(256) not null
);

-- user1@example.com
insert into calendar_user_authorities(calendar_user, authority)
select id, 'ROLE_USER'
from calendar_users
where email = 'user1@example.com';

-- admin1@example.com
insert into calendar_user_authorities(calendar_user, authority)
select id, 'ROLE_ADMIN'
```

```
from calendar_users
where email = 'admin1@example.com';
insert into calendar_user_authorities(calendar_user, authority)
select id, 'ROLE_USER'
from calendar_users
where email = 'admin1@example.com';

-- user2@example.com
insert into calendar_user_authorities(calendar_user, authority)
select id, 'ROLE_USER'
from calendar_users
where email = 'user2@example.com';
```

> **Important note**
>
> Notice that we use the id as the foreign key, which is better than utilizing the username as a foreign key (as Spring Security does). By using the id as the foreign key, we can allow users to easily change their username.

Inserting custom authorities

We need to update `DefaultCalendarService` to insert the authorities for the user using our custom schema when we add a new `CalendarUser` class. This is because while we reused the schema for the user definition, we did not define custom authorities in our existing application. Update `DefaultCalendarService`, as follows:

```
//src/main/java/com/packtpub/springsecurity/service/
DefaultCalendarService. java
@Repository
public class DefaultCalendarService implements CalendarService {
    private final EventDao eventDao;

    private final CalendarUserDao userDao;

    private final JdbcOperations jdbcOperations;
...
    public int createUser(CalendarUser user) {
        int userId = userDao.createUser(user);
        jdbcOperations.update("insert into calendar_user_
authorities(calendar_user,authority) values (?,?)", userId,
            "ROLE_USER");
        return userId;
```

```
        }
    }
```

> **Important note**
>
> You may have noticed the JdbcOperations interface that is used for inserting our user.
> This is a convenient template provided by Spring that helps manage boilerplate code, such
> as connection and transaction handling. For more details, refer to the *Appendix, Additional
> Reference Material* of this book to find the Spring Reference.

Configuring JdbcUserDetailsManager to use custom SQL queries

In order to use custom SQL queries for our non-standard schema, we'll simply update our
userDetailsService() method to include new queries. This is quite similar to how we enabled
support for GBAC, except instead of using the default SQL, we will use our modified SQL. Notice that
we remove our old setGroupAuthoritiesByUsernameQuery() method call, since we will
not be using it in this example, in order to keep things simple:

```
//src/main/java/com/packtpub/springsecurity/configuration/
SecurityConfig.ja va
private static String CUSTOM_USERS_BY_USERNAME_QUERY = "select email,
password, true " +
        "from calendar_users where email = ?";


private static String CUSTOM_AUTHORITIES_BY_USERNAME_QUERY = "select
cua.id, cua.authority " +
        "from calendar_users cu, calendar_user_authorities " +
        "cua where cu.email = ? " +
        "and cu.id = cua.calendar_user";



@Bean
public UserDetailsManager userDetailsService(DataSource dataSource) {
    JdbcUserDetailsManager jdbcUserDetailsManager = new
JdbcUserDetailsManager(dataSource);
    jdbcUserDetailsManager.setUsersByUsernameQuery(CUSTOM_USERS_BY_
USERNAME_QUERY);
    jdbcUserDetailsManager.setAuthoritiesByUsernameQuery(CUSTOM_
AUTHORITIES_BY_USERNAME_QUERY);
    return jdbcUserDetailsManager;
}
```

This is the only configuration required to use Spring Security to read settings from an existing,
non-default schema! Start up the application and ensure that everything is working properly.

> **Important note**
>
> Your code should now look like this: `calendar04.03-calendar`.

Keep in mind that the utilization of an existing schema commonly requires an extension of `JdbcUserDetailsManager` to support the changing of passwords, the renaming of user accounts, and other user-management functions.

If you are using `JdbcUserDetailsManager` to perform user-management tasks, then there are over 20 SQL queries utilized by the class that are accessible through the configuration. However, only the three covered are available through the namespace configuration. Please refer to the Javadoc or source code to review the defaults for the queries used by `JdbcUserDetailsManager`.

Configuring secure passwords

You might recall from the security audit in *Chapter 1, Anatomy of an Unsafe Application*, that the security of passwords stored in cleartext was a top priority of the auditors. In fact, in any secured system, password security is a critical aspect of trust and authoritativeness of an authenticated principal. Designers of a fully secured system must ensure that passwords are stored in a way in which malicious users would have an impractically difficult time compromising them.

The following general rules should be applied to passwords stored in a database:

- Passwords must not be stored in cleartext (plaintext)
- Passwords supplied by the user must be compared to the recorded passwords in the database
- A user's password should not be supplied to the user upon demand (even if the user forgets it)

For the purposes of most applications, the best fit for these requirements involves one-way encoding, known as the **hashing** of the passwords. Using a cryptographic hash provides properties such as security and uniqueness that are important to properly authenticate users, with the added bonus that once it is hashed, the password cannot be extracted from the value that is stored.

In most secure application designs, it is neither required nor desirable to ever retrieve the user's actual password upon request, as providing the user's password to them without the proper additional credentials could present a major security risk. Instead, most applications provide the user the ability to reset their password, either by presenting additional credentials (such as their social security number, date of birth, tax ID, or other personal information), or through an email-based system.

> **Storing other types of sensitive information**
>
> Many of the guidelines listed that apply to passwords apply equally to other types of sensitive information, including social security numbers and credit card information (although, depending on the application, some of these may require the ability to decrypt). Storing this type of information to represent it in multiple ways, for example, a customer's full 16-digit credit card number, would be stored in a highly encrypted form, but the last four digits might be stored in cleartext. For reference, think of any internet commerce site that displays XXXX XXXX XXXX 1234 to help you identify your stored credit cards.

You may already be thinking ahead and wondering, given our admittedly unrealistic approach of using SQL to populate our H2 database with users, how do we encode the passwords? H2, or most other databases for that matter, don't offer encryption methods as built-in database functions.

Typically, the bootstrap process (populating a system with initial users and data) is handled through a combination of SQL loads and Java code. Depending on the complexity of your application, this process can get very complicated.

For the JBCP calendar application, we'll retain the `dataSource()` bean declaration and `DataSource` is a name in code in the corresponding SQL, and then add some SQL that will modify the passwords to their hashed values.

We have seen in this section the best practices for configuring secure password.

In the following section, we will deep dive into the different options to configure secured passwords using the `PasswordEncoder` interface.

Exploring the PasswordEncoder interface

Password hashing in Spring Security is encapsulated and defined by implementations of the `o.s.s.authentication.encoding.PasswordEncoder` interface. The simple configuration of a password encoder is possible through the `createDelegatingPasswordEncoder()` method within the `PasswordEncoderFactories` element, as follows:

```
//src/main/java/com/packtpub/springsecurity/configuration/
SecurityConfig.java
@Bean
public PasswordEncoder encoder() {
    return PasswordEncoderFactories.createDelegatingPasswordEncoder();
}
```

You'll be happy to learn that Spring Security ships with a number of implementations of `passwordEncoder`, which are applicable for different needs and security requirements.

The following table provides a list of the out-of-the-box implementation classes and their benefits.

We can find the complete list of supported encoders in spring security in the `Password EncoderFactories` class. If one of these matches our requirement, we don't need to rewrite it.

Note that all implementations reside in the `o.s.s.crypto` package:

Encoder	Algorithm	Usage
`Pbkdf2PasswordEncoder`	PBKDF2	Provides key strengthening with configurable iteration count, suitable for password hashing. Suitable for password storage.
`SCryptPasswordEncoder`	Scrypt	Memory-hard key derivation function, making it resistant to brute-force attacks. Suitable for password storage.
`StandardPasswordEncoder`	SHA-256	Uses a standard SHA-256 algorithm. Note that SHA-256 alone is not recommended for password hashing due to its speed. Suitable for legacy systems but not recommended for new applications.
`NoOpPasswordEncoder`	No operation	No hashing or encoding; passwords are stored as plain text. Not recommended for production. Useful for testing and development.
`LdapShaPasswordEncoder`	SHA-1	Performs SHA-1 hashing with optional salt. Suitable for compatibility with LDAP directories. Suitable for integration with LDAP-based systems.
`BCryptPasswordEncoder`	BCrypt	One-way hash function with adaptive hashing, suitable for password hashing. Recommended for password storage.
`MessageDigest PasswordEncoder`	Configurable (e.g., MD5, SHA-256, SHA-512)	Uses various message digest algorithms, but the choice of algorithm is crucial for security. Depends on the selected algorithm. Not recommended for new applications due to weaknesses in some algorithms.

Table 4.3 – Main PasswordEncoder implementation

As with many other areas of Spring Security, it's also possible to reference a bean definition by implementing `PasswordEncoder` to provide more precise configuration and

allowing `PasswordEncoder` to be wired into other beans through the dependency injection. For the JBCP calendar application, we'll need to use this bean reference method in order to hash the passwords of the newly created users.

The DelegatingPasswordEncoder implementation

Prior to Spring Security 5.0, the default `PasswordEncoder` was `NoOpPasswordEncoder`, which required plain-text passwords. Based on the Password History section, you might expect that the default `PasswordEncoder` would now be something like `BCryptPasswordEncoder`. However, this ignores three real world problems:

- Many applications use old password encodings that cannot easily migrate.
- The best practice for password storage will change again.
- As a framework, Spring Security cannot make breaking changes frequently.

Instead, Spring Security introduces `DelegatingPasswordEncoder`, which solves all of the problems by:

- Ensuring that passwords are encoded by using the current password storage recommendations.
- Allowing for validating passwords in modern and legacy formats.
- Allowing for upgrading the encoding in the future.

You can easily construct an instance of `DelegatingPasswordEncoder` by using `Password EncoderFactories`:

```
PasswordEncoder passwordEncoder =
PasswordEncoderFactories.createDelegatingPasswordEncoder();
```

Let's walk through the process of configuring basic password encoding for the JBCP calendar application.

Configuring password encoding

Configuring basic password encoding involves two steps: hashing the passwords we load into the database after the SQL script executes and ensuring that Spring Security is configured to work with `PasswordEncoder`.

Configuring the PasswordEncoder method

First, we'll declare an instance of `PasswordEncoder` as a normal Spring bean, as follows:

```
//src/main/java/com/packtpub/springsecurity/configuration/
SecurityConfig.ja va
@Bean
public PasswordEncoder passwordEncoder() {
    String idForEncode = "SHA-256";
    Map<String, PasswordEncoder> encoders = new HashMap<>();
    encoders.put("SHA-256", new org.springframework.security.crypto.
password.MessageDigestPasswordEncoder("SHA-256"));
    return new DelegatingPasswordEncoder(idForEncode, encoders);
}
```

Making Spring Security aware of the PasswordEncoder method

We'll need to configure Spring Security to have a reference to a Bean of type `PasswordEncoder`, so that it can encode and compare the presented password during user login.

If you were to try the application at this point, you'd notice that what were previously valid login credentials would now be rejected. This is because the passwords stored in the database (loaded with the `calendar-users.sql` script) are not stored as a `hash` that matches the password encoder. We'll need to update the stored passwords to be hashed values.

Hashing the stored passwords

As illustrated in the following diagram, when a user submits a password, Spring Security hashes the submitted password and then compares that against the unhashed password in the database:

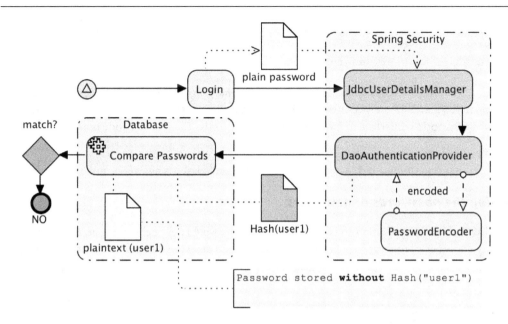

Figure 4.2 – Hashing the stored passwords workflow

This means that users cannot log in to our application. To fix this, we will update the SQL that is loaded at startup time to update the passwords to be the hashed values. Update the `DataSourceConfig.java` file, as follows:

```
//src/main/java/com/packtpub/springsecurity/configuration/
DataSourceConfig. java
@Bean
public DataSource dataSource() {
    return new EmbeddedDatabaseBuilder()
        .setName("dataSource")
        .setType(EmbeddedDatabaseType.H2)
        .addScript("/database/h2/calendar-schema.sql")
        .addScript("/database/h2/calendar-data.sql")
        .addScript("/database/h2/calendar-authorities.sql")
        .addScript("/database/h2/calendar-sha256.sql")
        .build();
}
```

The `calendar-sha256.sql` file simply updates the existing passwords to their expected hashed values, as follows:

```
-- original password was: user1
update calendar_users
```

```
set password = '{SHA-256}0a041b9462caa4a31bac3567e0b6e6fd9100787db2ab4
33d96f6d178cabfce90'
where email = 'user1@example.com';
```

How did we know what value to update the password to? We have provided o.s.s.authentication. encoding.Sha256PasswordEncoderMain to demonstrate how to use the configured PasswordEncoder interface to hash the existing passwords. The relevant code is as follows:

```
ShaPasswordEncoder encoder = new ShaPasswordEncoder(256);
String encodedPassword = encoder.encodePassword(password, null);
```

Hashing the passwords of new users

If we tried running the application and creating a new user, we would not be able to log in. This is because the newly-created user's password would not be hashed. We need to update DefaultCalendarService to hash the password. Make the following updates to ensure that the newly-created users passwords are hashed:

```
//src/main/java/com/packtpub/springsecurity/service/
DefaultCalendarService.java
public class DefaultCalendarService implements CalendarService {
    private final EventDao eventDao;

    private final CalendarUserDao userDao;

    private final JdbcOperations jdbcOperations;

    private final PasswordEncoder passwordEncoder;
...
    public int createUser(CalendarUser user) {
        String encodedPassword = passwordEncoder.encode(user.
getPassword());
        user.setPassword(encodedPassword);
        int userId = userDao.createUser(user);
        jdbcOperations.update("insert into calendar_user_
authorities(calendar_user,authority) values (?,?)", userId,
                "ROLE_USER");
        return userId;
    }
}
```

Not quite secure

Go ahead and start the application. Try creating a new user with user1 as the password. Log out of the application, then use the instructions on the **Welcome** page to open the **H2 console** and view all of the users passwords. Did you notice that the hashed values for the newly created user and `user1@ example.com` are the same value? The fact that we have now figured out another user's password is a little disturbing. We will solve this with a technique known as **salting**.

> **Important note**
>
> Your code should now look like this: `calendar04.04-calendar`.

Would you like some salt with that password? If the security auditor were to examine the encoded passwords in the database, he'd find something that would still make him concerned about the website's security. Let's examine the following stored username and password values for a few of our users:

Username	Plaintext password	Hashed password
admin1@example. com	admin1	{SHA-256}25f43b1486ad95a1 398e3eeb3d83bc4010015fcc9 bedb35b432e00298d5021f7
user1@example. com	user1	{SHA-256}0a041b9462caa4a3 1bac3567e0b6e6fd9100787db 2ab433d96f6d178cabfce90

Table 4.4 – Hashed users passwords

This looks very secure—the encrypted passwords obviously bear no resemblance to the original passwords. What could the auditor be concerned about? What if we add a new user who happens to have the same password as our `user1@example.com` user?

Username	Plaintext password	Hashed password
hacker@example. com	user1	{SHA-256}0a041b9462caa4 a31bac3567e0b6e6fd91007 87db2ab433d96f6d178cabfce90

Table 4.5 – Hacked hashed user password

Now, note that the encrypted password of the `hacker@example.com` user is exactly the same as the real user! Thus, a hacker who had somehow gained the ability to read the encrypted passwords in the database could compare their known password's encrypted representation with the unknown one for the user account, and see they are the same! If the hacker had access to an automated tool to perform this analysis, they could likely compromise the user's account within a matter of hours.

While it is difficult to guess a single password, hackers can calculate all the hashes ahead of time and store a mapping of the hash to the original password. Then, figuring out the original password is a matter of looking up the password by its hashed value in constant time. This is a hacking technique known as **rainbow tables**.

One common and effective method of adding another layer of security to encrypted passwords is to incorporate a **salt**. A salt is a second plaintext component, which is concatenated with the plaintext password prior to performing the hash, in order to ensure that two factors must be used to generate (and thus compare) the hashed password values. Properly selected salts can guarantee that no two passwords will ever have the same hashed value, thus preventing the scenario that concerned our auditor, and avoiding many common types of brute force password cracking techniques.

Best practice salts generally fall into one of the following three categories:

- They are algorithmically generated from some pieces of data associated with the user, for example, the timestamp that the user created

- They are randomly generated and stored in some form

- They are plaintext or two-way encrypted along with the user's password record

Remember that because the **salt** is added to the plaintext password, it can't be one-way encrypted—the application needs to be able to look up or derive the appropriate `salt` value for a given user's record in order to calculate the `hash` of the password, and to compare it with the stored `hash` of the user when performing authentication.

Using salt in Spring Security

Spring Security provides a cryptography module that are included in the `spring-security-core` module and are available separately in `spring-security-crypto`. The **crypto** module contains its own `o.s.s.crypto.password.PasswordEncoder` interface. In fact, using this interface is the preferred method for encoding passwords, because it will salt passwords using a random `salt`. At the time of this writing, there are the following three implementations of `o.s.s.crypto.password.PasswordEncoder`:

Class	Description
`o.s.s.crypto.bcrypt.BCryptPasswordEncoder`	This class uses the `bcrypt` hashing function. It supports `salt` and the ability to slow down to perform over time as technology improves. This helps protect against brute-force search attacks.
`o.s.s.crypto.password.NoOpPasswordEncoder`	This class does no encoding (it returns the password in its plaintext form). Provided for legacy and testing purposes only and is not considered secure.
`o.s.s.crypto.password.StandardPasswordEncoder`	This class uses `SHA-256` with multiple iterations and a random `salt` value. Provided for legacy and testing purposes only and is not considered secure.

Table 4.6 – Common PasswordEncoder implementations

Important note

For those who are familiar with `Spring Security`, The Spring Security Crypto module provides support for symmetric encryption, key generation, and password encoding. The classes above are part of the core module and has no dependencies on any other Spring Security (or Spring) code.

Updating the Spring Security configuration

This can be done by updating the Spring Security configuration. Remove the old `ShaPasswordEncoder` encoder and add the new `StandardPasswordEncoder` encoder, as follows:

```
//src/main/java/com/packtpub/springsecurity/configuration/
SecurityConfig.ja va
@Bean
public PasswordEncoder passwordEncoder() {
    return new StandardPasswordEncoder();
}
```

Migrating existing passwords

Let's take a look at the following steps and learn about migrating existing passwords:

1. We need to update our existing passwords to use the values produced by the new `PasswordEncoder` class. If you would like to generate your own passwords, you can use the following code snippet:

```
StandardPasswordEncoder encoder = new StandardPasswordEncoder();
String encodedPassword = encoder.encode("password");
```

2. Remove the previously used `calendar-sha256.sql` file, and add the provided `saltedsha256.sql` file as follows:

```
//src/main/java/com/packtpub/springsecurity/configuration/
DataSourceConfig.java
@Bean
public DataSource dataSource() {
    return new EmbeddedDatabaseBuilder()
            .setName("dataSource")
            .setType(EmbeddedDatabaseType.H2)
            .addScript("/database/h2/calendar-schema.sql")
            .addScript("/database/h2/calendar-data.sql")
            .addScript("/database/h2/calendar-authorities.sql")
            .addScript("/database/h2/calendar-saltedsha256.sql")
            .build();
}
```

Updating DefaultCalendarUserService

The `passwordEncoder()` method we defined previously is smart enough to handle the new password encoder interface. However, `DefaultCalendarUserService` needs to update to the new interface. Make the following updates to the `DefaultCalendarUserService` class:

```
//src/main/java/com/packtpub/springsecurity/service/
DefaultCalendarService.java
@Repository
public class DefaultCalendarService implements CalendarService {
    private final EventDao eventDao;

    private final CalendarUserDao userDao;

    private final JdbcOperations jdbcOperations;

    private final PasswordEncoder passwordEncoder;
```

```
    public int createUser(CalendarUser user) {
        String encodedPassword = passwordEncoder.encode(user.
getPassword());
        user.setPassword(encodedPassword);
        int userId = userDao.createUser(user);
        jdbcOperations.update("insert into calendar_user_
authorities(calendar_user,authority) values (?,?)", userId,
            "ROLE_USER");
        return userId;
    }
}
```

With the preceding code implementation, we have been able to configure Salt SHA256 in Spring Security. The `DefaultCalendarService` uses this Salt `PasswordEncoder` to insert user's password.

In the next section, we will explore another option to use `Salt` with `Bcrypt` algorithm.

Trying out salted passwords

Start up the application and try creating another user with the password `user1`. Use the H2 console to compare the new user's password and observe that they are different.

> **Important note**
> Your code should now look like this: `calendar04.05-calendar`.

Spring Security now generates a random `salt` and combines this with the password before hashing our password. It then adds the random `salt` to the beginning of the password in plaintext, so that passwords can be checked. The stored password can be summarized as follows:

```
salt = randomsalt()
hash = hash(salt+originalPassword)
storedPassword = salt + hash
```

This is the pseudocode for hashing a newly created password.

To authenticate a user, `salt` and `hash` can be extracted from the stored password, since both `salt` and `hash` are fixed lengths. Then, the extracted `hash` can be compared against a new `hash`, computed with extracted `salt` and the inputted password:

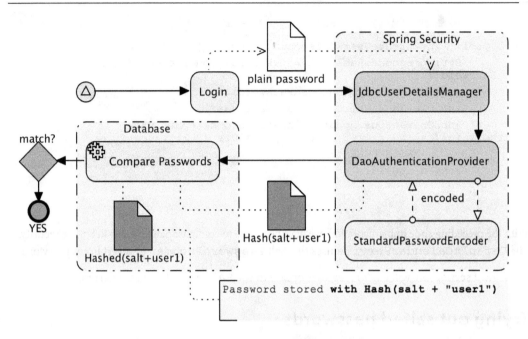

Figure 4.3 – Salting the stored passwords workflow

The following is the pseudocode for validating a salted password:

```
storedPassword = datasource.lookupPassword(username) salt,
expectedHash = extractSaltAndHash(storedPassword) actualHash =
hash(salt+inputedPassword)
authenticated = (expectedHash == actualHash)Trying out salted
passwords with StandardPasswordEncoder
```

BCryptPasswordEncoder is another salt implementation uses the widely supported bcrypt algorithm to hash the passwords. Bcrypt uses a random 16-byte salt value and is a deliberately slow algorithm, to hinder password crackers. You can tune the amount of work it does by using the strength parameter, which takes a value from 4 to 31. The higher the value, the more work has to be done to calculate the hash. The default value is 10. You can change this value in your deployed system without affecting existing passwords, as the value is also stored in the encoded hash. The following example uses the BCryptPasswordEncoder with strength parameter value equal to 4:

```
//src/main/java/com/packtpub/springsecurity/configuration/
SecurityConfig.java
@Bean
public PasswordEncoder passwordEncoder() {
    return new BCryptPasswordEncoder(4);
}
```

In addition, we had to use add the provided `calendar-bcrypt.sql` file as follows:

```
//src/main/java/com/packtpub/springsecurity/configuration/
DataSourceConfig.java
@Bean
public DataSource dataSource() {
    return new EmbeddedDatabaseBuilder()
            .setName("dataSource")
            .setType(EmbeddedDatabaseType.H2)
            .addScript("/database/h2/calendar-schema.sql")
            .addScript("/database/h2/calendar-data.sql")
            .addScript("/database/h2/calendar-authorities.sql")
            .addScript("/database/h2/calendar-bcrypt.sql")
            .build();
}
```

Start up the application and try logging to the application with username `user1@example.com` and password `user1`.

> **Important note**
>
> Your code should now look like this: `calendar04.06-calendar`.

Summary

In this chapter, we learned how to use Spring Security's built-in *JDBC support*. Specifically, we have learned that Spring Security provides a default schema for new applications. We also explored how to implement *GBAC* and how it can make managing users easier.

We also learned how to integrate Spring Security's JDBC support with an existing database and also how to secure our passwords by hashing them and using a randomly generated *salt*.

In the next chapter, we will explore the *Spring Data* project and how to configure Spring Security to use **object-relational mapping (ORM)** to connect to an RDBMS, as well as a document database.

5

Authentication with Spring Data

In the previous chapter, we covered how to leverage Spring Security's built-in **Java DataBase Connectivity (JDBC)** support. In this chapter, we will look at the Spring Data project and how to leverage **Jakarta Persistence API (JPA)** to perform authentication against a relational database. We will also explore how to perform authentication against a document database using **MongoDB**. This chapter's sample code is based on the Spring Security setup from *Chapter 4, JDBC-based Authentication*, and it has been updated to refactor the need for SQL and to use ORM for all database interactions.

During the course of this chapter, we will cover the following topics:

- Some of the basic concepts related to the Spring Data project
- Utilizing Spring Data JPA to authenticate against a relational database
- Utilizing Spring Data MongoDB to authenticate against a document database
- How to customize Spring Security for more flexibility when dealing with Spring Data integration
- Understanding the Spring Data project

The Spring Data project's mission is to provide a familiar and consistent Spring-based programming model for data access, while still retaining the special traits of the underlying data provider.

The following are just a few of the powerful features of this Spring Data project:

- Powerful repository and custom object-mapping abstractions
- Dynamic query derivation from repository method names
- Implementation of domain base classes, providing basic properties
- Support for transparent auditing (created and last changed)
- The ability to integrate custom repository code

- Easy Spring integration via Java-based configuration and custom XML namespaces
- Advanced integration with Spring MVC controllers
- Experimental support for cross-store persistence

This project simplifies the use of data access technologies, relational and non-relational databases, `MapReduce` frameworks, and cloud-based data services. This umbrella project contains many subprojects that are specific to a given database. These projects were developed by working together with many of the companies and developers that are behind these exciting technologies. There are also many community-maintained modules and other related modules, including *JDBC support* and *Apache Hadoop*.

The following table describes the main modules that make up the Spring Data project:

Module	Description
Spring Data Commons	Applies core Spring concepts to all Spring Data projects
Spring Data Gemfire	Provides easy configuration and access to Gemfire from Spring applications
Spring Data JPA	Makes it easy to implement JPA-based repositories
Spring Data Key Value	Map-based repositories and SPIs, which can easily build a Spring Data module for key-value stores
Spring Data LDAP	Provides Spring Data repository support for Spring LDAP
Spring Data MongoDB	Spring-based, object-document support and repositories for MongoDB
Spring Data REST	Exports Spring Data repositories as hypermedia-driven RESTful resources
Spring Data Redis	Provides easy configuration and access to Redis from Spring applications
Spring Data for Apache Cassandra	A Spring Data module for Apache Cassandra
Spring Data for Apache Solr	A Spring Data module for Apache Solr

Table 5.1 – Main modules of the Spring Data project

After exploring the core modules of the Spring Data project, let's now delve into the primary features of Spring Data JPA.

This chapter's code in action link is here: `https://packt.link/omOQK`.

Spring Data JPA

The Spring Data JPA project aims to significantly improve the ORM implementation of data access layers by reducing the effort to the amount that's actually needed. A developer only needs to write repository interfaces, including custom finder methods, and Spring will provide the implementation automatically.

The following are just a few of the powerful features specific to the Spring Data JPA Project:

- Sophisticated support for building repositories based on Spring and JPA

- Support for QueryDSL predicates and, thus, type-safe JPA queries

- Transparent auditing of domain classes

- Pagination support, dynamic query execution, and the ability to integrate custom data access code

- Validation of @Query-annotated queries at bootstrap time

- Support for XML based entity mapping

- The JavaConfig based repository configuration by introducing @EnableJpaRepositories

Updating our dependencies

We have already included all the dependencies you need for this chapter, so you will not need to make any updates to your build.gradle file. However, if you are just adding Spring Data JPA support to your application, you need to add spring-boot- starter-data-jpa as a dependency in the build.gradle file, as follows:

```
//build.gradle
dependencies {
    // JPA / ORM / Hibernate:
    implementation 'org.springframework.boot:spring-boot-starter-data-
jpa'
...
}
```

Notice haven't removed the spring-boot-starter-jdbc dependency.

The spring-boot-starter-data-jpa dependency will contain all the dependencies needed to wire our domain objects to our embedded database with JPA.

Reconfiguring the database configuration

Firstly, we will convert the current JBCP calendar project. Let's begin by reconfiguring the database.

We can begin by removing the DataSourceConfig.java file, as we will be leveraging Spring Boot's built-in support for an embedded H2 database.

Initializing the database

We can now remove the `src/main/resources/database` directory and all contents in that directory. This directory contains several `.sql` files.

Now, we need to create a `data.sql` file that will contain our seed data, as follows:

- Take a look at the following SQL statement, depicting the password for **user1**:

```
//src/main/resources/data.sql
insert into calendar_users(id,email,password,first_
name,last_name) values (0, 'user1@example.
com','$2a$04$qr7RWyqOnWWC1nwotUW1nOe1RD5.
mKJVHK16WZy6v49pymu1WDHmi','User','1');
```

- Take a look at the following SQL statement, depicting the password for **admin1**:

```
insert into calendar_users(id,email,password,first_
name,last_name) values (1,'admin1@example.com','$2a$04$0CF/
Gsquxlel3fWq5Ic/ZOGDCaXbMfXYiXsviTNMQofWRXhvJH3IK','Admin','1');
```

- Take a look at the following SQL statement, depicting the password for **user2**:

```
insert into calendar_users(id,email,password,first_
name,last_name) values (2,'user2@example.
com','$2a$04$PiVhNPAxunf0Q4IMbVeNIuH4M4ecySWHihyrclxW..
PLArjLbg8CC','User2','2');
```

- Take a look at the following SQL statement, depicting the user roles:

```
insert into role(id, name) values (0, 'ROLE_USER');
insert into role(id, name) values (1, 'ROLE_ADMIN');
```

- Here, **user1** has one role:

```
insert into user_role(user_id,role_id) values (0, 0);
```

- Here, **admin1** has two roles:

```
insert into user_role(user_id,role_id) values (1, 0);
insert into user_role(user_id,role_id) values (1, 1);
```

- Take a look at the following SQL statement, depicting events:

```
insert into events (id,date_
when,summary,description,owner,attendee) values (100,'2023-
07-03 20:30:00','Birthday Party','This is going to be a great
birthday',0,1);
insert into events (id,date_
when,summary,description,owner,attendee) values (101,'2023-12-23
13:00:00','Conference Call','Call with the client',2,0);
```

```
insert into events (id,date_
when,summary,description,owner,attendee) values (102,'2023-09-14
11:30:00','Vacation','Paragliding in Greece',1,2);
```

Now, we can update the application properties to define our embedded database properties in the `src/main/resources/application.yml` file, as follows:

```
datasource:
  url: jdbc:h2:mem:dataSource;DB_CLOSE_DELAY=-1;DB_CLOSE_ON_EXIT=FALSE
  driverClassName: org.h2.Driver
  username: sa
  password:

jpa:
  database-platform: org.hibernate.dialect.H2Dialect
  show-sql: true
  hibernate:
    ddl-auto: create-drop
```

At this point, we have removed the old database configuration and added the new configuration. The application will not work at this point, but this can still be considered a marker point before we continue to the next steps of conversion.

> **Important note**
> Your code should now look like this: `calendar05.01-calendar`.

Refactoring from SQL to ORM

Refactoring from an SQL to an ORM implementation is simpler than you might think. Most of the refactoring involves the removal of excess code in the form of an SQL. In this next section, we will refactor our SQL implementation to a JPA implementation.

For JPA to map our domain objects to our database, we need to perform some mapping on our domain objects.

Mapping domain objects using JPA

Take a look at the following steps to learn about mapping the domain objects:

1. Let's begin by mapping our `Event.java` file so that all the domain objects will use JPA, as follows:

    ```
    //src/main/java/com/packtpub/springsecurity/domain/Event.java
    @Entity
    ```

```
@Table(name = "events")
public class Event implements Serializable{

    @Id
    @GeneratedValue(strategy = GenerationType.IDENTITY)
    private Integer id;
    @NotEmpty(message = "Summary is required")
    private String summary;
    @NotEmpty(message = "Description is required")
    private String description;
    @NotNull(message = "When is required")
    private Calendar dateWhen;



    @NotNull(message = "Owner is required")
    @ManyToOne(fetch = FetchType.LAZY)
    @JoinColumn(name="owner", referencedColumnName="id")
    private CalendarUser owner;
    @ManyToOne(fetch = FetchType.LAZY)
    @JoinColumn(name="attendee", referencedColumnName="id")
    private CalendarUser attendee;

    ...

}
```

2. We need to create a Role.java file with the following contents:

```
//src/main/java/com/packtpub/springsecurity/domain/Role.java
@Entity
@Table(name = "role")
public class Role implements Serializable {

    @Id
    @GeneratedValue(strategy = GenerationType.AUTO)
    private Integer id;
    private String name;

    @ManyToMany(fetch = FetchType.EAGER, mappedBy = "roles")
    private Set<CalendarUser> users;

    ...

}
```

3. The Role object will be used to map authorities to our CalendarUser table. Let's map our CalendarUser.java file, now that we have a Role.java file:

```
//src/main/java/com/packtpub/springsecurity/domain/CalendarUser.
java

@Entity
@Table(name = "calendar_users")
public class CalendarUser implements Principal, Serializable {

    private static final long serialVersionUID =
8433999509932007961L;

    @Id
    @SequenceGenerator(name = "user_id_seq", initialValue =
1000)
    @GeneratedValue(generator = "user_id_seq")
    private Integer id;

    private String firstName;

    private String lastName;

    private String email;

    private String password;

    @ManyToMany(fetch = FetchType.EAGER)
    @JoinTable(name = "user_role",
            joinColumns = @JoinColumn(name = "user_id"),
            inverseJoinColumns = @JoinColumn(name = "role_id"))
    private Set<Role> roles;
    ...
}
```

At this point, we have mapped our domain objects with the required JPA annotation, including @Entity and @Table to define the **Relational Database Management System (RDBMS)** location, as well as structural, reference, and association mapping annotations.

At this stage, you can also **remove** the following dependency:

```
//build.gradle
dependencies {
...
implementation 'org.springframework.boot:spring-boot-starter-data-
jdbc'
...
}
```

The application will not work at this point, but this can still be considered a marker point before we continue to the next steps of conversion.

> **Important note**
>
> Your code should now look like this: `calendar05.02-calendar`.

Spring Data repositories

We will now add the required interfaces for Spring Data to map our required **Create, Read, Update, and Delete (CRUD)** operations to our embedded database, by performing the following steps:

1. We begin by adding a new interface to a new package, which will be `com.packtpub.springsecurity.repository`. The new file will be called `CalendarUserRepository.java`, as follows:

    ```
    //com/packtpub/springsecurity/repository/CalendarUserRepository.
    java
    public interface CalendarUserRepository extends
    JpaRepository<CalendarUser, Integer> {

        CalendarUser findByEmail(String email);

    }
    ```

2. We can now continue by adding a new interface to the same repository package, which will be `com.packtpub.springsecurity.repository`, and the new file will be called `EventRepository.java`:

    ```
    //com/packtpub/springsecurity/repository/EventRepository.java
    public interface EventRepository extends JpaRepository<Event,
    Integer> {

    }
    ```

This will allow for standard CRUD operations such as find(), save(), and delete() on our Event objects.

3. Finally, we will add a new interface to the same repository package, which will be com.packtpub. springsecurity.repository, and the new file will be called RoleRepository. java. This CrudRepository interface will be used to manage the Role object for our security roles associated with a given CalendarUser:

```
//com/packtpub/springsecurity/repository/RoleRepository.java

public interface RoleRepository extends JpaRepository<Role, Integer> {

}
```

This will allow for standard CRUD operations such as find(), save(), and delete() on our Role objects.

Data access objects

We need to refactor the JdbcEventDao.java file with a new name, JpaEventDao.java, so that we can replace the JDBC SQL code with our new Spring Data code. Let's take a look at the following steps:

1. Specifically, we need to add the new EventRepository interface and replace the SQL code with the new ORM repository, as shown in the following code:

```
//com/packtpub/springsecurity/dataaccess/JpaEventDao.java
@Repository
public class JpaEventDao implements EventDao {

    // --- members ---

    private EventRepository repository;

    // --- constructors ---

    public JpaEventDao(EventRepository repository) {
        if (repository == null) {
            throw new IllegalArgumentException("repository
    cannot be null");
        }
        this.repository = repository;
    }

    // --- EventService ---
```

```java
    @Override
    @Transactional(readOnly = true)
    public Event getEvent(int eventId) {
        return repository.findById(eventId).orElse(null);
    }

    @Override
    public int createEvent(final Event event) {
        if (event == null) {
            throw new IllegalArgumentException("event cannot be
null");
        }
        if (event.getId() != null) {
            throw new IllegalArgumentException("event.getId()
must be null when creating a new Message");
        }
        final CalendarUser owner = event.getOwner();
        if (owner == null) {
            throw new IllegalArgumentException("event.getOwner()
cannot be null");
        }
        final CalendarUser attendee = event.getAttendee();
        if (attendee == null) {
            throw new IllegalArgumentException("attendee.
getOwner() cannot be null");
        }
        final Calendar when = event.getDateWhen();
        if(when == null) {
            throw new IllegalArgumentException("event.getWhen()
cannot be null");
        }
        Event newEvent = repository.save(event);
        return newEvent.getId();
    }

    @Override
    @Transactional(readOnly = true)
    public List<Event> findForUser(final int userId) {
        Event example = new Event();
        CalendarUser cu = new CalendarUser();
        cu.setId(userId);
        example.setOwner(cu);

        return repository.findAll(Example.of(example));
```

```
        }

        @Override
        @Transactional(readOnly = true)
        public List<Event> getEvents() {
            return repository.findAll();
        }

    }
```

2. At this point, we need to refactor the DAO classes to support the new `CrudRepository` interfaces we have created. Let's begin by refactoring the `JdbcCalendarUserDao.java` file. First, we can rename the file `JpaCalendarUserDao.java` to indicate that this uses JPA, not standard JDBC:

```java
//com/packtpub/springsecurity/dataaccess/JpaCalendarUserDao.java
@Repository
public class JpaCalendarUserDao implements CalendarUserDao {

    private static final Logger logger = LoggerFactory
            .getLogger(JpaCalendarUserDao.class);

    // --- members ---

    private CalendarUserRepository userRepository;
    private RoleRepository roleRepository;

    // --- constructors ---

    public JpaCalendarUserDao(final CalendarUserRepository repository,
                              final RoleRepository roleRepository) {
        if (repository == null) {
            throw new IllegalArgumentException("repository
cannot be null");
        }
        if (roleRepository == null) {
            throw new IllegalArgumentException("roleRepository
cannot be null");
        }

        this.userRepository = repository;
        this.roleRepository = roleRepository;
```

```
    }

    // --- CalendarUserDao methods ---

    @Override
    @Transactional(readOnly = true)
    public CalendarUser getUser(final int id) {
        return userRepository.findById(id).orElse(null);
    }

    @Override
    @Transactional(readOnly = true)
    public CalendarUser findUserByEmail(final String email) {
        if (email == null) {
            throw new IllegalArgumentException("email cannot be
null");
        }
        try {
            return userRepository.findByEmail(email);
        } catch (EmptyResultDataAccessException notFound) {
            return null;
        }
    }

    @Override
    @Transactional(readOnly = true)
    public List<CalendarUser> findUsersByEmail(final String
email) {
        if (email == null) {
            throw new IllegalArgumentException("email cannot be
null");
        }
        if ("".equals(email)) {
            throw new IllegalArgumentException("email cannot be
empty string");
        }
        return userRepository.findAll();
    }

    @Override
    public int createUser(final CalendarUser userToAdd) {
        if (userToAdd == null) {
            throw new IllegalArgumentException("userToAdd cannot
be null");
```

```
        }
        if (userToAdd.getId() != null) {
            throw new IllegalArgumentException("userToAdd.
getId() must be null when creating a "+CalendarUser.class.
getName());
        }

        Set<Role> roles = new HashSet<>();
        roles.add(roleRepository.findById(0).orElse(null));
        userToAdd.setRoles(roles);

        CalendarUser result = userRepository.save(userToAdd);
        userRepository.flush();

        return result.getId();
    }

}
```

In the preceding code, the update fragments to leverage the JPA repositories have been placed in bold, so now the Event and CalendarUser objects are mapped to our underlying RDBMS.

The application will not work at this point, but this can still be considered a marker point before we continue to the next steps of conversion.

> **Important note**
> At this point, your source code should look the same as chapter05.03- calendar.

Application services

The only thing left to do is configure Spring Security to use the new artifacts.

We need to edit the DefaultCalendarService.java file and only remove the remaining code that was used to add USER_ROLE to any new User object that was created, as follows:

```
//com/packtpub/springsecurity/service/DefaultCalendarService.java
@Repository
public class DefaultCalendarService implements CalendarService {
... omitted for brevity ...
  public int createUser(CalendarUser user) {
    String encodedPassword = passwordEncoder.encode(user.
getPassword());
    user.setPassword(encodedPassword);
    int userId = userDao.createUser(user);
```

```
        return userId;
    }
}
```

The UserDetailsService object

Let's take a look at the following steps to add the UserDetailsService object:

1. Now, we need to add a new implementation of the UserDetailsService object; we will use our CalendarUserRepository interface to authenticate and authorize users again, with the same underlying RDBMS, but using our new JPA implementation, as follows:

    ```
    //com/packtpub/springsecurity/service/
    CalendarUserDetailsService.java

    @Component
    public class CalendarUserDetailsService implements
    UserDetailsService {
    @Override
    public UserDetails loadUserByUsername(String username) throws
    UsernameNotFoundException {
        CalendarUser user = calendarUserDao.
    findUserByEmail(username);
        if (user == null) {
            throw new UsernameNotFoundException("Invalid username/
    password.");
        }
        return new CalendarUserDetails(user);
      }
    }
    ```

2. Now, we have to configure Spring Security to use our custom UserDetailsService object, as follows:

    ```
    //com/packtpub/springsecurity/configuration/SecurityConfig.java
    @Configuration
    @EnableWebSecurity
    public class SecurityConfig {
    ... omitted for brevity ...
        @Bean
        public AuthenticationManager authManager(HttpSecurity http)
    throws Exception {
            AuthenticationManagerBuilder
    authenticationManagerBuilder =
                http.getSharedObject(AuthenticationManagerBuilder.
    class);
    ```

```
            return authenticationManagerBuilder.build();
        }
    }
    ...
    }
```

3. Start the application and try logging in to the application. Any of the configured users can now log in and create new events. You can also create a new user and will be able to log in as this new user immediately.

> **Important note**
>
> Your code should now look like `calendar05.04-calendar`.

Refactoring from an RDBMS to a document database

Luckily, with the Spring Data project, once we have a Spring Data implementation, we have most of the difficult work completed. Now, there are only a few implementation-specific changes that need to be refactored.

Document database implementation with MongoDB

We are now going to work on refactoring our RDBMS implementation—with JPA as our ORM provider—to a document database implementation, using MongoDB as our underlying database provider. MongoDB is a free and open-source cross-platform document-oriented database program. Classified as a NoSQL database program, MongoDB uses JSON-like documents with schemas. MongoDB is developed by MongoDB Inc. and is located at `https://github.com/mongodb/mongo`.

Updating our dependencies

We have already included all of the dependencies you need for this chapter, so you will not need to make any updates to your `build.gradle` file. However, if you are just adding Spring Data JPA support to your own application, you will need to add `spring-boot-starter-data-jpa` as a dependency to the `build.gradle` file, as follows:

```
//build.gradle
dependencies {
// MondgoDB
implementation 'org.springframework.boot:spring-boot-starter-data-
mongodb'
```

```
implementation 'de.flapdoodle.embed:de.flapdoodle.embed.mongo.
spring30x:4.9.2'


}
```

Notice that we removed the `spring-boot-starter-jpa` dependency. The `spring-boot-starter-data-mongodb` dependency will contain all the dependencies needed to wire our domain objects to our embedded MongoDB database, with a mix of Spring and MongoDB annotations.

We also added the `Flapdoodle-embedded` MongoDB database, but this is only meant for testing and demonstration purposes. Embedded MongoDB will provide a platform-neutral way to run MongoDB in unit tests. This embedded database is located at `https://github.com/flapdoodle-oss/de.flapdoodle.embed.mongo`.

Reconfiguring the database configuration in MongoDB

First, we will begin to convert the current JBCP calendar project. Let's begin by reconfiguring the database to use the Flapdoodle-embedded MongoDB database. Previously, when we updated the dependencies for this project, we added a Flapdoodle dependency that gave the project an embedded MongoDB database, which we could automatically use instead of installing a full version of MongoDB installation. To stay consistent with the JBCP application, we need to change the name of our database. With Spring Data, we can change the MongoDB configuration using the YAML configuration, as follows:

```
//src/main/resources/application.yml
spring:
  ## Thymeleaf configuration:
  thymeleaf:
    cache: false
    mode: HTML

  # MongoDB
  data:
    mongodb:
      host: localhost
      database: dataSource

de:
  flapdoodle:
    mongodb:
      embedded:
        version: 7.0.0
```

The most important configuration for our current requirements is changing the database name to `dataSource`, which is the same name we used throughout this book.

Initializing the MongoDB database

With the JPA implementation, we used the `data.sql` file to initialize the data in our database. For MongoDB implementation, we can remove the `data.sql` file and replace it with a Java configuration file, which we will call `MongoDataInitializer.java`:

```
//src/main/java/com/packtpub/springsecurity/configuration/
MongoDataInitializer.java
@Configuration
public class MongoDataInitializer {

    private static final Logger logger = LoggerFactory
            .getLogger(MongoDataInitializer.class);

    private RoleRepository roleRepository;

    private CalendarUserRepository calendarUserRepository;

    private EventRepository eventRepository;

    public MongoDataInitializer(RoleRepository roleRepository,
    CalendarUserRepository calendarUserRepository, EventRepository
    eventRepository) {
        this.roleRepository = roleRepository;
        this.calendarUserRepository = calendarUserRepository;
        this.eventRepository = eventRepository;
    }

    @PostConstruct
    public void setUp() {
    }

    CalendarUser user, admin, user2;

    // CalendarUsers
    {
        user = new CalendarUser(0, "user1@
    example.com","$2a$04$qr7RWyqOnWWC1nwotUW1nOe1RD5.
    mKJVHK16WZy6v49pymu1WDHmi","User","1");
        admin = new CalendarUser(1,"admin1@example.com","$2a$04$0CF/
    Gsquxlel3fWq5Ic/ZOGDCaXbMfXYiXsviTNMQofWRXhvJH3IK","Admin","1");
        user2 = new CalendarUser(2,"user2@example.
    com","$2a$04$PiVhNPAxunf0Q4IMbVeNIuH4M4ecySWHihyrclxW..
    PLArjLbg8CC","User2","2");

    }
```

```
Role user_role, admin_role;

private void seedRoles(){
    user_role = new Role(0, "ROLE_USER");
    user_role = roleRepository.save(user_role);

    admin_role = new Role(1, "ROLE_ADMIN");
    admin_role = roleRepository.save(admin_role);
}

private void seedEvents(){

    // Event 1
    Event event1 = new Event(
            100,
            "Birthday Party",
            "This is going to be a great birthday",
        LocalDateTime.of(2023, 6,3,6,36,00),
            user,
            admin
            );

    // Event 2
    Event event2 = new Event(
            101,
            "Conference Call",
            "Call with the client",
        LocalDateTime.of(2023, 11,23,13,00,00),
            user2,
            user
            );

    // Event 3
    Event event3 = new Event(
            102,
            "Vacation",
            "Paragliding in Greece",
        LocalDateTime.of(2023, 8,14,11,30,00),
            admin,
            user2
            );
```

```
        // save Event
        eventRepository.save(event1);
        eventRepository.save(event2);
        eventRepository.save(event3);

        List<Event> events = eventRepository.findAll();

        logger.info("Events: {}", events);

    }

    private void seedCalendarUsers(){

        // user1
        user.addRole(user_role);

        // admin2
        admin.addRole(user_role);
        admin.addRole(admin_role);

        // user2
        user2.addRole(user_role);

        // CalendarUser
        calendarUserRepository.save(user);
        calendarUserRepository.save(admin);
        calendarUserRepository.save(user2);

        List<CalendarUser> users = calendarUserRepository.findAll();

        logger.info("CalendarUsers: {}", users);
    }

}
```

This will be executed at load time and will seed the same data into our MongoDB as we did with our H2 database.

Mapping domain objects with MongoDB

Let's begin by mapping our Event.java file so that each of the domain objects is saved as a document in our MongoDB database. This can be done by performing the following steps:

1. With a document database, domain object mapping is a little different, but the same ORM concepts hold true. Let's begin with the Event JPA implementation, and then we can transform our Entity to document mapping:

```java
//src/main/java/com/packtpub/springsecurity/domain/Event.java
@Entity
@Table(name = "events")
public class Event implements Serializable{

    @Id
    @GeneratedValue(strategy = GenerationType.AUTO)
    private Integer id;
    @NotEmpty(message = "Summary is required")
    private String summary;
    @NotEmpty(message = "Description is required")
    private String description;
    @NotNull(message = "When is required")
    private Calendar dateWhen;


    @NotNull(message = "Owner is required")
    @ManyToOne(fetch = FetchType.LAZY)
    @JoinColumn(name="owner", referencedColumnName="id")
    private CalendarUser owner;
    @ManyToOne(fetch = FetchType.LAZY)
    @JoinColumn(name="attendee", referencedColumnName="id")
    private CalendarUser attendee;
    ...
}
```

2. In entity-based JPA mapping, we needed to use six different annotations to create the required mapping. Now, with document-based MongoDB mapping, we need to change all the previous mapping annotations. Here is a fully refactored example of our Event.java file:

```java
//src/main/java/com/packtpub/springsecurity/domain/Event.java
@Document(collection="events")
public class Event implements Persistable<Integer>,
Serializable{

    @Id
```

```
        private Integer id;
        @NotEmpty(message = "Summary is required")
        private String summary;
        @NotEmpty(message = "Description is required")
        private String description;
        @NotNull(message = "When is required")
        private LocalDateTime dateWhen;


        @NotNull(message = "Owner is required")
        @DBRef
        private CalendarUser owner;
        @DBRef
        private CalendarUser attendee;
    ...
    }
```

In the preceding code, we can see the following notable changes.

3. First, we declare the class to be of type `@o.s.d.mongodb.core.mapping.Document` and provide a collection name for these documents.

4. Next, the Event class must implement the `o.s.d.domain.Persistable` interface, providing the primary key type (`Integer`) for our document.

5. Now, we change the annotation for our domain ID to `@o.s.d.annotation.Id`, to define the domain primary key.

6. Previously, we had to map our owner and attendee `CalendarUser` object to two different mapping annotations.

    Now, we only have to define the two types to be of type `@o.s.d.mongodb.core.mapping.DBRef` and allow Spring Data to take care of the underlying references.

7. The final annotation we have to add defines a specific constructor to be used for new documents to be added to our document, by using the `@o.s.d.annotation.PersistenceConstructor` annotation.

8. Now that we have reviewed the changes needed to refactor from JPA to MongoDB, let's refactor the other domain object, starting with the `Role.java` file, as follows:

```
//src/main/java/com/packtpub/springsecurity/domain/Role.java
@Document(collection="role")
public class Role  implements Persistable<Integer>, Serializable
{

    @Id
    private Integer id;
```

```
            private String name;
    ...
    }
```

9. The final domain object that we need to refactor is our `CalendarUser.java` file. After all, this is the most complex domain object we have in this application:

```
//src/main/java/com/packtpub/springsecurity/domain/CalendarUser.
java
@Document(collection="calendar_users")
public class CalendarUser implements Persistable<Integer>,
Serializable {

    @Id
    private Integer id;

    private String firstName;
    private String lastName;
    private String email;
    private String password;

    @DBRef(lazy = false)
    private Set<Role> roles = new HashSet<>(5);
    ...
}
```

As you can see, the effort to refactor our domain objects from JPA to MongoDB is fairly simple, requiring less annotation configuration than the JPA configuration.

Spring Data repositories of MongoDB

We now have only a few changes to make to refactor from a JPA implementation to a MongoDB implementation. We will begin by refactoring our `CalendarUserRepository.java` file by changing the interface that our repository extends, as follows:

```
//com/packtpub/springsecurity/repository/CalendarUserRepository.java
public interface CalendarUserRepository extends
MongoRepository<CalendarUser, Integer> {

    CalendarUser findByEmail(String email);

}
...
```

sw `RoleRepository.java` files accordingly.

Important note

If you need help with any of these changes, remember the source for `chapter05.05` will have the completed code available for your reference.

Data access objects in MongoDB

In our `EventDao` interface, we are required to create a new `Event` object. With JPA, we can automatically generate our object ID. With MongoDB, there are several ways to assign primary key identifiers, but for the sake of this demonstration, we are just going to use an atomic counter, as follows:

```java
//src/main/java/com/packtpub/springsecurity/dataaccess/MongoEventDao.
java
@Repository
public class MongoEventDao implements EventDao {

private EventRepository repository;

// Simple Primary Key Generator
private AtomicInteger eventPK = new AtomicInteger(102);

    @Override
      public int createEvent(final Event event) {
...

        // Get the next PK instance
        event.setId(eventPK.incrementAndGet());
        Event newEvent = repository.save(event);
        return newEvent.getId();
    }

...

}
```

There was technically no change to our `CalendarUserDao` object, but for consistency in this book, we renamed the implementation file to denote the use of `Mongo`:

```java
@Repository
public class MongoCalendarUserDao implements CalendarUserDao {
```

There are no other **Data Access Object (DAO)** changes required for this refactoring example.

Go ahead and start the application, and it will behave just as before.

Try to log in as **user1** and **admin1**. Then test the application to ensure that both users can add new events to the system, ensuring the mapping is correct for the entire application.

> **Important note**
> You should start with the source from `chapter05.05-calendar`.

Summary

We have looked at the power and flexibility of the Spring Data project and explored several aspects related to application development, as well as its integration with Spring Security. In this chapter, we covered the Spring Data project and a few of its capabilities. We also saw the refactoring process to convert from legacy JDBC code using SQL to ORM with JPA, and from a JPA implementation with Spring Data to a MongoDB implementation using Spring Data. We also covered configuring Spring Security to leverage an `ORM Entity` in a relational database and a document database.

In the next chapter, we will explore Spring Security's built-in support for *LDAP-based authentication*.

LDAP Directory Services

In this chapter, we will review the **Lightweight Directory Access Protocol** (**LDAP**) and learn how it can be integrated into a Spring Security-enabled application to provide authentication, authorization, and user information.

During the course of this chapter, we will cover the following topics:

- Learning some of the basic concepts related to the LDAP protocol and server implementations

- Configuring a self-contained LDAP server within Spring Security

- Enabling LDAP authentication and authorization

- Understanding the model behind LDAP search and user matching

- Retrieving additional user details from standard LDAP structures

- Differentiating between LDAP authentication methods and evaluating the pros and cons of each type

- Explicitly configuring Spring Security LDAP using Spring bean declarations

- Connecting to external LDAP directories

- Exploring the built-in support for Microsoft AD

- We will also explore how to customize Spring Security for more flexibility when dealing with custom AD deployments

This chapter's code in action link is here: `https://packt.link/f2tf1`.

Understanding LDAP

LDAP has its roots in logical directory models dating back over 30 years, conceptually akin to a combination of an organizational chart and an address book. Today, LDAP is used more and more to centralize corporate user information, partition thousands of users into logical groups, and allow unified sharing of user information between many disparate systems.

For security purposes, LDAP is quite commonly used to facilitate centralized username and password authentication—user credentials are stored in the LDAP directory, and authentication requests can be made against the directory on the user's behalf. This eases management for administrators, as user credentials—login ID, password, and other details—are stored in a single location in the LDAP directory. Additionally, organizational information, such as group or team assignments, geographic location, and corporate hierarchy membership, are defined based on the user's location in the directory.

LDAP

At this point, if you have never used LDAP before, you may be wondering what it is. We'll illustrate a sample LDAP schema with a screenshot from the Apache Directory Server example directory:

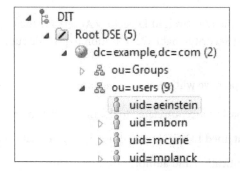

Figure 6.1 – Example of LDAP directory structure

Starting at a particular user entry for `uid=admin1@example.com` (highlighted in the preceding screenshot), we can infer the organizational membership of `admin1` by starting at this node in the tree and moving upward. We can see that the user `aeinstein` is a member of the `users` organizational unit (`ou=users`), which itself is a part of the `example.com` domain (the abbreviation `dc` shown in the preceding screenshot stands for **domain component**).

Preceding this are the organizational elements (`DIT` and `Root DSE`) of the LDAP tree itself, which don't concern us in the context of Spring Security. The position of the user `aeinstein` in the LDAP hierarchy is semantically and definitively meaningful—you can imagine a much more complex hierarchy easily illustrating the organizational and departmental boundaries of a huge organization.

The complete top-to-bottom path formed by walking down the tree to an individual leaf node forms a string composed of all intervening nodes along the way, as with the node path of `admin1`, as follows:

```
uid=admin1,ou=users,dc=example,dc=com
```

The preceding node path is unique and is known as a node's **Distinguished Name** (DN). The DN is akin to a database primary key, allowing a node to be uniquely identified and located in a complex tree structure. We'll see a node's DN used extensively throughout the authentication and searching process with Spring Security LDAP integration.

Note that there are several other users listed at the same level of organization as `admin1`. All of these users are assumed to be within the same organizational position as `admin1`. Although this example organization is relatively simple and flat, the structure of LDAP is arbitrarily flexible, with many levels of nesting and logical organization possible.

Spring Security LDAP support is assisted by the Spring LDAP module (`https://spring.io/projects/spring-ldap`), which is a separate project from the core Spring Framework and Spring Security projects. It's considered to be stable and provides a helpful set of wrappers around the standard Java LDAP functionality.

Common LDAP attribute names

Each entry in the tree is defined by one or more object classes. An object class is a logical unit of organization, grouping a set of semantically related attributes. By declaring an entry in the tree as an instance of a particular object class, such as a person, the organizer of the LDAP directory can provide users of the directory with a clear indication of what each element of the directory represents.

LDAP has a rich set of standard schemas covering the available LDAP object classes and their applicable attributes (along with gobs of other information). If you are planning on doing extensive work with LDAP, it's highly advised that you review a good reference guide, such as the appendix of the book *Zytrax OpenLDAP* (`https://www.zytrax.com/books/ldap/ape/`).

In the previous section, we were introduced to the fact that each entry in an LDAP tree has a DN, which uniquely identifies it in the tree. The DN is composed of a series of attributes, one (or more) of which is used to uniquely identify the path down the tree of the entry represented by the DN. As each segment of the path described by the DN represents an LDAP attribute, you could refer to the available, well-defined LDAP schemas and object classes to determine what each of the attributes in any given DN means.

We've included some of the common attributes and their meanings in the following table. These attributes tend to be organizing attributes—meaning that they are typically used to define the organizational structure of the LDAP tree—and are ordered from top to bottom in the structure that you're likely to see in a typical LDAP installation:

Attribute name	Description	Example
dc	Domain component: Generally, the highest level of organization in an LDAP hierarchy.	dc=jbcpcalendar,dc=com
c	Country: Some LDAP hierarchies are structured at a high level by country.	c=US
o	Organization name: This is a parent business organization used for classifying LDAP resources.	o=Oracle Corporation

Attribute name	Description	Example
ou	Organizational unit: This is a divisional business organization that is generally within an organization.	ou=Product Development
cn	Common name: This is a common name or a unique or human-readable name for the object. For humans, this is usually the person's full name, while for other resources in LDAP (computers, and so on), it's typically the hostname.	cn=Super Visor cn=Jim Bob
uid	User ID: Although not organizational in nature, the uid attribute is generally what Spring looks for during user authentication and search.	uid=svisor
userPassword	User password: This attribute stores the password for the person object to which this attribute is associated. It is typically one-way hashed using SHA or something similar.	userPassword=plaintext userPassword={SHA} cryptval

Table 6.1 – Example of LDAP directory structure

The attributes in the preceding table do, however, tend to be organizing attributes on the directory tree and, as such, will probably form various search expressions or mappings that you will use to configure Spring Security to interact with the LDAP server.

> **Important note**
>
> Remember that there are hundreds of standard LDAP attributes—these represent a very small fraction of those you are likely to see when integrating with a fully populated LDAP server.

Updating our dependencies

We have already included all of the dependencies you need for this chapter, so you will not need to make any updates to your build.gradle file. However, if you were just adding LDAP support to your own application, you would need to add spring-security-ldap as a dependency in build.gradle, as follows:

```
//build.gradle
dependencies {
...
// LDAP
    implementation 'org.springframework.security:spring-security-ldap'
```

```
. . .
}
```

> **Important note**
>
> Remember that there are hundreds of standard LDAP attributes—these represent a very small
> fraction of those you are likely to see when integrating with a fully populated LDAP server.

As mentioned previously, Spring Security's LDAP support is built on top of Spring LDAP. Gradle will
automatically bring this dependency in as a transitive dependency, so there is no need to explicitly list it.

Configuring embedded LDAP integration

Let's now enable the JBCP calendar application to support LDAP-based authentication. Fortunately, this
is a relatively simple exercise, using the embedded LDAP server and a sample **LDAP Data Interchange
Format** (LDIF) file. For this exercise, we will be using an LDIF file created for this book that's intended
to capture many of the common configuration scenarios with LDAP and Spring Security. We have
included several more sample LDIF files for an embedded UnboundID server. This is done by adding
an unboundid-ldapsdk dependency in build.gradle, as follows:

```
//build.gradle
dependencies {
. . .
// LDAP
    implementation 'com.unboundid:unboundid-ldapsdk'
. . .
}
```

Configuring an LDAP server reference

The first step is to configure the embedded LDAP server. Spring Boot will automatically configure an
embedded LDAP server, but we will need to tweak the configuration a bit. Make the following updates
to your application.yml file:

```
spring:
  ldap:
    base: dc=jbcpcalendar,dc=com
    embedded:
      ldif: classpath:/ldif/calendar.ldif
      baseDn: ${spring.ldap.base}
      port: 33389
```

> **Important note**
>
> You should be starting with the source from `chapter06.00-calendar`.

We are loading the `calendar.ldif` file from `classpath` and using it to populate the LDAP server. The `root` attribute declares the root of the LDAP directory using the specified DN. This should correspond to the logical root DN in the LDIF file we're using.

> **Tip**
>
> Be aware that for embedded LDAP servers, the `base-dn` attribute is required. If it is not specified or is specified incorrectly, you may receive several odd errors upon initialization. Also, be aware that the `ldif` resource should only load a single `ldif`, otherwise the server will fail to start up. Spring Security requires a single resource, since using something such as `classpath*:calendar.ldif` does not provide the deterministic ordering that is required.

We'll reuse the bean ID defined here later, in the Spring Security configuration files, when we declare the LDAP user service and other configuration elements. All other attributes on the `<ldap-server>` declaration are optional when using the embedded LDAP mode.

Enabling the LDAP AuthenticationManager interface

Next, we'll need to configure another `AuthenticationManager` interface that checks user credentials against the LDAP provider. Simply update the Spring Security configuration to use an `o.s.s.ldap.authentication.AuthenticationManager` reference, as follows:

```
//src/main/java/com/packtpub/springsecurity/configuration/
SecurityConfig.java
@Bean
AuthenticationManager authenticationManager(BaseLdapPathContextSource
contextSource, LdapAuthoritiesPopulator authorities) {
    LdapBindAuthenticationManagerFactory factory = new
LdapBindAuthenticationManagerFactory(contextSource);
    factory.setUserSearchBase("");
    factory.setUserSearchFilter("(uid={0})");
    factory.setLdapAuthoritiesPopulator(authorities);
    return factory.createAuthenticationManager();
}
```

Configuring the LdapAuthoritiesPopulator interface

Spring Security's LdapAuthoritiesPopulator is used to determine what authorities are returned for the user. The following example shows how to configure LdapAuthoritiesPopulator:

```
//src/main/java/com/packtpub/springsecurity/configuration/
SecurityConfig.java

@Bean
LdapAuthoritiesPopulator authorities(BaseLdapPathContextSource
contextSource) {
    String groupSearchBase = "ou=Groups";
    DefaultLdapAuthoritiesPopulator authorities =
            new DefaultLdapAuthoritiesPopulator(contextSource,
groupSearchBase);
    authorities.setGroupSearchFilter("(uniqueMember={0})");
    return authorities;
}
```

In addition, we have deleted all the references to PasswordEncoder Bean and the class CalendarUserDetailsService.

We'll discuss these attributes a bit more later. For now, get the application back up and running, and try logging in with admin1@example.com as the username and admin1 as the password. You should be logged in!

> **Important note**
> You should be starting with the source from chapter06.01-calendar.

Troubleshooting embedded LDAP

It is quite possible that you will run into hard-to-debug problems with embedded LDAP. If you are getting a 404 error when trying to access the application in your browser, there is a good chance that things did not start up properly. Some things to double-check if you can't get this simple example running are as follows:

- Ensure the baseDn attribute is set in your configuration file, and make sure it matches the root defined in the LDIF file that's loaded at startup. If you get errors referencing missing partitions, it's likely that either the root attribute was missed or doesn't match your LDIF file.

- Be aware that a failure starting up the embedded LDAP server is not a fatal failure. In order to diagnose errors loading LDIF files, you will need to ensure that the appropriate log settings, including logging for the LDAP server, are enabled, at least at the error level.

- If the application server shuts down non-gracefully, you may be required to delete some files in your temporary directory (`%TEMP%` on Windows systems or `/tmp` on Linux-based systems) in order to start the server again. The error messages regarding this are (fortunately) fairly clear. Unfortunately, embedded LDAP isn't as seamless and easy to use as the embedded H2 database, but it is still quite a bit easier than trying to download and configure many of the freely available external LDAP servers.

An excellent tool for troubleshooting or accessing LDAP servers in general is the Apache Directory Studio project, which offers standalone and Eclipse plugin versions. The free download is available at `http://jxplorer.org/`.

Understanding how Spring LDAP authentication works

We saw that we were able to log in using a user-defined in the LDAP directory. But what exactly happens when a user issues a login request for a user in LDAP? There are the following three basic steps to the LDAP authentication process:

1. Authenticate the credentials supplied by the user against the LDAP directory.

2. Determine the `GrantedAuthority` object that the user has, based on their information in LDAP.

3. Pre-load information from the LDAP entry for the user into a custom `UserDetails` object for further use by the application.

Authenticating user credentials

For the first step, authentication against the LDAP directory, a custom authentication provider is wired into `AuthenticationManager`. The `o.s.s.ldap.authentication.LdapAuthenticationProvider` interface takes the user's provided credentials and verifies them against the LDAP directory, as illustrated in the following diagram:

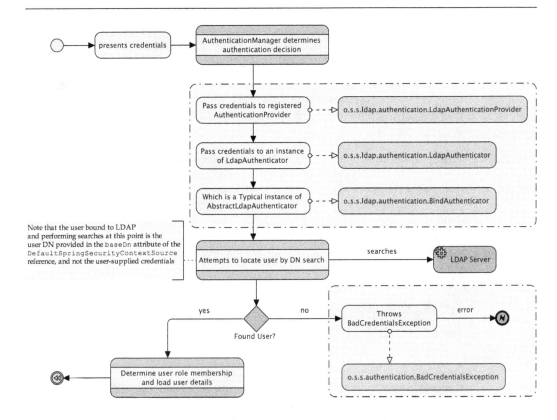

Figure 6.2 – Spring Security LDAP authentication workflow

We can see that the o.s.s.ldap.authentication.LdapAuthenticator interface defines a delegate to allow the provider to make the authentication request in a customizable way. The implementation that we've implicitly configured to this point, o.s.s.ldap.authentication. BindAuthenticator, attempts to use the user's credentials to bind (log in) to the LDAP server as if it were the user themselves making a connection. For an embedded server, this is sufficient for our authentication needs; however, external LDAP servers may be stricter, and in these, users may not be allowed to bind to the LDAP directory. Fortunately, an alternative method of authentication exists, which we will explore later in this chapter.

As noted in the preceding diagram, keep in mind that the search is performed under an LDAP context created by the credentials specified in the DefaultSpringSecurityContextSource reference's baseDn attribute. With an embedded server, we don't use this information, but with an external server reference, unless baseDn is supplied, anonymous binding is used. Retaining some control over the public availability of information in the directory is very common for organizations that require valid credentials to search an LDAP directory, and as such, baseDn will be almost always required in real-world scenarios. The baseDn attribute represents the full DN of a user with valid access to bind the directory and perform searches.

Demonstrating authentication with JXplorer

We are going to demonstrate how the authentication process works by using JXplorer to connect to our embedded LDAP instance and perform the same steps that Spring Security is performing. We will use user1@example.com throughout the simulation. These steps will help to ensure a firm grasp of what is happening behind the scenes and will help if you are having difficulty figuring out the correct configuration.

Ensure that the calendar application is started up and working. Next, start Jxplorer.

Binding anonymously to LDAP

The first step is to bind anonymously to LDAP. The bind is done anonymously because we did not specify the baseDn and password attributes on our DefaultSpringSecurityContextSource object. Within Jxplorer, create a connection using the following steps:

1. Click on **File** | **Connect**.

2. Enter the following information:

   - Hostname: localhost

   - Port: 33389

3. We did not specify baseDn, so select **No Authentication** as the **Authentication Method**.

4. Click on **OK**.

   You can safely ignore the message indicating no default schema information is present.

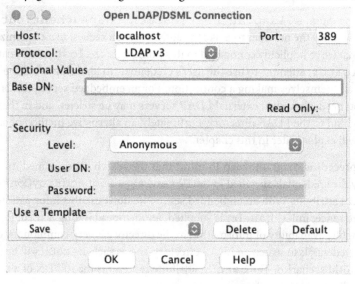

Figure 6.3 – Binding anonymously to LDAP

You should now see that you are connected to the embedded LDAP instance.

Searching for the user

Now that we have a connection, we can use it to look up the user's DN that we wish to bind to, by performing the following steps:

1. Right-click on **World** and select **Search**.

2. Enter a search base of `dc=jbcpcalendar,dc=com`. This corresponds to the `baseDn` attribute of our `spring.ldap.base` property that we specified.

3. Enter a filter of `uid=user1@example.com`. This corresponds to the value we specified for the `userSearchFilter` method of `AuthenticationManagerBuilder`.

4. Click on **Search**.

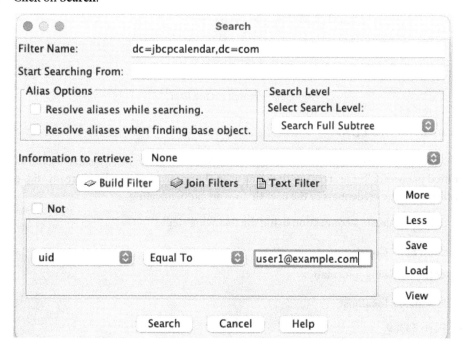

Figure 6.4 – Search for the user

5.   Click on **Copy DN** of the single result returned by our search. You can now see that our LDAP user is displayed. Note that this DN matches the value we searched for. Remember this DN, as it will be used in our next step.

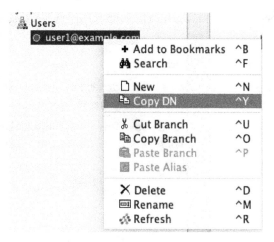

Figure 6.5 – Search for the user

Binding as a user to LDAP

Now that we have found the full DN of our user, we need to try to bind to LDAP as that user to validate the submitted password. These steps are the same as in the anonymous bind we already did, except that we will specify the credentials of the user that we are authenticating.

Within `Jxplorer`, create a connection using the following steps:

1.   Click on **File | Connect**.

2.   Enter the following information:

   •   Hostname: **localhost**

   •   Port: **33389**

3.   Set **Security Level** to **User + Password**.

4.   Enter the DN from our search result as **User DN**. The value should be `uid=admin1@example.com,ou=Administrators,ou=Users,dc=jbcpcalendar,dc=com`.

5.   The password should be the password that was submitted at the time of login. In our case, we want to use **admin1** to successfully authenticate. If the wrong password was entered, we would fail to connect and Spring Security would report an error.

6. Click on **OK**.

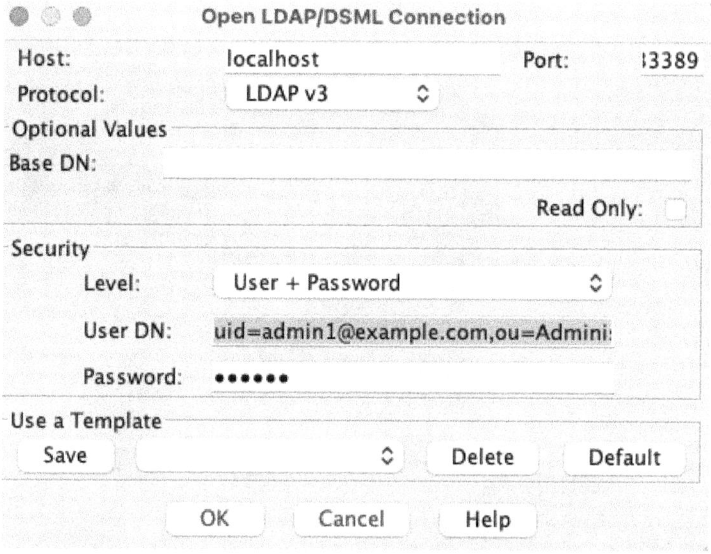

Figure 6.6 – Binding as a user to LDAP

Spring Security will determine that the username and password were correct for this user when it is able to successfully bind with the provided username and password (similar to how we were able to create a connection). Spring Security will then proceed with determining the user's role membership.

Determining the user's role membership

After the user has been successfully authenticated against the LDAP server, authorization information must be determined next. Authorization is defined by a principal's list of roles, and an LDAP-authenticated user's role membership is determined as illustrated in the following diagram:

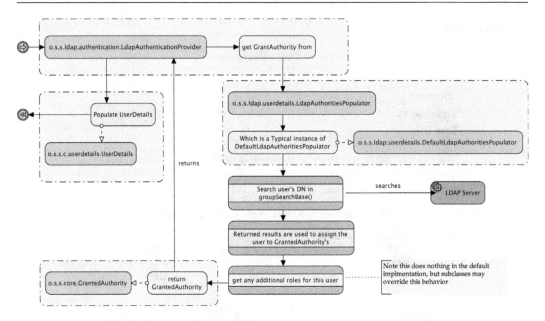

Figure 6.7 – User role membership

We can see that after authenticating the user against LDAP, `LdapAuthenticationProvider` delegates to `LdapAuthoritiesPopulator`. The `DefaultLdapAuthoritiesPopulator` interface will attempt to locate the authenticated user's DN in an attribute located at or below another entry in the LDAP hierarchy. The DN of the location searched for user role assignments is defined in the `groupSearchBase` method; in our sample, we set this to `groupSearchBase("ou=Groups")`. When the user's DN is located within an LDAP entry below the DN of `groupSearchBase`, an attribute on the entry in which their DN is found is used to confer a role to them.

How Spring Security roles are associated with LDAP users can be a little confusing, so let's look at the JBCP calendar LDAP repository and see how the association of a user with a role works. The `DefaultLdapAuthoritiesPopulator` interface uses several methods of the `AuthenticationManagerBuilder` declaration to govern searching for roles for the user. These attributes are used approximately in the following order:

1. `groupSearchBase`: This defines the base DN under which the LDAP integration should look for one or more matches for the user's DN. The default value performs a search from the LDAP root, which may be expensive.

2. `groupSearchFilter`: This defines the LDAP search filter used to match the user's DN to an attribute of an entry located under `groupSearchBase`. This search filter is parameterized with two parameters—the first (`{0}`) being the user's DN, and the second (`{1}`) being the user's username. The default value is `uniqueMember={0}`.

3. `groupRoleAttribute`: This defines the attribute of the matching entries, which will be used to compose the user's `GrantedAuthority` object. The default value is `cn`.

4. `rolePrefix`: This is the prefix that will be prepended to the value found in `groupRoleAttribute`, to make a Spring Security `GrantedAuthority` object. The default value is ROLE_.

This can be a little abstract and hard for new developers to follow because it's very different from anything we've seen so far with our JDBC and JPA-based `UserDetailsService` implementations. Let's continue walking through the login process with our `user1@example.com` user in the JBCP calendar LDAP directory.

Determining roles with Jxplorer

We will now try to determine the roles for our user with `Jxplorer`. Using the connection, we created previously, perform the following steps:

1. Right-click on **World** and select **Search**.

2. Enter a search base of `ou=Groups,dc=jbcpcalendar,dc=com`. This corresponds to the `baseDn` attribute of the `DefaultSpringSecurityContextSource` object we specified, plus the `groupSearchBase` attribute we specified for the `AuthenticationManagerBuilder` object.

3. Enter a text filter of `uniqueMember=uid=user1@example.com,ou=Users,dc=jbcpcalendar,dc= com`. This corresponds to the default `groupSearchFilter` attribute of `(uniqueMember={0})`. Notice that we have substituted the full DN of the user we found in our previous exercise for the `{0}` value.

4. Click on **Search**.

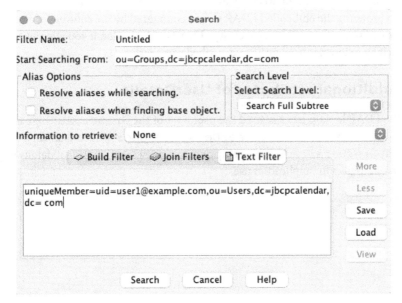

Figure 6.8 – Role search

5. You will observe that the **User** group is the only group returned in our search results. Click on copy DN of the single result returned by our search. You can now see the **User** group displayed in `Jxplorer`. Note that the group has a `uniqueMember` attribute with the full DN of our user and other users.

Spring Security now creates the `GrantedAuthority` object for each result by forcing the name of the group that was found into uppercase and prepending `ROLE_` to the group name. The pseudocode would look similar to the following code snippet:

```
foreach group in groups:
authority = ("ROLE_"+group).upperCase()
grantedAuthority = new GrantedAuthority(authority)
```

> **Tip**
>
> Spring LDAP is as flexible as your gray matter. Keep in mind that, although this is one way to organize an LDAP directory to be compatible with Spring Security, typical usage scenarios are exactly the opposite—an LDAP directory already exists that Spring Security needs to be wired into. In many cases, you will be able to reconfigure Spring Security to deal with the hierarchy of the LDAP server; however, it's key that you plan effectively and understand how Spring works with LDAP when it's querying. Use your brain, map out the user search and group search, and come up with the most optimal plan you can think of—keep the scope of searches as minimal and as precise as possible.

Can you describe how the results of the login process would differ for our `admin1@example.com` user? If you are confused at this point, we'd suggest that you take a breather and try using `Jxplorer` to work through browsing the embedded LDAP server, configured by the running of an application. It can be easier to grasp the flow of Spring Security's LDAP configuration if you attempt to search the directory yourself by following the algorithm described previously.

Mapping additional attributes of UserDetails

Finally, once the LDAP lookup has assigned the user a set of the `GrantedAuthority` objects, `o.s.s.ldap.userdetails.LdapUserDetailsMapper` will consult `o.s.s.ldap.userdetails.UserDetailsContextMapper` to retrieve any additional details to populate the `UserDetails` object for application use.

Using `AuthenticationManagerBuilder`, we've configured up until this point that `LdapUserDetailsMapper` will be used to populate a `UserDetails` object with information gleaned from the user's entry in the LDAP directory:

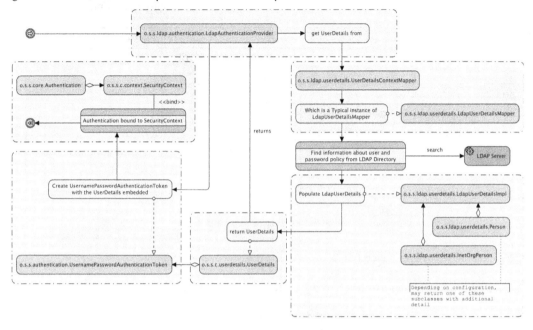

Figure 6.9 – Mapping additional attributes of UserDetails

We'll see in a moment how `UserDetailsContextMapper` can be configured to pull a wealth of information from the standard LDAP `person` and `inetOrgPerson` objects. With the baseline `LdapUserDetailsMapper`, little more than `username`, `password`, and `GrantedAuthority` are stored.

Although there is more machinery involved behind the scenes in LDAP user authentication and detail retrieval, you'll notice that the overall process seems somewhat similar to the JDBC authentication that we studied in *Chapter 4, JDBC-based Authentication* (authenticating the user and populating `GrantedAuthority`). As with JDBC authentication, there is the ability to perform advanced configuration of LDAP integration. Let's dive deeper and see what's possible!

Advanced LDAP configuration

Once we get beyond the basics of LDAP integration, there's a plethora of additional configuration capabilities in the Spring Security LDAP module that are still within the security `SecurityFilterChain` bean. These include retrieval of user personal information, additional options for user authentication, and the use of LDAP as the `UserDetailsService` interface in conjunction with a standard `DaoAuthenticationProvider` class.

Sample JBCP LDAP users

We've supplied a number of different users in the JBCP calendar LDIF file. The following quick reference chart may help you with the advanced configuration exercises, or with self-exploration:

Username/password	Role(s)	Password encoding
admin1@example.com/admin1	ROLE_ADMIN, ROLE_USER	Plaintext
user1@example.com/user1	ROLE_USER	Plaintext
shauser@example.com/shauser	ROLE_USER	{sha}
sshauser@example.com/sshauser	ROLE_USER	{ssha}
hasphone@example.com/hasphone	ROLE_USER	Plaintext (in the telephoneNumber attribute)

Table 6.2 – List of LDAP users

We'll explain why password encoding matters in the next section.

Password comparison versus bind authentication

Some LDAP servers will be configured so that certain individual users are not allowed to bind directly to the server, or so that anonymous binding (what we have been using for user search up until this point) is disabled. This tends to occur in very large organizations that want a restricted set of users to be able to read information from the directory.

In these cases, the standard Spring Security LDAP authentication strategy will not work, and an alternative strategy must be used, implemented by o.s.s.ldap.authentication. PasswordComparisonAuthenticator (a sibling class of BindAuthenticator):

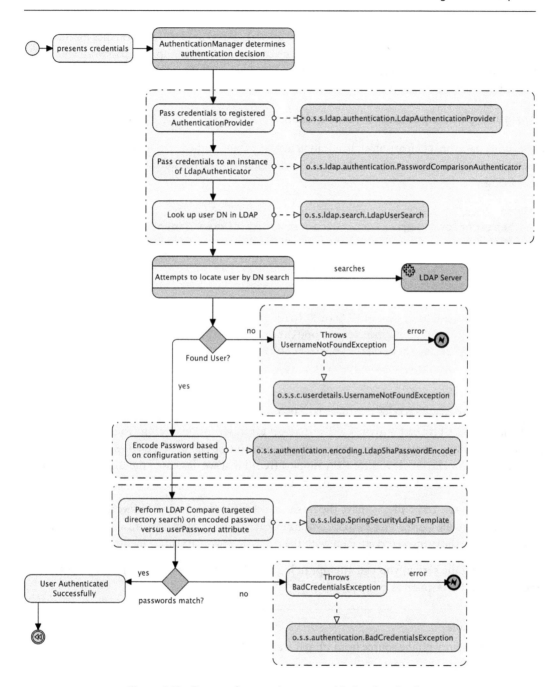

Figure 6.10 – Password comparison versus bind authentication

The `PasswordComparisonAuthenticator` interface binds to LDAP and searches for the DN matching the username provided by the user. It then compares the user-supplied password with the `userPassword` attribute stored on the matching LDAP entry. If the encoded password matches, the user is authenticated and the flow proceeds, as with `BindAuthenticator`.

Configuring basic password comparison

Configuring password comparison authentication instead of bind authentication is as simple as adding a method to the `AuthenticationManager` declaration. Update the `SecurityConfig.java` file as follows:

```
@Bean
AuthenticationManager authenticationManager(BaseLdapPathContextSource
contextSource, LdapAuthoritiesPopulator authorities) {
    LdapPasswordComparisonAuthenticationManagerFactory factory = new
LdapPasswordComparisonAuthenticationManagerFactory(
        contextSource, new LdapShaPasswordEncoder());
    factory.setUserSearchBase("");
    factory.setUserSearchFilter("(uid={0})");
    factory.setLdapAuthoritiesPopulator(authorities);
    factory.setPasswordAttribute("userPassword");
    return factory.createAuthenticationManager();
}
```

The `PasswordCompareConfigurer` class, that is used by declaring the `passwordCompare` method, uses `PlaintextPasswordEncoder` for password encoding. To use the `SHA-1` password algorithm, we need to set a password encoder, and we can use `o.s.s.a.encoding.LdapShaPasswordEncoder` for SHA support (recall that we discussed the `SHA-1` password algorithm extensively in *Chapter 4, JDBC-based Authentication*).

In our `calendar.ldif` file, we have the `password` field set to `userPassword`. The default password attribute for the `PasswordCompareConfigurer` class is `password`. So, we also need to override the `password` attribute with the `passwordAttribute` method.

After restarting the server, you can attempt to log in using `shauser@example.com` as the username and `shauser` as password.

> **Important note**
> You should start with the source from `chapter06.02-calendar`.

LDAP password encoding and storage

LDAP has general support for a variety of password encoding algorithms, ranging from plaintext to one-way hash algorithms—similar to those we explored in the previous chapter—with database-backed authentication. The most common storage formats for LDAP passwords are SHA (SHA-1 one-way hashed) and SSHA (SHA-1 one-way hashed with a salt value). Other password formats often supported by many LDAP implementations are thoroughly documented in *RFC 2307, An Approach to Using LDAP as a Network Information Service* (http://tools.ietf.org/html/rfc2307). The designers of *RFC 2307* did a very clever thing with regard to password storage. Passwords retained in the directory are, of course, encoded with whatever algorithm is appropriate (SHA and so on), but then, they are prefixed with the algorithm used to encode the password. This makes it very easy for the LDAP server to support multiple algorithms for password encoding. For example, an SHA encoded password is stored in the directory as follows:

```
{SHA}5baa61e4c9b93f3f0682250b6cf8331b7ee68fd8
```

We can see that the password storage algorithm is very clearly indicated with the {SHA} notation and stored along with the password.

The SSHA notation is an attempt to combine the strong SHA-1 hash algorithm with password salting to prevent dictionary attacks. As with password salting, which we reviewed in the previous chapter, the salt is added to the password prior to calculating the hash. When the hashed password is stored in the directory, the salt value is appended to the hashed password. The password is prepended with {SSHA} so that the LDAP directory knows that the user-supplied password needs to be compared differently. Most modern LDAP servers utilize SSHA as their default password storage algorithm.

The drawbacks of a password comparison authenticator

Now that you know a bit about how LDAP uses passwords, and we have PasswordComparisonAuthenticator set up, what do you think will happen if you log in using our sshauser@example.com user with their password, stored in SSHA format?

Go ahead, put the book aside and try it, and then come back. Your login was denied, right? And yet you were still able to log in as the user with the SHA-encoded password. Why? The password encoding and storage didn't matter when we were using bind authentication. Why do you think that is?

The reason it didn't matter with bind authentication was that the LDAP server was taking care of the authentication and validation of the user's password. With password compare authentication, Spring Security LDAP is responsible for encoding the password in the format expected by the directory and then matching it against the directory to validate the authentication.

For security purposes, password comparison authentication can't actually read the password from the directory (reading directory passwords is often denied by the security policy). Instead, PasswordComparisonAuthenticator performs an LDAP search, rooted at the user's directory entry, attempting to match with a password attribute and value as determined by the password that's been encoded by Spring Security.

So, when we try to log in with sshauser@example.com, PasswordComparisonAuthenticator encodes the password using the configured SHA algorithm and attempts to do a simple match, which fails, as the directory password for this user is stored in the SSHA format.

Our current configuration, using LdapShaPasswordEncoder, already supports SHA and SSHA, so currently, it still doesn't work. Let's think why that might be. Remember that SSHA uses a salted password, with the salt value stored in the LDAP directory along with the password. However, PasswordComparisonAuthenticator is coded so that it cannot read anything from the LDAP server (this typically violates the security policy with companies that don't allow binding). Thus, when PasswordComparisonAuthenticator computes the hashed password, it has no way to determine what salt value to use.

In conclusion, PasswordComparisonAuthenticator is valuable in certain limited circumstances where the security of the directory itself is a concern, but it will never be as flexible as straight bind authentication.

Configuring the UserDetailsContextMapper object

As we noted earlier, an instance of the o.s.s.ldap.userdetails.UserDetailsContextMapper interface is used to map a user's entry into the LDAP server to a UserDetails object in memory. The default UserDetailsContextMapper object behaves similarly to JpaDaoImpl, given the level of detail that is populated on the returned UserDetails object—that is to say, not a lot of information is returned besides the username and password.

However, an LDAP directory potentially contains many more details about individual users than usernames, passwords, and roles. Spring Security ships with two additional methods of pulling more user data from two of the standard LDAP object schemas—person and inetOrgPerson.

Implicit configuration of UserDetailsContextMapper

In order to configure a different UserDetailsContextMapper implementation than the default, we simply need to declare which LdapUserDetails class we want LdapAuthenticationProvider to return. The security namespace parser will be smart enough to instantiate the correct UserDetailsContextMapper implementation based on the type of the LdapUserDetails interface requested.

Let's reconfigure our `SecurityConfig.java` file to use the `inetOrgPerson` version of the mapper. Update the `SecurityConfig.java` file, as illustrated in the following code:

```
//src/main/java/com/packtpub/springsecurity/configuration/
SecurityConfig.java
@Bean
LdapAuthoritiesPopulator authorities(BaseLdapPathContextSource
contextSource) {
    String groupSearchBase = "ou=Groups";
    DefaultLdapAuthoritiesPopulator authorities =
        new DefaultLdapAuthoritiesPopulator(contextSource,
groupSearchBase);
    authorities.setGroupSearchFilter("(uniqueMember={0})");
    return authorities;
}
```

> **Important note**
>
> If we remove the `passwordEncoder` method, then the LDAP users that are using SHA passwords will fail to authenticate.

If you were to restart the application and attempt to log in as an LDAP user, you would see that nothing changed. In fact, `UserDetailsContextMapper` has changed behind the scenes to read the additional details in the case where attributes from the `inetOrgPerson` schema are available in the user's directory entry.

Viewing additional user details

To assist you in this area, we'll add the ability to view the current account to the JBCP calendar application. We'll use this page to illustrate how the richer person and the `inetOrgPerson` LDAP schemas can provide additional (optional) information to your LDAP-enabled application.

You may have noticed that this chapter came with an additional controller named `AccountController`. You can see the relevant code, as follows:

```
//src/main/java/com/packtpub/springsecurity/web/controllers/
AccountControll er.java
@Controller
public class AccountController {

    @RequestMapping("/accounts/my")
    public String view(Model model) {
        Authentication authentication = SecurityContextHolder.
getContext().getAuthentication();
        if(authentication == null) {
```

```
                    throw new IllegalStateException("authentication cannot be
null. Make sure you are logged in.");
        }
        Object principal = authentication.getPrincipal();
        model.addAttribute("user", principal);
        model.addAttribute("isLdapUserDetails", principal instanceof
LdapUserDetails);
        model.addAttribute("isLdapPerson", principal instanceof
Person);
        model.addAttribute("isLdapInetOrgPerson", principal instanceof
InetOrgPerson);
        return "accounts/show";
    }
}
```

The preceding code will retrieve the `UserDetails` object (principal) stored in the `Authentication` object by `LdapAuthenticationProvider` and determine what type of `LdapUserDetailsImpl` interface it is. The page code itself will then display various details depending on the type of `UserDetails` object that has been bound to the user's authentication information, as we see in the following JSP code. We have already included JSP as well:

```
//src/main/resources/templates/accounts/show.html
<dl>
    <dt>Username</dt>
    <dd id="username" th:text="${user.username}">ChuckNorris</dd>
    <dd> </dd>
    <dt>DN</dt>
    <dd id="dn" th:text="${user.dn}"></dd>
    <dd> </dd>

    <span th:if="${isLdapPerson}">

        <dt>Description</dt>
        <dd id="description" th:text="${user.description}"></dd>
        <dd> </dd>
        <dt>Telephone</dt>
        <dd id="telephoneNumber" th:text="${user.telephoneNumber}"></
dd>
        <dd> </dd>
        <dt>Full Name(s)</dt>

        <span th:each="cn : ${user.cn}">
            <dd th:text="${cn}"></dd>
        </span>
        <dd> </dd>
```

```
    </span>

    <span th:if="${isLdapInetOrgPerson}">
        <dt>Email</dt>
        <dd id="email" th:text="${user.mail}"></dd>
        <dd> </dd>
        <dt>Street</dt>
        <dd id="street" th:text="${user.street}"></dd>
        <dd> </dd>
    </span>
</dl>
```

The only work that actually needs to be done is to add a link in our header.html file, as shown in the following code snippet:

```
//src/main/resources/templates/fragments/header.html
<li class="nav-item">
    <a class="nav-link" th:href="@{/accounts/my}">Welcome <span
class="navbar-text"
                                     th:text="${#authentication.
name}"></span></a>
</li>
```

We've added the following two more users that you can use to examine the differences in the available data elements:

Username	Password	Type
shainet@example.com	shainet	inetOrgPerson
shaperson@example.com	shaperson	person

Table 6.3 – List of newly added LDAP users

> **Important note**
> Your code should look like chapter06.03-calendar.

Restart the server and examine the Account Details page for each of the types of users by clicking on username in the upper-right corner. You'll note that when UserDetails class is configured to use inetOrgPerson, although o.s.s.ldap.userdetails.InetOrgPerson is what is returned, the fields may or may not be populated depending on the available attributes in the directory entry.

In fact, `inetOrgPerson` has many more attributes that we've illustrated on this simple page. You can review the full list in *RFC 2798, Definition of the inetOrgPerson LDAP Object Class* (`http://tools.ietf.org/html/rfc2798`).

One thing you may notice is that there is no facility to support additional attributes that may be specified on an object entry, but don't fall into a standard schema. The standard `UserDetailsContextMapper` interfaces don't support arbitrary lists of attributes, but it is possible nonetheless to customize it with a reference to your own `UserDetailsContextMapper` interface using the `userDetailsContextMapper` method.

Using an alternate password attribute

In some cases, it may be necessary to use an alternate LDAP attribute instead of `userPassword`, for authentication purposes. This can happen during occasions when companies have deployed custom LDAP schemas or don't have the requirement for strong password management (arguably, this is never a good idea, but it definitely does occur in the real world).

The `PasswordComparisonAuthenticator` interface also supports the ability to verify the user's password against an alternate LDAP entry attribute instead of the standard `userPassword` attribute. This is very easy to configure, and we can demonstrate a simple example using the plaintext `telephoneNumber` attribute. Update the `SecurityConfig.java` as follows:

```
//src/main/java/com/packtpub/springsecurity/configuration/
SecurityConfig.java
@Bean
AuthenticationManager authenticationManager(BaseLdapPathContextSource
contextSource, LdapAuthoritiesPopulator authorities) {
    LdapPasswordComparisonAuthenticationManagerFactory factory = new
LdapPasswordComparisonAuthenticationManagerFactory(
            contextSource, new LdapShaPasswordEncoder());
    factory.setUserSearchBase("");
    factory.setUserDetailsContextMapper(new
InetOrgPersonContextMapper());
    factory.setUserSearchFilter("(uid={0})");
    factory.setLdapAuthoritiesPopulator(authorities);
    factory.setPasswordAttribute("telephoneNumber");
    return factory.createAuthenticationManager();
}
```

We can restart the server and attempt to log in with `hasphone@example.com` as the username attribute and `0123456789` as the `password` (telephone number) attribute.

> **Important note**
> Your code should look like `chapter06.04-calendar`.

Of course, this type of authentication has all of the perils we discussed earlier regarding authentication based on `PasswordComparisonAuthenticator`; however, it's good to be aware of it on the off-chance that it comes up with an LDAP implementation.

Using LDAP as UserDetailsService

One thing to note is that LDAP may also be used as `UserDetailsService`. As we will discuss later in the book, `UserDetailsService` is required to enable various other bits of functionality in the Spring Security infrastructure, including the remember-me and OpenID authentication features.

We will modify our `AccountController` object to use the `LdapUserDetailsService` interface to obtain the user. Before doing this, make sure to remove the `passwordCompare` method, as shown in the following code snippet:

```
//src/main/java/com/packtpub/springsecurity/configuration/
SecurityConfig.java
@Bean
AuthenticationManager authenticationManager(BaseLdapPathContextSource
contextSource, LdapAuthoritiesPopulator authorities) {
    LdapBindAuthenticationManagerFactory factory = new
LdapBindAuthenticationManagerFactory(contextSource);
    factory.setUserSearchBase("");
    factory.setUserSearchFilter("(uid={0})");
    factory.setLdapAuthoritiesPopulator(authorities);
    factory.setUserDetailsContextMapper(new
InetOrgPersonContextMapper());
    return factory.createAuthenticationManager();
}
```

Configuring LdapUserDetailsService

The configuration of LDAP as a `UserDetailsService` function is very similar to the configuration of an LDAP `AuthenticationProvider`. Like the JDBC `UserDetailsService`, an LDAP `UserDetailsService` interface is configured as a sibling to the `<http>` declaration. Make the following updates to the `SecurityConfig.java` file:

```
//src/main/java/com/packtpub/springsecurity/configuration/
SecurityConfig.java

@Bean
public UserDetailsService userDetailsService(BaseLdapPathContextSource
contextSource, LdapAuthoritiesPopulator authorities) {
    return new LdapUserDetailsService(new
FilterBasedLdapUserSearch("", "(uid={0})", contextSource),
authorities);
}
```

Functionally, `o.s.s.ldap.userdetails.LdapUserDetailsService` is configured in almost exactly the same way as `LdapAuthenticationProvider`, with the exception that there is no attempt to use the principal's username to bind to LDAP. Instead, the credentials are supplied by the `DefaultSpringSecurityContextSource` reference and are used to perform the user lookup.

> **Important note**
>
> Do not make the very common mistake of configuring `AuthenticationManagerBuilder` with the `UserDetailsService` referring to `LdapUserDetailsService` if you intend to authenticate the user against LDAP itself! As discussed previously, the `password` attribute often cannot be retrieved from LDAP due to security reasons, which makes `UserDetailsService` useless for authenticating. As noted previously, `LdapUserDetailsService` uses the baseDn attribute supplied with the `DefaultSpringSecurityContextSource` declaration to get its information—this means that it does not attempt to bind the user to LDAP and, as such, may not behave as you expect.

Updating AccountController to use LdapUserDetailsService

We will now update the `AccountController` object to use the `LdapDetailsUserDetailsService` interface to look up the user that it displays:

```
//src/main/java/com/packtpub/springsecurity/web/controllers/
AccountControll er.java
@Controller
public class AccountController {

    private final UserDetailsService userDetailsService;

    public AccountController(UserDetailsService userDetailsService) {
        if (userDetailsService == null) {
            throw new IllegalArgumentException("userDetailsService
cannot be null");
        }
        this.userDetailsService = userDetailsService;
    }

    @RequestMapping("/accounts/my")
    public String view(Model model) {
        Authentication authentication = SecurityContextHolder.
getContext().getAuthentication();
        if(authentication == null) {
            throw new IllegalStateException("authentication cannot be
null. Make sure you are logged in.");
```

```
    }
    Object principal = userDetailsService.
loadUserByUsername(authentication.getName());
    model.addAttribute("user", principal);
    model.addAttribute("isLdapUserDetails", principal instanceof
LdapUserDetails);
    model.addAttribute("isLdapPerson", principal instanceof
Person);
    model.addAttribute("isLdapInetOrgPerson", principal instanceof
InetOrgPerson);
    return "accounts/show";
    }
}
```

Obviously, this example is a bit silly, but it demonstrates the use of `LdapUserDetailsService`. Go ahead and restart the application and give this a try with the `username` as `admin1@example.com` and the `password` as `admin1`. Can you figure out how to modify the controller to display an arbitrary user's information?

Can you figure out how you should modify the security settings to restrict access to an administrator?

> **Important note**
> Your code should look like `chapter06.05-calendar`.

Integrating Spring Security with an external LDAP server

It is likely that once you test basic integration with the embedded LDAP server, you will want to interact with an external LDAP server. Fortunately, this is very straightforward and can be done using a slightly different syntax along with the same

Update the Spring Security configuration to connect to an external LDAP server on port 33389, as follows:

```
//src/main/java/com/packtpub/springsecurity/configuration/
SecurityConfig.java
@Bean
AuthenticationManager authenticationManager(LdapAuthoritiesPopulator
authorities) {
    BaseLdapPathContextSource contextSource=  new
DefaultSpringSecurityContextSource(
        List.of("ldap://localhost:" + LDAP_PORT + "/"),
"dc=jbcpcalendar,dc=com"){{
        setUserDn("uid=admin,ou=system");
        setPassword("secret");
    }};
```

```
    LdapPasswordComparisonAuthenticationManagerFactory factory = new
LdapPasswordComparisonAuthenticationManagerFactory(
        contextSource, new LdapShaPasswordEncoder());
    factory.setUserSearchBase("");
    factory.setUserDetailsContextMapper(new
InetOrgPersonContextMapper());
    factory.setUserSearchFilter("(uid={0})");
    factory.setLdapAuthoritiesPopulator(authorities);
    factory.setPasswordAttribute("userPassword");
    return factory.createAuthenticationManager();
}
```

The notable differences here (aside from the LDAP URL) are that the DN and password for an account are provided. The account (which is actually optional) should be allowed to bind to the directory and perform searches across all relevant DNs for user and group information. The binding resulting from the application of these credentials against the LDAP server URL is used for the remaining LDAP operations across the LDAP-secured system.

Be aware that many LDAP servers also support **SSL-encrypted LDAP** (**LDAPS**)—this is, of course, preferred for security purposes and is supported by the Spring LDAP stack. Simply use ldaps:// at the beginning of the LDAP server URL. LDAPS typically runs on TCP port 636. Note that there are many commercial and non-commercial implementations of LDAP.

The exact configuration parameters that you will use for connectivity, user binding, and the population of GrantedAuthoritys will wholly depend on both the vendor and the structure of the directory. We will cover one very common LDAP implementation, Microsoft AD, in the next section.

If you do not have an LDAP server handy and would like to give this a try, go ahead and add the following code to your SecurityConfig.java file, which starts up the embedded LDAP server we have been using:

```
//src/main/java/com/packtpub/springsecurity/configuration/
SecurityConfig.java
private BaseLdapPathContextSource
getDefaultSpringSecurityContextSource () {
    DefaultSpringSecurityContextSource
defaultSpringSecurityContextSource =      new
DefaultSpringSecurityContextSource(
        List.of("ldap://localhost:" + LDAP_PORT),
"dc=jbcpcalendar,dc=com");
    defaultSpringSecurityContextSource.
setUserDn("uid=admin,ou=system");
    defaultSpringSecurityContextSource.setPassword("secret");
    defaultSpringSecurityContextSource.afterPropertiesSet();
    return defaultSpringSecurityContextSource;
}
```

```
@Bean
LdapAuthoritiesPopulator authorities() {
    String groupSearchBase = "ou=Groups";
    DefaultLdapAuthoritiesPopulator authorities =
        new DefaultLdapAuthoritiesPopulator(this.
getDefaultSpringSecurityContextSource(), groupSearchBase);
    authorities.setGroupSearchFilter("(uniqueMember={0})");
    return authorities;
}

@Bean
AuthenticationManager authenticationManager(LdapAuthoritiesPopulator
authorities) {
    LdapPasswordComparisonAuthenticationManagerFactory factory = new
LdapPasswordComparisonAuthenticationManagerFactory(
        this.getDefaultSpringSecurityContextSource(), new
LdapShaPasswordEncoder());
    factory.setUserSearchBase("");
    factory.setUserDetailsContextMapper(new
InetOrgPersonContextMapper());
    factory.setUserSearchFilter("(uid={0})");
    factory.setLdapAuthoritiesPopulator(authorities);
    factory.setPasswordAttribute("userPassword");
    return factory.createAuthenticationManager();
}

@Bean
public UserDetailsService userDetailsService(LdapAuthoritiesPopulator
authorities) {
    return new LdapUserDetailsService(new
FilterBasedLdapUserSearch("", "(uid={0})", this.
getDefaultSpringSecurityContextSource()), authorities);
}
```

If this isn't convincing, start up your LDAP server import `calendar.ldif` into it. You can then connect to the external LDAP server. Go ahead and restart the application and give this a try with the username as `shauser@example.com` and the `password` as `shauser`.

> **Important note**
>
> Your code should look like `chapter06.06-calendar`.

Explicit LDAP bean configuration

In this section, we'll lead you through the set of bean configurations required to explicitly configure both a connection to an external LDAP server and the `LdapAuthenticationProvider` interface required to support authentication against an external server. As with other explicit bean-based configurations, you really want to avoid doing this unless you find yourself in a situation where the capabilities of the security namespace style of configuration will not support your business or your technical requirements, in which case, read on!

Configuring an external LDAP server reference

To implement this configuration, we'll assume that we have a local LDAP server running on port 33389, with the same configuration corresponding to the `DefaultSpringSecurityContextSource` interface example provided in the previous section. The required bean definition is provided in the `SecurityConfig.java` file. In fact, to keep things simple, we have provided the entire `SecurityConfig.java` file. Review the LDAP server reference in the following code snippet:

```
//src/main/java/com/packtpub/springsecurity/configuration/
SecurityConfig.java
@Bean
public DefaultSpringSecurityContextSource contextSource() {
    DefaultSpringSecurityContextSource
defaultSpringSecurityContextSource =        new
DefaultSpringSecurityContextSource(
        List.of("ldap://localhost:" + LDAP_PORT),
"dc=jbcpcalendar,dc=com");
    defaultSpringSecurityContextSource.
setUserDn("uid=admin,ou=system");
    defaultSpringSecurityContextSource.setPassword("secret");
    return defaultSpringSecurityContextSource;
}
```

Next, we'll explore how we can perform a search to locate the user in the LDAP directory.

Performing a search to locate the user in the LDAP directory

If you've read and understood the explanations throughout this chapter describing how Spring Security LDAP authentication works behind the scenes, this bean configuration will be perfectly understandable, with the following characteristics:

- User credential binding authentication (not password comparison)
- Use of `InetOrgPerson` in `UserDetailsContextMapper`

Take a look at the following steps:

1. The first bean provided for us is `BindAuthenticator`, and the supporting `FilterBased LdapUserSearch` bean is used to locate the user's DN in the LDAP directory prior to binding, as follows:

```java
//src/main/java/com/packtpub/springsecurity/configuration/
SecurityConfig.java
@Bean
public BindAuthenticator
bindAuthenticator(FilterBasedLdapUserSearch userSearch,
BaseLdapPathContextSource contextSource){
    BindAuthenticator bindAuthenticator = new
BindAuthenticator(contextSource);
    bindAuthenticator.setUserSearch(userSearch);
    return bindAuthenticator;
}
@Bean
public FilterBasedLdapUserSearch
filterBasedLdapUserSearch(BaseLdapPathContextSource
contextSource){
    return new FilterBasedLdapUserSearch("", //user-search-base
        "(uid={0})", //user-search-filter
        contextSource); //ldapServer
}
```

2. Second, `LdapAuthoritiesPopulator` and `UserDetailsContextMapper` perform the roles we examined earlier in the chapter:

```java
//src/main/java/com/packtpub/springsecurity/configuration/
SecurityConfig.java
@Bean
public LdapAuthoritiesPopulator
authoritiesPopulator(BaseLdapPathContextSource contextSource){
    DefaultLdapAuthoritiesPopulator
defaultLdapAuthoritiesPopulator = new
DefaultLdapAuthoritiesPopulator(contextSource,"ou=Groups");
    defaultLdapAuthoritiesPopulator.
setGroupSearchFilter("(uniqueMember={0})");
    return defaultLdapAuthoritiesPopulator;
}

@Bean
public UserDetailsContextMapper userDetailsContextMapper(){
    return new InetOrgPersonContextMapper();
}
```

3. Finally, we must update Spring Security to utilize our explicitly configured `UserDetailsService` bean as follows:

```
//src/main/java/com/packtpub/springsecurity/configuration/
SecurityConfig.java
@Bean
public UserDetailsService
userDetailsService(FilterBasedLdapUserSearch
filterBasedLdapUserSearch,
        LdapAuthoritiesPopulator authoritiesPopulator,
UserDetailsContextMapper userDetailsContextMapper) {
    LdapUserDetailsService ldapUserDetailsService =
new LdapUserDetailsService(filterBasedLdapUserSearch,
authoritiesPopulator);
    ldapUserDetailsService.
setUserDetailsMapper(userDetailsContextMapper);
    return ldapUserDetailsService;
}
```

4. At this point, we have fully configured LDAP authentication with explicit Spring bean notation. Employing this technique in the LDAP integration is useful in a few cases, such as when the security namespace does not expose certain configuration attributes, or when custom implementation classes are required to provide functionality tailored to a particular business scenario. We'll explore one such scenario later in this chapter when we examine how to connect to Microsoft AD via LDAP.

5. Go ahead and start the application and give the configuration a try with the `username` as `shauser@example.com` and the `password` as `shauser`.

    Assuming you have an external LDAP server running, or you have kept the configured in-memory `DefaultSpringSecurityContextSource` object, everything should still be working.

> **Important note**
> Your code should look like `chapter06.07-calendar`.

Delegating role discovery to UserDetailsService

One technique for populating user roles that are available to use with explicit bean configuration is implementing the support for looking up a user by username in `UserDetailsService` and getting the `GrantedAuthority` objects from this source.

The configuration is as simple as replacing the bean with the `ldapAuthoritiesPopulator` ID bean with an updated `UserDetailsService` and `LdapAuthoritiesPopulator` object, with a reference to `UserDetailsService`. Make the following updates to the `SecurityConfig.java` file, ensuring you remove the previous `ldapAuthoritiesPopulator` bean definition:

```
//src/main/java/com/packtpub/springsecurity/configuration/
SecurityConfig.java
@Bean
public BindAuthenticator bindAuthenticator(FilterBasedLdapUserSearch
userSearch, BaseLdapPathContextSource contextSource){
    BindAuthenticator bindAuthenticator = new
BindAuthenticator(contextSource);
    bindAuthenticator.setUserSearch(userSearch);
    return bindAuthenticator;
}


@Bean
public FilterBasedLdapUserSearch
filterBasedLdapUserSearch(BaseLdapPathContextSource contextSource){
    return new FilterBasedLdapUserSearch("", //user-search-base
            "(uid={0})", //user-search-filter
            contextSource); //ldapServer
}


@Bean
public LdapAuthoritiesPopulator
authoritiesPopulator(UserDetailsService userDetailsService){
    return new
UserDetailsServiceLdapAuthoritiesPopulator(userDetailsService);
}
```

We will also need to ensure that we have defined `userDetailsService`. To keep things simple, add an in-memory `UserDetailsService` interface, as follows:

```
//src/main/java/com/packtpub/springsecurity/configuration/
SecurityConfig.java
@Bean
public UserDetailsManager userDetailsService() {
    InMemoryUserDetailsManager manager = new
InMemoryUserDetailsManager();
    manager.createUser(User.withUsername("user1@example.com").
password("user1").roles("USER").build());
    manager.createUser(User.withUsername("admin1@example.com").
password("admin1").roles("USER", "ADMIN").build());
    return manager;
}
```

At the end, we configured a custom `LdapAuthenticationProvider` interface, as follows:

```
//src/main/java/com/packtpub/springsecurity/configuration/
SecurityConfig.java
@Bean
public LdapAuthenticationProvider
authenticationProvider(BindAuthenticator ba,
        LdapAuthoritiesPopulator lap,
        UserDetailsContextMapper cm){
    LdapAuthenticationProvider ldapAuthenticationProvider =  new
LdapAuthenticationProvider(ba, lap);
    ldapAuthenticationProvider.setUserDetailsContextMapper(cm);
    return ldapAuthenticationProvider;
}
```

If you have it, you will want to remove the references to `UserDetailsService` from `AccountController`, as follows:

```
//src/main/java/com/packtpub/springsecurity/web/controllers/
AccountController.java
@Controller
public class AccountController {

    @RequestMapping("/accounts/my")
    public String view(Model model) {
        Authentication authentication = SecurityContextHolder.
getContext().getAuthentication();
        if(authentication == null) {
            throw new IllegalStateException("authentication cannot be
null. Make sure you are logged in.");
        }
        Object principal = authentication.getPrincipal();
        model.addAttribute("user", principal);
        model.addAttribute("isLdapUserDetails", principal instanceof
LdapUserDetails);
        model.addAttribute("isLdapPerson", principal instanceof
Person);
        model.addAttribute("isLdapInetOrgPerson", principal instanceof
InetOrgPerson);
        return "accounts/show";
    }
}
```

You should now be able to authenticate with `admin1@example.com` as the `username` and `admin1` as the `password`. Naturally, we could also substitute this in-memory `UserDetailsService` interface for the JDBC or JPA-based one we discussed in *Chapter 4, JDBC-based Authentication*, and in *Chapter 5, Authentication with Spring Data*.

> **Important note**
> Your code should look like `chapter06.08-calendar`.

The logistical and managerial problem you may notice with this is that the usernames and roles must be managed both in the LDAP server and the repository used by `UserDetailsService`—this is probably not a scalable model for a large user base.

The more common use of this scenario is when LDAP authentication is required to ensure that users of the secured application are valid corporate users, but the application itself wants to store authorization information. This keeps potentially application-specific data out of the LDAP directory, which can be a beneficial separation of concerns.

Integrating with Microsoft Active Directory via LDAP

One of the convenient features of Microsoft AD is not only its seamless integration with Microsoft Windows-based network architectures, but also that it can be configured to expose the contents of AD using the LDAP protocol. If you are working in a company that is heavily leveraging Microsoft Windows, it is probable that any LDAP integration you do will be against your AD instance.

Depending on your configuration of Microsoft AD (and the directory administrator's willingness to configure it to support Spring Security LDAP), you may have a difficult time, not with the authentication and binding process, but with the mapping of AD information to the user's `GrantedAuthority` objects within the Spring Security system.

The sample AD LDAP tree for JBCP calendar corporate within our LDAP browser looks like the following screenshot:

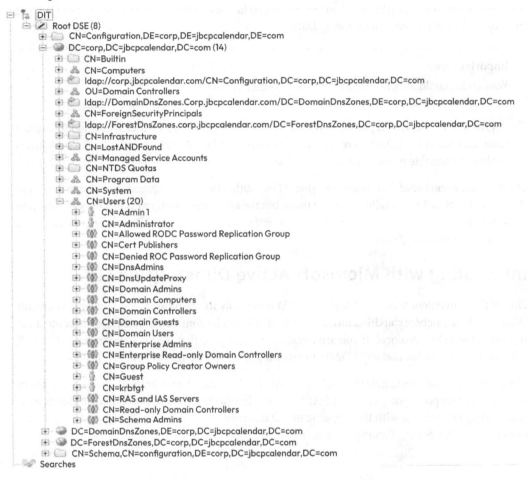

Figure 6.11 – Example of a Microsoft Active Directory structure

What you do not see here is ou=Groups, which we saw in our sample LDAP structure earlier; this is because AD stores group membership as attributes on the LDAP entries of the users themselves.

We need to alter our configuration to support our AD structure. Assuming we are starting with the bean configuration detailed in the previous section, make the following updates:

```
//src/main/java/com/packtpub/springsecurity/configuration/
SecurityConfig.java
@Bean
public AuthenticationProvider authenticationProvider(){
    ActiveDirectoryLdapAuthenticationProvider ap = new
```

```
ActiveDirectoryLdapAuthenticationProvider(
        "corp.jbcpcalendar.com",
        "ldap://corp.jbcpcalendar.com");
    ap.setConvertSubErrorCodesToExceptions(true);
    return ap;
}


@Bean
public DefaultSpringSecurityContextSource contextSource() {
    DefaultSpringSecurityContextSource
defaultSpringSecurityContextSource =      new
DefaultSpringSecurityContextSource(
        List.of("ldap://corp.jbcpcalendar.com"),
"dc=corp,dc=jbcpcalendar,dc=com");
    defaultSpringSecurityContextSource.
setUserDn("CN=bnl,CN=Users,DC=corp,DC=jbcpcalendar,DC=com");
    defaultSpringSecurityContextSource.setPassword("admin123!");
    return defaultSpringSecurityContextSource;
}


@Bean
public FilterBasedLdapUserSearch
filterBasedLdapUserSearch(BaseLdapPathContextSource contextSource) {
    return new FilterBasedLdapUserSearch("CN=Users", //user-search-
base
        "(sAMAccountName={0})", //user-search-filter
        contextSource); //ldapServer
}
```

If you have it defined, you will want to remove the `UserDetailsService` declaration in the `SecurityConfig.java` file. Finally, you will want to remove the references to `UserDetailsService` from `AccountController`.

The `sAMAccountName` attribute is the AD equivalent of the `uid` attribute we use in a standard LDAP entry. Although most AD LDAP integrations are likely to be more complex than this example, this should give you a starting point to jump off and explore your conceptual understanding of the inner workings of Spring Security LDAP integration; supporting even a complex integration will be much easier.

> **Important note**
>
> If you want to run this sample, you will need an instance of AD up and running that matches the schema displayed in the screenshot. The alternative is to adjust the configuration to match your AD schema. A simple way to play around with AD is to install `Active Directory Lightweight Directory Services`, which can be found at `https://www.microsoft.com/fr-FR/download/details.aspx?id=1451`. Your code should look like `chapter06.09-calendar`.

Built-in AD support in Spring Security 6.1

Active Directory supports its own non-standard authentication options, and the normal usage pattern does not fit too cleanly with the standard `LdapAuthenticationProvider`. Typically, authentication is performed by using the domain username (in the form of `user@domain`), rather than using an LDAP distinguished name. To make this easier, Spring Security has an authentication provider that is customized for a typical Active Directory setup.

Configuring `ActiveDirectoryLdapAuthenticationProvider` is quite straightforward. You need only supply the domain name and an LDAP URL that supplies the address of the server as we have covered in the previous sections. A reminder of what the configuration looks like can be found in the following code snippet:

```
//src/main/java/com/packtpub/springsecurity/configuration/
SecurityConfig.ja
@Bean
public AuthenticationProvider authenticationProvider(){
    ActiveDirectoryLdapAuthenticationProvider ap = new
ActiveDirectoryLdapAuthenticationProvider(
            "corp.jbcpcalendar.com",
            "ldap://corp.jbcpcalendar.com");
    ap.setConvertSubErrorCodesToExceptions(true);
    return ap;
}
```

There are a few things that should be noted about the provided `ActiveDirectory LdapAuthenticationProvider` class, as follows:

- The users that need to be authenticated must be able to bind to AD (there is no manager user)
- The default method for populating user authorities is to search the users `memberOf` attributes
- Users must contain an attribute named `userPrincipalName`, which is in the `username@<domain>` format

Due to the complex LDAP deployments that occur in the real world, the built-in support will most likely provide a guide to as how you can integrate with your custom LDAP schema.

Summary

We have seen that LDAP servers can be relied on to provide authentication and authorization information, as well as rich user profile information when requested. In this chapter, we covered the LDAP terminology and concepts, and how LDAP directories might be commonly organized to work with Spring Security. We also explored the configuration of both standalone (embedded) and external LDAP servers from a Spring Security configuration file.

We covered the authentication and authorization of users against LDAP repositories, and their subsequent mapping to Spring Security actors. We also saw the differences in authentication schemes, password storage, and security mechanisms in LDAP, and how they are treated in Spring Security. We also learned to map user detail attributes from the LDAP directory to the `UserDetails` object for rich information exchange between LDAP and the Spring-enabled application. We also explained bean configuration for LDAP and the pros and cons of this approach. Lastly, we also covered integration with Microsoft AD.

In the next chapter, we will discuss Spring Security's `remember-me` feature, which allows a user's session to securely persist even after closing the browser.

7

Remember-me Services

In this chapter, we'll add the ability for an application to remember a user even after their session has expired and the browser is closed. The following topics will be covered in this chapter:

- Discussing what **remember-me** is

- Learning how to use the token-based remember-me feature

- Discussing how secure remember-me is, and various ways of making it more secure

- Enabling the persistent-based remember-me feature, and how to handle additional considerations for using it

- Presenting the overall remember-me architecture

- Learning how to create a custom remember-me implementation that is restricted to the user's IP address

This chapter's code in action link is here: `https://packt.link/WEEx2`.

What is remember-me?

A convenient feature to offer frequent users of a website is the remember-me feature. This feature allows a user to elect to be remembered even after their browser is closed. In Spring Security, this is implemented through the use of a remember-me **cookie** that is stored in the user's browser. If Spring Security recognizes that the user is presenting a remember-me cookie, then the user will automatically be logged into the application, and will not need to enter a username or password.

> **What is a cookie?**
>
> A cookie is a way for a client (that is, a web browser) to persist the state. For more information about cookies, refer to additional online resources, such as Wikipedia (`https://en.wikipedia.org/wiki/HTTP_cookie`).

Spring Security provides the following two different strategies that we will discuss in this chapter:

- The first is the token-based remember-me feature, which relies on a cryptographic signature
- The second method, the `persistent-based remember-me` feature, requires a datastore (a database)

As we previously mentioned, we will discuss these strategies in much greater detail throughout this chapter. The remember-me feature must be explicitly configured in order to enable it. Let's start by trying the token-based remember-me feature and see how it affects the flow of the login experience.

Dependencies

The token-based remember-me section does not need any additional dependencies other than the basic setup from *Chapter 2, Getting Started with Spring Security*. However, you will want to ensure you include the following additional dependencies in your `build.gradle` file if you are leveraging the persistent-based remember-me feature. We have already included these dependencies in the chapter's sample, so there is no need to update the sample application:

```
//build.gradle
dependencies {
...
    // JPA / ORM / Hibernate:
    implementation 'org.springframework.boot:spring-boot-starter-data-
jpa'
    // H2 db
    implementation 'com.h2database:h2'
...
}
```

The token-based remember-me feature

Spring Security provides two different implementations of the remember-me feature. We will start by exploring how to set up token-based remember-me services.

Configuring the token-based remember-me feature

Completing this exercise will allow us to provide a simple and secure method to keep users logged in for extended periods of time. To start, perform the following steps:

1. Modify the `SecurityConfig.java` configuration file and add the `rememberMe` method.

    Take a look at the following code snippet:

    ```
    //src/main/java/com/packtpub/springsecurity/configuration/
    SecurityConfig.java
    ```

```
@Bean
public SecurityFilterChain filterChain(HttpSecurity http) throws
Exception {
...
    // Remember Me
    http.rememberMe(httpSecurityRememberMeConfigurer ->
        httpSecurityRememberMeConfigurer.key("jbcpCalendar"));
...
}
```

> **Important note**
>
> You should start with the source from `chapter07.00-calendar`.

2. If we try running the application now, we'll see nothing different in the flow. This is because we also need to add a field to the login form that allows the user to opt for this functionality. Edit the `login.html` file and add a checkbox, as shown in the following code snippet:

```
//src/main/resources/templates/login.html
<div class="mb-3">
    <label class="form-label" for="password">Password</label>
    <input class="form-control" id="password" name="password"
        type="password"/>
</div>

<div class="mb-3">
    <label for="remember-me">Remember Me?</label>
    <input type="checkbox" id="remember-me" name="remember-me"
th:checked="true" />
</div>

<div class="mb-3">
    <input class="btn btn-primary" id="submit" name="submit"
type="submit"
        value="Login"/>
</div>
```

> **Important note**
>
> You should start with the source from `chapter07.01-calendar`.

3. When we next log in, if the remember-me box is selected, a remember-me cookie is set in the user's browser.

    Spring Security understands that it should remember the user by inspecting the `remember-me` HTTP parameter.

> **Important note**
>
> In Spring Security 4.x and after, the default remember-me form field is `remember-me`. This can be overridden with the `rememberMeParameter` method.

4. If the user then closes their browser and reopens it to an authenticated page on the JBCP calendar website, they won't be presented with the login page a second time. Try it yourself now—log in with the remember-me option selected, bookmark the home page, then restart the browser and access the home page. You'll see that you're immediately logged in successfully without needing to supply your login credentials again. If this appears to be happening to you, it means that your browser or a browser plugin is restoring the session.

> **Tip**
>
> Try closing the tab first and then close the browser.

One more effective solution is to use Chrome developer tools to remove the `JSESSIONID` cookie. This can often save time and annoyance during the development and verification of this type of feature on your site.

Figure 7.1 – Exploring the remember-me cookie

After logging in and selecting `remember-me`, you should see two cookies have been set, `JSESSIONID` and `remember-me`, as shown in the screenshot.

How the token-based remember-me feature works

The remember-me feature sets a cookie in the user's browser containing a Base64-encoded string with the following pieces:

- The username
- An expiration date/time

- An SHA-256 hash value of the `expiration` date/time, `username`, `password`, and the `key` attribute of the `rememberMe` method

These are combined into a single cookie value that is stored in the browser for later use. The cookie is composed as follows:

```
base64(username + ":" + expirationTime + ":" + algorithmName + ":"
algorithmHex(username + ":" + expirationTime + ":" password + ":" +
key))
```

```
username:           As identifiable to the UserDetailsService
password:           That matches the one in the retrieved UserDetails
expirationTime:     The date and time when the remember-me token
expires, expressed in milliseconds
key:                A private key to prevent modification of the
remember-me token
algorithmName:      The algorithm used to generate and to verify the
remember-me token
```

In the upcoming section, we'll explore the SHA-256 algorithm in conjunction with Spring Security.

SHA-256 Algorithm

By default, this implementation uses the SHA-256 algorithm to encode the token signature. To verify the token signature, the algorithm retrieved from `algorithmName` is parsed and used. If `algorithmName` is not present, the default matching algorithm will be used, which is `SHA-256`. You can specify different algorithms for signature encoding and for signature matching; this allows users to safely upgrade to a different encoding algorithm while still being able to verify old ones if `algorithmName` is not present. To do that, you can specify your customized `TokenBasedRememberMeServices` as a bean and use it in the configuration:

```
@Bean
SecurityFilterChain securityFilterChain(HttpSecurity http,
RememberMeServices rememberMeServices) throws Exception {
    http
        .authorizeHttpRequests((authorize) -> authorize
            .anyRequest().authenticated()
        )
        .rememberMe((remember) -> remember
            .rememberMeServices(rememberMeServices)
        );
    return http.build();
}

@Bean
```

```
RememberMeServices rememberMeServices(UserDetailsService
userDetailsService) {
    RememberMeTokenAlgorithm encodingAlgorithm =
RememberMeTokenAlgorithm.SHA256;
    TokenBasedRememberMeServices rememberMe = new
TokenBasedRememberMeServices(myKey, userDetailsService,
encodingAlgorithm);
    rememberMe.setMatchingAlgorithm(RememberMeTokenAlgorithm.MD5);
    return rememberMe;
}
```

To recap, we've covered the SHA-256 algorithm, and in the next section, we'll delve into the remember-me signature.

Remember-me signature

We can see how SHA-256 can ensure that we have downloaded the correct file, but how does this apply to Spring Security's remember-me service? Much like the file we downloaded, the cookie is untrusted, but we can trust it if we can validate the signature that originated from our application. When a request comes in with the remember-me cookie, its contents are extracted, and the expected signature is compared to the signature found in the cookie.

The steps in calculating the expected signature are illustrated in the following diagram:

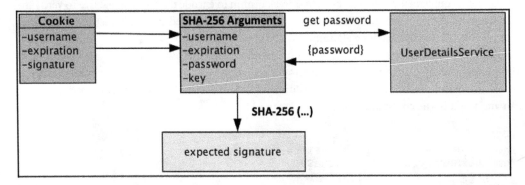

Figure 7.2 – SHA-256 Hash-Based Token Approach

The remember-me cookie contains the username, expiration, and signature. Spring Security will extract the username and expiration from the cookie. It will then utilize the username from the cookie to look up the password using UserDetailsService. The key is already known because it was provided using the rememberMe method. Now that all of the arguments are known, Spring Security can calculate the expected signature using the username, expiration, password, and key. It then compares the expected signature against the cookie's signature.

If the two signatures match, we can trust that the `username` and `expiration` date are valid. Forging a `signature` is next to impossible without knowing the remember-me key (which only the application knows) and the user's password (which only this user knows). This means that if the signatures match and the token is not expired, the user can be logged in.

> **Important note**
>
> You have anticipated that if the user changes their username or password, any remember-me token set will no longer be valid. Make sure that you provide appropriate messaging to users if you allow them to change these bits of their account. Later in this chapter, we will look at an alternative remember-me implementation that is reliant only on the username and not on the password.

Note that it is still possible to differentiate between users who have been authenticated with a remember-me cookie and users who have presented the username and password (or equivalent) credentials. We'll experiment with this shortly when we investigate the security of the remember-me feature.

Token-based remember-me configuration directives

The following two configuration changes are commonly made to alter the default behavior of the remember-me functionality:

Attribute	Description
`key`	This defines a unique key used when producing the remember-me cookie's signature.
`tokenValiditySeconds`	This defines the length of time (in seconds). The remember-me cookie will be considered valid for authentication. It is also used to set the cookie expiration timestamp.

Table 7.3 – The main configuration of the remember-me cookie

As you may infer from the discussion of how the cookie contents are hashed, the `key` attribute is critical for the security of the remember-me feature. Make sure that the key you choose is likely to be unique to your application and long enough so that it can't be easily guessed.

Keeping in mind the purpose of this book, we've kept the `key` values relatively simple, but if you're using remember-me in your application, it's suggested that your key contains the unique name of your application and is at least 36 random characters long. Password generator tools (search `online password generator` on Google) are a great way to get a pseudo-random mix of alphanumeric and special characters to compose your remember-me key. For applications that exist in multiple environments (such as development, test, and production), the remember-me cookie value should include this fact as well. This will prevent remember-me cookies from inadvertently being used in the wrong environment during testing!

An example key value in a production application might be similar to the following:

```
prodJbcpCalendar-rmkey- YWRtaW4xJTQwZXhhbXBsZS5jb206MTY5ODc2MTM
2ODgwNjpTSEEyNTY6YzE5ZjE2YzliN2U2ZjA
xZGMyMjdkMWJmN2J1YWQzNGRhYWJiMGFmND1iMDE0ZGY5MTg4YjIzYzM1YjQzZmMzNw
```

The `tokenValiditySeconds` method is used to set the number of seconds after which the remember-me token will not be accepted for the automatic login function, even if it is otherwise a valid token. The same attribute is also used to set the maximum lifetime of the remember-me cookie on the user's browser.

Configuration of the remember-me session cookies

If `tokenValiditySeconds` is set to `-1`, the login cookie will be set to a session cookie, which does not persist after the browser is closed by the user. The token will be valid (assuming the user doesn't close the browser) for a non-configurable length of two weeks. Don't confuse this with the cookie that stores your user's session ID—they're two different things with similar names!

You may have noticed that we listed very few attributes. Don't worry, we will spend time covering some of the other configuration attributes throughout this chapter.

Is remember-me secure?

Any feature related to security that has been added for user convenience has the potential to expose our carefully protected site to a security risk. The remember-me feature, in its default form, runs the risk of the user's cookie being intercepted and reused by a malicious user. The following diagram illustrates how this might happen:

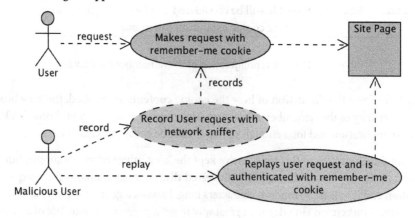

Figure 7.3 –Remember-me session cookie replay attack

The use of **Secure Sockets Layer (SSL)** (covered in the *Appendix, Additional Reference Material*) and other network security techniques can mitigate this type of attack, but be aware that there are other techniques, such as **Cross-Site Scripting (XSS)**, that can steal or compromise a remembered user session. While convenient for the user, we don't want to risk financial or other personal information being inadvertently changed or possibly stolen if the remembered session is misused.

> **Important note**
>
> Although we don't cover malicious user behavior in detail in this book, when implementing any secured system, it is important to understand the techniques employed by users who may be trying to hack your customers or employees. XSS is one such technique, but many others exist. It's highly recommended that you review the *OWASP Top Ten* article (`https://owasp.org/www-project-top-ten/`) for a good list, and also pick up a web application security reference book in which many of the techniques demonstrated are illustrated to apply to any technology.

One common approach for maintaining the balance between convenience and security is identifying the functional locations on the site where personal or sensitive information could be present. You can then use the `fullyAuthenticated` expression to ensure these locations are protected using an authorization that checks not just the user's role, but that they have been authenticated with a full username and password. We will explore this feature in greater detail in the next section.

Authorization rules for remember-me

We'll fully explore the advanced authorization techniques later in *Chapter 11, Fine-Grained Access Control*, however, it's important to realize that it's possible to differentiate access rules based on whether or not an authenticated session was remembered.

Let's assume we want to limit users trying to access the H2 admin console to administrators who have been authenticated using a username and password. This is similar to the behavior found in other major consumer-focused commerce sites, which restrict access to the elevated portions of the site until a password is entered. Keep in mind that every site is different, so don't blindly apply such rules to your secure site. For our sample application, we'll concentrate on protecting the H2 database console. Update the `SecurityConfig.java` file to use the `fullyAuthenticated` keyword, which ensures that remembered users who try to access the H2 database are denied access. This is shown in the following code snippet:

```
//src/main/java/com/packtpub/springsecurity/configuration/
SecurityConfig.java

...

@Bean
public SecurityFilterChain filterChain(HttpSecurity http,
PersistentTokenRepository persistentTokenRepository) throws Exception
{
```

```
    http.authorizeHttpRequests( authz -> authz
            .requestMatchers("/webjars/**").permitAll()
            .requestMatchers("/css/**").permitAll()
            .requestMatchers("/favicon.ico").permitAll()

            // H2 console:
            .requestMatchers("/admin/h2/**")
            .access(new
WebExpressionAuthorizationManager("isFullyAuthenticated() and
hasRole('ADMIN')"))
...
    // Remember Me
    http.rememberMe(httpSecurityRememberMeConfigurer ->
httpSecurityRememberMeConfigurer
        .key("jbcpCalendar").
tokenRepository(persistentTokenRepository));
...
}
```

The existing rules remain unchanged. We've added a rule that requires requests for account information to have the appropriate GrantedAuthority of ROLE_ADMIN, and that the user is fully authenticated; that is, during this authenticated session, they have presented a username and password or other suitable credentials. Note the syntax of the **Spring Expression Language (SpEL)** logical operators here— AND, OR, and NOT are used for logical operators in SpEL. This was thoughtful of the SpEL designers, as the && operator would be awkward to represent in XML, even though the preceding example uses Java-based configuration!

> **Important note**
> You should start with the source from chapter07.02-calendar.

Go ahead and log in with the username admin1@example.com and the password admin1, ensuring you select the remember-me feature. Access the H2 database console and you will see that the access is granted. Now, delete the JSESSIONID cookie (or close the tab and then all of the browser instances), and ensure that access is still granted to the **All Events** page.

Now, navigate to the H2 console and observe that the access is denied.

This approach combines the usability enhancements of the remember-me feature with an additional level of security by requiring a user to present a full set of credentials to access sensitive information. Throughout the rest of the chapter, we will explore other ways of making the remember-me feature more secure.

Persistent remember-me

Spring Security provides the capability to alter the method for validating the remember-me cookie by leveraging different implementations of the `RememberMeServices` interface. In this section, we will discuss how we can use persistent remember-me tokens using a database, and how this can increase the security of our application.

Using the persistent-based remember-me feature

Modifying our remember-me configuration at this point to persist to the database is surprisingly trivial. The Spring Security configuration parser will recognize a new `tokenRepository` method on the `rememberMe` method, and simply switch implementation classes for `RememberMeServices`. Let's now review the steps required to accomplish this.

Adding SQL to create the remember-me schema

We have placed the SQL file containing the expected schema in our `src/main/resources` folder in the same place we did in *Chapter 3, Custom Authentication*. You can view the schema definition in the following code snippet:

```
//src/main/resources/schema.sql
create table persistent_logins
(
    username   varchar_ignorecase(50) not null,
    series     varchar(64) primary key,
    token      varchar(64) not null,
    last_used timestamp    not null
);
```

Initializing the data source with the remember-me schema

Spring Data will automatically initialize the embedded database with `schema.sql`, as described in the preceding section. Note, however, that with **Jakarta Persistence API (JPA)**, in order for the schema to be created and the `data.sql` file used to seed the database, we must ensure data source initialization is deferred as follows:

```
//src/main/resources/application.yml
spring:
  jpa:
    database-platform: org.hibernate.dialect.H2Dialect
    show-sql: false
    hibernate:
      ddl-auto: create-drop
    defer-datasource-initialization: true
```

After reviewing the persistent-based remember-me functionality, specifically using a database, the next section will cover the configuration of this feature with JPA.

Configuring the persistent-based remember-me feature

Finally, we'll need to make some brief configuration changes to the rememberMe declaration to point it to the data source we're using, as shown in the following code snippet:

```
//src/main/java/com/packtpub/springsecurity/configuration/SecurityC
onfig.java
@Bean
public SecurityFilterChain filterChain(HttpSecurity http,
PersistentTokenRepository persistentTokenRepository) throws Exception
{
    http.authorizeRequests( authz -> authz
    ...
    // Remember Me
    http.rememberMe(httpSecurityRememberMeConfigurer ->
httpSecurityRememberMeConfigurer
        .key("jbcpCalendar").
tokenRepository(persistentTokenRepository));
    return http.build();
}


@Bean
public PersistentTokenRepository persistentTokenRepository(DataSource
dataSource) {
    JdbcTokenRepositoryImpl db = new JdbcTokenRepositoryImpl();
    db.setDataSource(dataSource);
    return db;
}
```

This is all we need to do to switch over to using persistent-based remember-me authentication. Go ahead and start up the application and give it a try. From a user standpoint, we do not notice any differences, but we know that the implementation backing this feature has changed.

> **Important note**
> You should start with the source from chapter07.03-calendar.

How does the persistent-based remember-me feature work?

Instead of validating a signature present in the cookie, the persistent-based remember-me service validates if the token exists in a database. Each persistent remember-me cookie consists of the following:

- **Series identifier**: This identifies the initial login of a user and remains consistent each time the user is automatically logged in to the original session

- **Token value**: A unique value that changes each time a user is authenticated using the remember-me feature

Take a look at the following diagram:

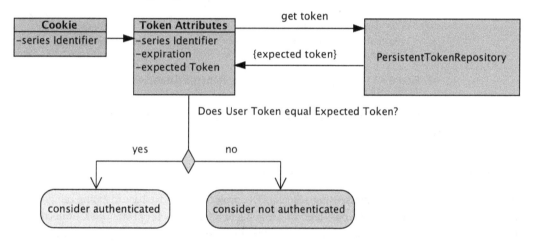

Figure 7.4 – Exploring the persistent-based remember-me feature

When the remember-me cookie is submitted, Spring Security will use an `o.s.s.web.authentication.rememberme.PersistentTokenRepository` implementation to look up the expected token value and expiration using the submitted series identifier. It will then compare the token value in the cookie to the expected token value. If the token is not expired and the two tokens match, the user is considered authenticated. A new remember-me cookie with the same series identifier, a new token value, and an updated expiration date will be generated.

If the series token submitted is found in the database, but the tokens do not match, it can be assumed that someone stole the remember-me cookie. In this case, Spring Security will terminate this series of remember-me tokens and warn the user that their login has been compromised.

The persisted tokens can be found in the database and viewed with the H2 console, as shown in the following screenshot:

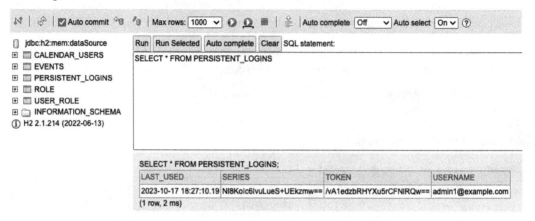

Figure 7.5 – Getting the persisted token from the database

After gaining an understanding of how the persistent-based remember-me feature operates in this chapter, we will delve into the JPA-based `PersistentTokenRepository` in the subsequent section.

JPA-based PersistentTokenRepository

As we have seen in the earlier chapters, using a Spring Data project for our database mapping can greatly simplify our work. So, to keep things consistent, we are going to refactor our **Java Database Connectivity (JDBC)** `PersistentTokenRepository` interface, which uses `JdbcTokenRepositoryImpl`, to one that is JPA-based. We will do so by performing the following steps:

1. First, let's create a domain object to hold the persistent logins, as shown in the following code snippet:

    ```
    //src/main/java/com/packtpub/springsecurity/domain/
    PersistentLogin.java
    @Entity
    @Table(name = "persistent_logins")
    public class PersistentLogin implements Serializable {

        @Id
        private String series;

        private String username;
        private String token;
        private Date lastUsed;
    ```

```
        public PersistentLogin(){}

        public PersistentLogin(PersistentRememberMeToken token){
            this.series = token.getSeries();
            this.username = token.getUsername();
            this.token = token.getTokenValue();
            this.lastUsed = token.getDate();
        }

    // getters/setters omitted for brevity
    }
```

2. Next, we need to create an `o.s.d.jpa.repository.JpaRepository` repository instance, as shown in the following code snippet:

```
//src/main/java/com/packtpub/springsecurity/repository/
RememberMeTokenRepository.java
import java.util.Date;
import java.util.List;

import com.packtpub.springsecurity.domain.PersistentLogin;

import org.springframework.data.jpa.repository.JpaRepository;

public interface RememberMeTokenRepository extends
JpaRepository<PersistentLogin, String> {

    PersistentLogin findBySeries(String series);
    List<PersistentLogin> findByUsername(String username);
    Iterable<PersistentLogin> findByLastUsedAfter(Date
expiration);

}
```

3. Now, we need to create a custom `PersistentTokenRepository` interface to replace the `Jdbc` implementation. We have four methods we must override, but the code should look fairly familiar as we will be using JPA for all of the operations:

```
//src/main/java/com/packtpub/springsecurity/web/authentication/
rememberme/JpaPersistentTokenRepository.java:
public class JpaPersistentTokenRepository implements
PersistentTokenRepository {

    private final RememberMeTokenRepository
rememberMeTokenRepository;
```

```
    public
JpaPersistentTokenRepository(RememberMeTokenRepository
rememberMeTokenRepository) {
        this.rememberMeTokenRepository =
rememberMeTokenRepository;
    }

    @Override
    public void createNewToken(PersistentRememberMeToken token)
{
        PersistentLogin newToken = new PersistentLogin(token);
        this.rememberMeTokenRepository.save(newToken);
    }

    @Override
    public void updateToken(String series, String tokenValue,
Date lastUsed) {
        PersistentLogin token = this.rememberMeTokenRepository.
findBySeries(series);
        if (token != null) {
            token.setToken(tokenValue);
            token.setLastUsed(lastUsed);
            this.rememberMeTokenRepository.save(token);
        }
    }

    @Override
    public PersistentRememberMeToken getTokenForSeries(String
seriesId) {
        PersistentLogin token = this.rememberMeTokenRepository.
findBySeries(seriesId);
        if(token == null){
            return null;
        } else {
            return new PersistentRememberMeToken(token.
getUsername(),
                    token.getSeries(),
                    token.getToken(),
                    token.getLastUsed());
        }
    }

    @Override
    public void removeUserTokens(String username) {
        List<PersistentLogin> tokens = this.
rememberMeTokenRepository.findByUsername(username);
```

```
                 this.rememberMeTokenRepository.deleteAll(tokens);
        }


    }
```

4. Now, we need to make a few changes in the `SecurityConfig.java` file to declare the new `PersistentTokenTokenRepository` interface, but the rest of the configuration from the last section does not change, as shown in the following code snippet:

```
/src/main/java/com/packtpub/springsecurity/configuration/
SecurityConfig.java
@Bean
public PersistentTokenRepository persistentTokenRepository(
        RememberMeTokenRepository rmtr) {
    return new JpaPersistentTokenRepository(rmtr);
}
```

5. This is all we need to do to switch JDBC to JPA persistent-based remember-me authentication. Go ahead and start up the application and give it a try. From a user standpoint, we do not notice any differences, but we know that the implementation backing this feature has changed.

> **Important note**
> You should start with the source from `chapter07.04-calendar`.

Custom RememberMeServices

Up to this point, we have used a fairly simple implementation of `PersistentTokenRepository`. We have used a JDBC-backed and JPA-backed implementation. This provided limited control over the cookie persistence; if we want more control, we wrap our `PersistentTokenRepository` interface in `RememberMeServices`. Spring Security has a slightly modified version, as previously described, called `PersistentTokenBasedRememberMeServices`, which we can wrap our custom `PersistentTokenRepository` interface in and use in our remember-me service.

In the following section, we are going to wrap our existing `PersistentTokenRepository` interface with `PersistentTokenBasedRememberMeServices` and use the `rememberMeServices` method to wire it into our remember-me declaration:

```
//src/main/java/com/packtpub/springsecurity/configuration/
SecurityConfig.java
@Bean
public SecurityFilterChain filterChain(HttpSecurity http,
        PersistentTokenRepository persistentTokenRepository,
RememberMeServices rememberMeServices) throws Exception {
    http.authorizeHttpRequests( authz -> authz
```

```
                    .requestMatchers("/webjars/**").permitAll()
    ...

    // Remember Me
    http.rememberMe(httpSecurityRememberMeConfigurer ->
httpSecurityRememberMeConfigurer
            .key("jbcpCalendar")
            .rememberMeServices(rememberMeServices)
            .tokenRepository(persistentTokenRepository));


    return http.build();
}
@Bean
public RememberMeServices rememberMeServices
(PersistentTokenRepository ptr,
UserDetailsService  userDetailsService){
    PersistentTokenBasedRememberMeServices rememberMeServices = new
            PersistentTokenBasedRememberMeServices("jbcpCalendar",
            userDetailsService, ptr);
    rememberMeServices.setAlwaysRemember(true);
    return rememberMeServices;
}
```

> **Important note**
>
> You should start with the source from `chapter07.05-calendar`.

Are database-backed persistent tokens more secure?

Just like `TokenBasedRememberMeServices`, persistent tokens may be compromised by cookie theft or other man-in-the-middle techniques. The use of SSL, as covered in the *Appendix, Additional Reference Material* can circumvent man-in-the-middle techniques. If you are using a **Servlet 5.0** environment (that is, Tomcat 10+), Spring Security will mark the cookie as `HttpOnly`, which will help to mitigate against the cookie being stolen in the event of an XSS vulnerability in the application. To learn more about the `HttpOnly` attribute, refer to the external resource on cookies provided earlier in the chapter.

One of the advantages of using the persistent-based remember-me feature is that we can detect whether the cookie is compromised. If the correct series token and an incorrect token are presented, we know that any remember-me feature using that series token should be considered compromised, and we should terminate any sessions associated with it. Since the validation is stateful, we can also terminate the specific remember-me feature without needing to change the user's password.

Cleaning up the expired remember-me sessions

The downside of using the persistent-based remember-me feature is that there is no built-in support for cleaning up the expired sessions. To do this, we need to implement a background process that cleans up the expired sessions. We have included code in the chapter's sample code to perform the cleanup.

For conciseness, we display a version that does not do validation or error handling in the following code snippet. You can view the full version in the sample code of this chapter:

```
//src/main/java/com/packtpub/springsecurity/web/authentication/
rememberme/ JpaTokenRepositoryCleaner.java
public class JpaTokenRepositoryCleaner implements Runnable {

    private Logger logger = LoggerFactory.getLogger(getClass());

    private final RememberMeTokenRepository rememberMeTokenRepository;

    private final long tokenValidityInMs;

    public JpaTokenRepositoryCleaner(RememberMeTokenRepository
rememberMeTokenRepository,
                                     long tokenValidityInMs) {
        if (rememberMeTokenRepository == null) {
            throw new IllegalArgumentException("jdbcOperations cannot
be null");
        }
        if (tokenValidityInMs < 1) {
            throw new IllegalArgumentException("tokenValidityInMs must
be greater than 0. Got " + tokenValidityInMs);
        }
        this.rememberMeTokenRepository = rememberMeTokenRepository;
        this.tokenValidityInMs = tokenValidityInMs;
    }

    public void run() {
        long expiredInMs = System.currentTimeMillis() -
tokenValidityInMs;

        logger.info("Searching for persistent logins older than {}ms",
tokenValidityInMs);

        try {
            Iterable<PersistentLogin> expired =
rememberMeTokenRepository.findByLastUsedAfter(new Date(expiredInMs));
            for(PersistentLogin pl: expired){
                logger.info("*** Removing persistent login for {}
```

```
***", pl.getUsername());
                remrmberMeTokenRepository.delete(pl);
        }
    } catch(Throwable t) {
        logger.error("**** Could not clean up expired persistent
remember me tokens. ***", t);
    }
  }
}
```

The sample code for this chapter also includes a simple Spring configuration that will execute the cleaner every ten minutes. If you are unfamiliar with Spring's task abstraction and want to learn it, then you may want to read more about it at `https://docs.spring.io/spring-framework/reference/integration/scheduling.html`. You can find the relevant configuration in the following code snippet. For clarity, we are putting this scheduler in the `JavaConfig.java` file:

```
//src/main/java/com/packtpub/springsecurity/configuration/ JavaConfig.
java@Configuration

@Configuration
@EnableScheduling
public class JavaConfig {

    private RememberMeTokenRepository rememberMeTokenRepository;

    public JavaConfig(RememberMeTokenRepository
rememberMeTokenRepository) {
        this.rememberMeTokenRepository = rememberMeTokenRepository;
    }

    @Scheduled(fixedRate = 600_000)
    public void tokenRepositoryCleaner(){
        Thread trct = new Thread(
            new JpaTokenRepositoryCleaner(
                rememberMeTokenRepository,
                100_000L));
        trct.start();
    }

}
```

> **Important note**
>
> Keep in mind that this configuration is not cluster aware. Therefore, if this is deployed to a cluster, the cleaner will execute once for every **Java Virtual Machine (JVM)** that the application is deployed to.

Start up the application and give the updates a try. The configuration that was provided will ensure that the cleaner is executed every ten minutes. You may want to change the cleaner task to run more frequently and clean up the more recently used remember-me tokens by modifying the @Scheduled declaration. You can then create a few remember-me tokens and see that they get deleted by querying for them in the H2 database console.

> **Important note**
> You should start with the source from chapter07.06-calendar.

The remember-me architecture

We have gone over the basic architecture of both TokenBasedRememberMeServices and PersistentTokenBasedRememberMeServices, but we have not described the overall architecture. Let's see how all of the remember-me pieces fit together.

The following diagram illustrates the different components involved in the process of validating a token-based remember-me token:

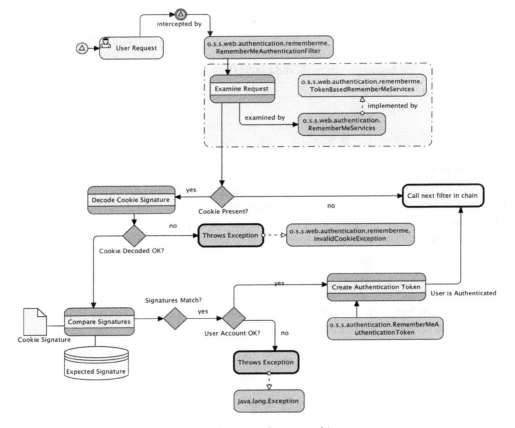

Figure 7.6 – The remember-me architecture

As with any of the Spring Security filters, RememberMeAuthenticationFilter is invoked from within FilterChainProxy. The job of RememberMeAuthenticationFilter is to inspect the request, and if it is of interest, an action is taken. The RememberMeAuthenticationFilter interface will use the RememberMeServices implementation to determine whether the user is already logged in. The RememberMeServices interface does this by inspecting the HTTP request for a remember-me cookie that is then validated using either the token-based validation or the persistent-based validation we previously discussed. If the token checks out, the user will be logged in.

Remember-me and the user life cycle

The implementation of RememberMeServices is invoked at several points in the user life cycle (the life cycle of an authenticated user's session). To assist you in your understanding of the remember-me functionality, it can be helpful to be aware of the points in time when remember-me services are informed of life cycle functions:

Action	What should happen?	The RememberMeServices method invoked
Successful login	The implementation sets a remember-me cookie (if the form parameter has been sent)	loginSuccess
Failed login	The implementation should cancel the cookie if it's present	loginFailed
User logout	The implementation should cancel the cookie if it's present	Logout

Table 7.8 – Remember-me life cycle events

> **Important note**
>
> The logout method is not present on the RememberMeServices interface. Instead, each RememberMeServices implementation also implements the LogoutHandler interface, which contains the logout method. By implementing the LogoutHandler interface, each RememberMeServices implementation can perform the necessary cleanup when the user logs out.

Knowing where and how RememberMeServices ties into the user's life cycle will be important when we begin to create custom authentication handlers because we need to ensure that any authentication processor treats RememberMeServices consistently to preserve the usefulness and security of this functionality.

Restricting the remember-me feature to an IP address

Let's put our understanding of the remember-me architecture to use. A common requirement is that any remember-me token should be tied to the IP address of the user who created it. This adds additional security to the remember-me feature. To do this, we only need to implement a custom `PersistentTokenRepository` interface. The configuration changes that we will make will illustrate how to configure a custom `RememberMeServices`. Throughout this section, we will take a look at `IpAwarePersistentTokenRepository`, which is included in the chapter's source code. The `IpAwarePersistenTokenRepository` interface ensures that the series identifier is internally combined with the current user's IP address, and the series identifier includes only the identifier externally. This means that whenever a token is looked up or saved, the current IP address is used to look up or persist the token. In the following code snippets, you can see how `IpAwarePersistentTokenRepository` works. If you want to dig in even deeper, we encourage you to view the source code included in the chapter.

The trick to looking up the IP address is using `RequestContextHolder` of Spring Security. The relevant code is as follows:

> **Important note**
>
> It should be noted that to use `RequestContextHolder`, you need to ensure you have set up your `web.xml` file to use `RequestContextListener`. We have already performed this setup for our sample code. However, this can be useful when utilizing the example code in an external application. Refer to the Javadoc of `IpAwarePersistentTokenRepository` for details on how to set this up.

Take a look at the following code snippet:

```
//src/main/java/com/packtpub/springsecurity/web/authentication/
rememberme/ IpAwarePersistentTokenRepository.java
private String ipSeries(String series) {
    ServletRequestAttributes attributes = (ServletRequestAttributes)
RequestContextHolder.getRequestAttributes();
    if (attributes == null) {
        throw new IllegalStateException("RequestContextHolder.
getRequestAttributes() cannot be null");
    }
    String remoteAddr = attributes.getRequest().getRemoteAddr();
    logger.debug("Remote address is {}", remoteAddr);

    return series + remoteAddr;
}
```

We can build on this method to force tokens that are saved to include the IP address in the series identifier, as follows:

```
@Override
public void createNewToken(PersistentRememberMeToken token) {
    String ipSeries = ipSeries(token.getSeries());
    PersistentRememberMeToken ipToken = tokenWithSeries(token,
ipSeries);
    this.delegateRepository.createNewToken(ipToken);
}
```

You can see that we first created a new series with the IP address concatenated to it.

The tokenWithSeries method is just a helper that creates a new token with all of the same values, except a new series. We then submit the new token with a series identifier, which includes the IP address, to delegateRepository, which is the original implementation of PersistentTokenRepository.

Whenever the tokens are looked up, we require that the current user's IP address is appended to the series identifier. This means that there is no way for a user to obtain a token for a user with a different IP address:

```
@Override
public PersistentRememberMeToken getTokenForSeries(String seriesId) {
    String ipSeries = ipSeries(seriesId);
    PersistentRememberMeToken ipToken = delegateRepository.
getTokenForSeries(ipSeries);
    return tokenWithSeries(ipToken, seriesId);
}
```

The remainder of the code is quite similar. Internally, we construct the series identifier to be appended to the IP address, and externally, we present only the original series identifier. By doing this, we enforce the constraint that only the user who created the remember-me token can use it.

Let's review the Spring configuration included in this chapter's sample code for IpAwarePersistent TokenRepository. In the following code snippet, we first create the IpAwarePersistent TokenRepository declaration that wraps a new JpaPersistentTokenRepository declaration. We then initialize a RequestContextFilter class by instantiating an OrderedRequestContextFilter interface:

```
//src/main/java/com/packtpub/springsecurity/configuration/
SecurityConfig.java
@Bean
public RememberMeServices rememberMeServices(PersistentTokenRepository
ptr, UserDetailsService userDetailsService) {
    PersistentTokenBasedRememberMeServices rememberMeServices = new
            PersistentTokenBasedRememberMeServices("jbcpCalendar",
```

```
        userDetailsService, ptr);
    rememberMeServices.setAlwaysRemember(true);
    return rememberMeServices;
}


@Bean
public IpAwarePersistentTokenRepository
tokenRepository(RememberMeTokenRepository rmtr) {
    return new IpAwarePersistentTokenRepository(new
JpaPersistentTokenRepository(rmtr));
}


@Bean
public OrderedRequestContextFilter requestContextFilter() {
    return new OrderedRequestContextFilter();
}
```

Now, go ahead and start up the application. You can use the second computer along with a plugin, such as Firebug, to manipulate your remember-me cookie. If you try to use the remember-me cookie from one computer on another computer, Spring Security will now ignore the remember-me request and delete the associated cookie.

> **Important note**
> You should start with the source from chapter07.07-calendar.

Note that the IP-based remember-me tokens may behave unexpectedly if the user is behind a shared or load-balanced network infrastructure, such as a multi **Wide-Area Network (WAN)** corporate environment. In most scenarios, however, the addition of an IP address to the remember-me function provides an additional, welcome layer of security to a helpful user feature.

Custom cookie and HTTP parameter names

Curious users may wonder whether the expected value of the remember-me form field checkbox to be remember-me, or the cookie name to be remember-me, can be changed to obscure the use of Spring Security. This change can be made in one of two locations. Take a look at the following steps:

1. We can simply define more properties to change the checkbox and cookie names in RememberMeServices bean, as follows:

    ```
    //src/main/java/com/packtpub/springsecurity/configuration/
    SecurityConfig.java
    @Bean
    public RememberMeServices
    ```

```
remememberMeServices(PersistentTokenRepository ptr,
UserDetailsService userDetailsService) {
    PersistentTokenBasedRememberMeServices rememberMeServices =
new
        PersistentTokenBasedRememberMeServices("jbcpCalendar",
        userDetailsService, ptr);
    rememberMeServices.setAlwaysRemember(true);
    rememberMeServices.setParameter("obscure-remember-me");
    rememberMeServices.setCookieName("obscure-remember-me");
    return rememberMeServices;
}
```

2. Don't forget to change the `login.html` page to set the name of the `checkbox form field` and to match the parameter value we declared. Go ahead and make the updates to `login.html`, as follows:

```
//src/main/resources/templates/login.html
<div class="mb-3">
    <label for="remember-me">Remember Me?</label>
    <input type="checkbox" id="remember-me" name="obscure-
remember-me" th:checked="true" />
</div>
```

3. We'd encourage you to experiment here to ensure you understand how these settings are related. Go ahead and start up the application and give it a try.

> **Important note**
> You should start with the source from `chapter07.08-calendar`.

Summary

This chapter explained and demonstrated the use of the remember-me feature in Spring Security. We started with the most basic setup and learned how to gradually make the feature more secure. Specifically, we learned about a token-based remember-me service and how to configure it. We also explore how persistent-based remember-me services can provide additional security, how they work, and the additional considerations necessary when using them.

We also covered the creation of a custom remember-me implementation that restricts the remember-me token to a specific IP address. We saw various other ways to make the remember-me feature more secure.

Up next is **certificate-based authentication**, and we will discuss how to use trusted client-side certificates to perform authentication.

8

Client Certificate Authentication with TLS

Although username and password authentication are extremely common, as we discussed in *Chapter 1, Anatomy of an Unsafe Application*, and in *Chapter 2, Getting Started with Spring Security*, forms of authentication exist that allow users to present different types of credentials. Spring Security caters to these requirements as well. In this chapter, we'll move beyond form-based authentication to explore authentication using trusted client-side certificates.

During the course of this chapter, we will cover the following topics:

- Learning how client certificate authentication is negotiated between the user's browser and a compliant server

- Configuring Spring Security to authenticate users with client certificates

- Understanding the architecture of client certificate authentication in Spring Security

- Exploring advanced configuration options related to client certificate authentication

- Reviewing pros, cons, and common troubleshooting steps when dealing with client certificate authentication

This chapter's code in action link is here: `https://packt.link/XgAQ7`.

How does client certificate authentication work?

Client certificate authentication requires a request for information from the server and a response from the browser to negotiate a trusted authentication relationship between the client (that is, a user's browser) and the server application. This trusted relationship is built through the use of the exchange of trusted and verifiable credentials, known as **certificates**.

Unlike much of what we have seen up to this point, with client certificate authentication, the Servlet container or application server itself is typically responsible for negotiating the trust relationship between the browser and server by requesting a certificate, evaluating it, and accepting it as valid.

Client certificate authentication is also known as **mutual authentication** and is part of the **Secure Sockets Layer (SSL)** protocol and its successor, **Transport Layer Security (TLS)**. As mutual authentication is part of the SSL and TLS protocols, it follows that an HTTPS connection (secured with SSL or TLS) is required in order to make use of client certificate authentication. For more details on SSL/TLS support in Spring Security, please refer to the *Generating a server certificate* section in the *Appendix, Additional Reference Material*. Setting up SSL/TLS is required to implement client certificate authentication.

The following sequence diagram illustrates the interaction between the client browser and the web server when negotiating an SSL connection and validating the trust of a client certificate used for mutual authentication:

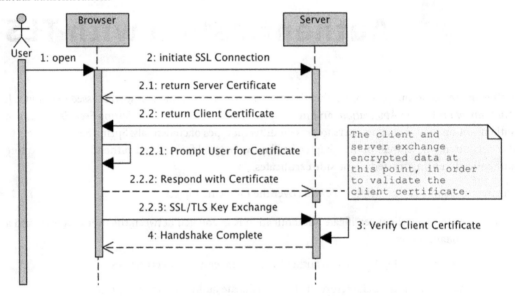

Figure 8.1 – Client certificate authentication

We can see that the exchange of two certificates, the server and client certificates, provides the authentication that both parties are known and can be trusted to continue their conversation securely. In the interest of clarity, we omit some details of the SSL handshake and trust the checking of the certificates themselves; however, you are encouraged to do further reading in the area of the SSL and TLS protocols, and certificates in general, as many good reference guides on these subjects exist. *RFC 8446, The Transport Layer Security (TLS) Protocol Version 1.3* (https://datatracker.ietf.org/doc/html/rfc8446), is a good place to begin reading about client certificate presentation, and if you'd like to get into more detail, *SL and TLS: Designing and Building Secure Systems, Eric Rescorla, Addison-Wesley* (https://www.amazon.com/SSL-TLS-Designing-Building-Systems/dp/0201615983) has an incredibly detailed review of the protocol and its implementation.

An alternative name for client certificate-based authentication is **X.509 authentication**. The term X.509 is derived from the X.509 standard, originally published by the **International Telecommunication Union Telecommunication (ITU-T)** organization for use in directories based on the X.500 standard (the origins of **Lightweight Directory Access Protocol (LDAP)**, as you may recall from *Chapter 6, LDAP Directory Services*). Later, this standard was adapted for use in securing internet communications.

We mention this here because many of the classes in Spring Security related to this subject refer to X.509. Remember that X.509 doesn't define the mutual authentication protocol itself, but defines the format and structure of the certificates and the encompassing trusted certificate authorities instead.

Setting up the client certificate authentication infrastructure

Unfortunately for you as an individual developer, being able to experiment with client certificate authentication requires some non-trivial configuration and setup prior to the relatively easy integration with Spring Security. As these setup steps tend to cause a lot of problems for first-time developers, we feel it is important to walk you through them.

We assume that you are using a local, self-signed server certificate, self-signed client certificates, and Apache Tomcat. This is typical of most development environments; however, it's possible that you may have access to a valid server certificate, a **certificate authority (CA)**, or another application server. If this is the case, you may use these setup instructions as guidelines and configure your environment in an analogous manner. Please refer to the SSL setup instructions in the *Appendix, Additional Reference Material*, for assistance in configuring Tomcat and Spring Security to work with SSL in a standalone environment.

Understanding the purpose of a public key infrastructure

This chapter focuses on setting up a self-contained development environment for the purposes of learning and education. However, in most cases where you are integrating Spring Security into an existing client certificate-secured environment, there will be a significant amount of infrastructure (usually a combination of hardware and software) in place to provide functionality, such as certificate granting and management, user self-service, and revocation. Environments of this type define a public key infrastructure—a combination of hardware, software, and security policies that result in a highly secure authentication-driven network ecosystem.

In addition to being used for web application authentication, certificates or hardware devices in these environments can be used for secure, non-repudiated email (using S/MIME), network authentication, and even physical building access (using PKCS 11-based hardware devices).

While the management overhead of such an environment can be high (and requires both IT and process excellence to implement well), it is arguably one of the most secure operating environments possible for technology professionals.

Creating a client certificate key pair

The self-signed client certificate is created in the same way as the self-signed server certificate is created—by generating a key pair using the `keytool` command. A client certificate key pair differs in that it requires the key store to be available to the web browser and requires the client's public key to be loaded into the server's trust store (we'll explain what this is in a moment).

If you do not wish to generate your own key right now, you may skip to the next section and use the sample certificates in the `./src/main/resources/keys` folder in the sample chapter. Otherwise, create the client key pair, as follows:

```
keytool -genkeypair -alias jbcpclient -keyalg RSA -validity 365
-keystore jbcp_clientauth.p12 -storetype PKCS12
```

> **Important note**
>
> You can find additional information about `keytool`, along with all of the configuration options, at Oracle's site, here: `https://docs.oracle.com/en/java/javase/17/docs/specs/man/keytool.html`.

Most of the arguments to `keytool` are fairly arbitrary for this use case. However, when prompted to set up the first and last name (the **Common Name (CN)**, the portion of the owner's **Distinguished Name (DN)** for the client certificate, ensure that the answer to the first prompt matches a user that we have set up in our Spring Security JDBC store. For example, `admin1@example.com` is an appropriate value since we have the `admin1@example.com` user setup in Spring Security. An example of the command-line interaction is as follows:

```
What is your first and last name? [Unknown]: admin1@example.com
... etc
Is CN=admin1@example.com, OU=JBCP Calendar, O=JBCP, L=Park City,
ST=UT, C=US correct?
[no]: yes
```

We'll see why this is important when we configure Spring Security to access the information from the certificate-authenticated user. We have one final step before we can set up certificate authentication within Tomcat, which is explained in the next section.

Configuring the Tomcat trust store

Recall that the definition of a key pair includes both a private and public key. Similar to SSL certificates verifying and securing server communication, the validity of the client certificate needs to be verified by the certifying authority that created it.

As we have created our own self-signed client certificate using the `keytool` command, the **Java Virtual Machine (JVM)** will not implicitly trust it as having been assigned by a trusted certificate authority.

Let's take a look at the following steps:

1. We will need to force Tomcat to recognize the certificate as a trusted certificate. We do this by exporting the public key from the key pair and adding it to the Tomcat trust store.

    Again, if you do not wish to perform this step now, you can use the existing trust store in `.src/main/resources/keys` and skip to where we configure `server.xml` later in this section.

2. We'll export the public key to a standard certificate file named `jbcp_clientauth.cer`, as follows:

    ```
    keytool -exportcert -alias jbcpclient -keystore jbcp_clientauth.
    p12 -storetype PKCS12 -storepass changeit -file jbcp_clientauth.
    cer
    ```

3. Next, we'll import the certificate into the trust store (this will create the trust store, but in a typical deployment scenario you'd probably already have some other certificates in the trust store):

    ```
    keytool -importcert -alias jbcpclient -keystore tomcat.
    truststore -file jbcp_clientauth.cer
    ```

    The preceding command will create the trust store called `tomcat.truststore` and prompt you for a password (we chose `changeit` as the password). You'll also see some information about the certificate and will finally be asked to confirm that you do trust the certificate, as follows:

    ```
    Owner: CN=admin1@example.com, OU=JBCP Calendar, O=JBCP, L=Park
    City, ST=UT, C=US
    Issuer: CN=admin1@example.com, OU=JBCP Calendar, O=JBCP, L=Park
    City, ST=UT, C=US
    Serial number: 464fc10c
    Valid from: Fri Jun 23 11:10:19 MDT 2017 until: Thu Feb 12
    10:10:19
    MST 2043
    //Certificate fingerprints:
    MD5: 8D:27:CE:F7:8B:C3:BD:BD:64:D6:F5:24:D8:A1:8B:50
    SHA1:
    C1:51:4A:47:EC:9D:01:5A:28:BB:59:F5:FC:10:87:EA:68:24:E3:1F
    SHA256: 2C:F6:2F:29:ED:09:48:FD:FE:A5:83:67:E0:A0:B9:DA:C5:3B:
    FD:CF:4F:95:50:3A:
    2C:B8:2B:BD:81:48:BB:EF
    Signature algorithm name: SHA256withRSA Version: 3
    //Extensions
    #1: ObjectId: 2.5.29.14 Criticality=false SubjectKeyIdentifier [
    KeyIdentifier [
    0000: 29 F3 A7 A1 8F D2 87 4B
    EA 74 AC 8A 4B BC 4B 5D
    )
    K.t..K.K]
    ```

```
0010: 7C 9B 44 4A
..DJ
]
]
Trust this certificate? [no]: yes
```

Remember the location of the new `tomcat.truststore` file, as we will need to reference it in our Tomcat configuration.

What's the difference between a key store and a trust store?

The **Java Secure Socket Extension** (**JSSE**) documentation defines a key store as a storage mechanism for private keys and their corresponding public keys. The key store (containing key pairs) is used to encrypt or decrypt secure messages, and so on. The trust store is intended to store only public keys for trusted communication partners when verifying an identity (similar to how the trust store is used in certificate authentication). In many common administration scenarios, however, the key store and trust store are combined into a single file (in Tomcat, this would be done through the use of the `keystoreFile` and `truststoreFile` attributes of the connector). The format of the files themselves can be exactly the same. Really, each file can be any JSSE-supported keystore format, including **Java KeyStore** (**JKS**), PKCS 12, and so on.

---

4. As previously mentioned, we assume you have already configured the SSL Connector, as outlined in the *Appendix, Additional Reference Material*. If you do not see the `keystoreFile` or `keystorePass` attributes in `server.xml`, it means you should visit the *Appendix, Additional Reference Material*, to get SSL set up.

5. Finally, we'll need to point Tomcat at the trust store and enable client certificate authentication. This is done by adding three additional attributes to the SSL connector in the Tomcat `server.xml` file, as follows:

```
//sever.xml
<Connector port="8443" protocol="HTTP/1.1" SSLEnabled="true"
maxThreads="150"
        scheme="https" secure="true" sslProtocol="TLS"
keystoreFile="<KEYSTORE_PATH>/tomcat.keystore"
        keystorePass="changeit" truststoreFile="<CERT_PATH>/
tomcat.truststore"
        truststorePass="changeit"  clientAuth="true" />
```

6. This should be the remaining configuration required to trigger Tomcat to request a client certificate when an SSL connection is made. Of course, you will want to ensure you replace both `<CERT_PATH>` and `<KEYSTORE_PATH>` with the full paths. For example, on a Unix-based **operating system** (**OS**), the path might look like this: `/home/packt/chapter8/keys/tomcat.keystore`.

7. Go ahead and try to start up Tomcat to ensure that the server starts up without any errors in the logs.

> **Important note**
>
> There's also a way to configure Tomcat to optionally use client certificate authentication—we'll enable this later in the chapter. For now, we require the use of client certificates to even connect to the Tomcat server in the first place. This makes it easier to diagnose whether or not you have set this up correctly!

Configuring Tomcat in Spring Boot

We can also configure the embedded Tomcat instance within Spring Boot, which is how we will be working with Tomcat for the rest of this chapter.

Configuring Spring Boot to use our newly created certificates is as straightforward as configuring the properties of the YAML entry, as shown in the following code snippet:

```
## Chapter 8 TLS over HTTP/1.1:
## https://localhost:8443
server:
  port: 8443
  ssl:
    key-store: classpath:keys/jbcp_clientauth.p12
    key-store-password: changeit
    keyStoreType: PKCS12
    keyAlias: jbcpclient
    protocol: TLS

    client-auth: need
    trust-store: classpath:keys/tomcat.truststore
    trust-store-password: changeit
```

The final step is to import the certificate into the client browser.

Importing the certificate key pair into a browser

Depending on what browser you are using, the process of importing a certificate may differ. We will provide instructions for installations of Firefox, Chrome, and Internet Explorer here, but if you are using another browser, please consult its help section or your favorite search engine for assistance.

Using Mozilla Firefox

Perform the following steps to import the key store containing the client certificate key pair in Firefox:

1. Click on **Edit | Preferences**.

2. Click on the **Advanced** button.

3. Click on the **Encryption** tab.

4. Click on the **View Certificates** button. The **Certificate Manager** window should open up.

5. Click on the **Your Certificates** tab.

6. Click on the **Import...** button.

7. Browse to the location where you saved the `jbcp_clientauth.p12` file and select it. You will need to enter the password (that is, `changeit`) that you used when you created the file.

The client certificate should be imported, and you should see it on the list.

Using Google Chrome

Perform the following steps to import the key store containing the client certificate key pair in Chrome:

1. Click on the wrench icon on the browser toolbar.

2. Select **Settings**.

3. Click on **Show advanced settings...**.

4. In the **HTTPS/SSL** section, click on the **Manage certificates...** button.

5. In the **Personal** tab click on the **Import...** button.

6. Browse to the location where you saved the `jbcp_clientauth.p12` file and select it (Ensure that you utilize the .p12 extension for the certificate).

7. You will need to enter the password (that is, `changeit`) that you used when you created the file.

8. Click on **OK**.

Using Microsoft Edge

Let's take a look at the steps of using Microsoft Edge with Windows OS:

1. Double-click on the `jbcp_clientauth.p12` file in Windows Explorer. The **Certificate Import Wizard** window should open (Ensure that you utilize the .p12 extension for the certificate).

2. Click on **Next** and accept the default values until you are prompted for the certificate password.

3. Enter the certificate password (that is, `changeit`) and click **Next**.

4. Accept the default **Automatically select the certificate store** option and click **Next**.

5. Click on **Finish**.

To verify that the certificate was installed correctly, you will need to perform another series of steps:

1. Open the **Settings** menu in Microsoft Edge.

2. Select **Privacy, search and services**

3. Scroll down to **Security** and

4. Click **Manage certificates**.

5. Click on the **Personal** tab, if it is not already selected. You should see the certificate listed here.

Wrapping up testing

You should now be able to connect to the **Jim Bob Circle Pants Online Calendar (JBCP)** calendar site using the client certificate. Navigate to `https://localhost:8443/`, taking care to use **HTTPS** and **8443**. If all is set up correctly, you should be prompted for a certificate when you attempt to access the site—in Chrome, the certificate is displayed as follows:

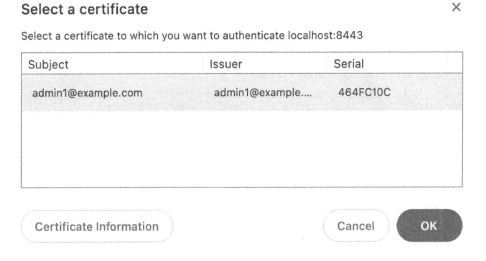

Figure 8.2 – Client certificate details in Chrome

You'll notice, however, that if you attempt to access a protected section of the site, such as the **My Events** section, you'll be redirected to the login page. This is because we haven't yet configured Spring Security to recognize the information in the certificate—at this point, the negotiation between the client and server has stopped at the Tomcat server itself.

Important note

You should start with the code from `chapter08.00-calendar`.

Troubleshooting client certificate authentication

Unfortunately, if we said that getting client certificate authentication configured correctly for the first time—without anything going wrong—was easy, we'd be lying to you. The fact is, although this is a great and very powerful security apparatus, it is poorly documented by both the browser and web server manufacturers, and the error messages, when present, can be confusing at best and misleading at worst.

Remember that, at this point, we have not involved Spring Security in the equation at all, so a debugger will most likely not help you (unless you have the Tomcat source code handy). There are some common errors and things to check.

You aren't prompted for a certificate when you access the site. There are many possible causes for this, and this can be the most puzzling problem to try to solve. Here are some things to check:

1. Ensure that the certificate has been installed in the browser client you are using. Sometimes, you need to restart the whole browser (close all windows) if you attempted to access the site previously and were rejected.

2. Ensure you are accessing the SSL port for the server (typically 8443 in a development setup), and have selected the HTTPS protocol in your URL. The client certificates are not presented in insecure browser connections. Make sure the browser also trusts the server SSL certificate, even if you have to force it to trust a self-signed certificate.

3. Ensure you have added the clientAuth directive to your Tomcat configuration (or the equivalent for whatever application server you are using).

4. If all else fails, use a network analyzer or packet sniffer, such as Wireshark (http://www.wireshark.org/) or Fiddler2 (http://www.fiddler2.com/), to review the traffic and SSL key exchange over the wire (check with your IT department first—many companies do not allow tools of this kind on their networks).

5. If you are using a self-signed client certificate, make sure the public key has been imported into the server's trust store. If you are using a CA-assigned certificate, make sure the CA is trusted by the **Java Virtual Machine (JVM)** or that the CA certificate is imported into the server's trust store.

6. Internet Explorer, in particular, does not report details of client certificate failures at all (it simply reports a generic **Page Cannot be Displayed** error). Use Firefox to diagnose whether an issue you are seeing is related to client certificates or not.

7. The following JVM option will enable SSL handshake level logging: -Djavax.net.debug=ssl:handshake. This debugging flag can produce a large amount of output but can be very helpful in diagnosing underlying SSL connectivity issues.

Configuring client certificate authentication in Spring Security

Unlike the authentication mechanisms that we have utilized thus far, the use of client certificate authentication results in the user's request being pre-authenticated by the server. As the server (Tomcat) has already established that the user has provided a valid and trustworthy certificate, Spring Security can simply trust this assertion of validity.

An important component of the secure login process is still missing, which is the authorization of the authenticated user. This is where our configuration of Spring Security comes in—we must add a component to Spring Security that will recognize the certificate authentication information from the user's HTTP session (populated by Tomcat), and then validate the presented credentials against the Spring Security UserDetailsService invocation. The invocation of UserDetailsService will result in the determination of whether the user declared in the certificate is known to Spring Security at all, and then it will assign GrantedAuthority as per the usual login rules.

Configuring client certificate authentication using the security namespace

With all of the complexity of LDAP configuration, configuring client certificate authentication is a welcome reprieve. If we are using the security namespace style of configuration, the addition of client certificate authentication is a simple one-line configuration change, added within the HttpSecurity declaration. Go ahead and make the following changes to the provided SecurityConfig. java configuration:

```
//src/main/java/com/packtpub/springsecurity/configuration/
SecurityConfig.java
// SSL / TLS x509 support
http.x509(httpSecurityX509Configurer -> httpSecurityX509Configurer
            .userDetailsService(userDetailsService));
```

> **Important note**
>
> Observe that the .x509() method references our existing userDetailsService() configuration. For simplicity, we use the UserDetailsServiceImpl implementation covered in *Chapter 5, Authentication with Spring Data*. However, we could easily swap this out with any other implementation (that is, the LDAP or JDBC-based implementation covered in *Chapter 4, JDBC-based Authentication*).

After restarting the application, you'll again be prompted for a client certificate, but this time, you should be able to access areas of the site requiring authorization. You can see from the logs (if you have them enabled) that you have been logged in as the admin1@example.com user.

> **Important note**
>
> Your code should look like `chapter08.01-calendar`.

How does Spring Security use certificate information?

As previously discussed, Spring Security's involvement in certificate exchange is to pick up information from the presented certificate and map the user's credentials to a user service. What we did not see in the use of the `.x509()` method was the magic that makes this happen. Recall that when we set the client certificate up, a DN similar to an LDAP DN was associated with the certificate:

```
Owner: CN=admin@example.com, OU=JBCP Calendar, O=JBCP, L=Park
City, ST=UT, C=US
```

Spring Security uses the information in this DN to determine the actual username of the principal and it will look for this information in `UserDetailsService`. In particular, it allows for the specification of a regular expression, which is used to match a portion of the DN established with the certificate, and the utilization of this portion of the DN as the principal name. The implicit, default configuration for the `.x509()` method would be as follows:

```
http.x509(httpSecurityX509Configurer -> httpSecurityX509Configurer
            .subjectPrincipalRegex("CN=(.*?)(?:,|$)")
            .userDetailsService(userDetailsService));
```

We can see that this regular expression would match the `admin1@example.com` value as the principal's name. This regular expression must contain a single matching group, but it can be configured to support the username and DN issuance requirements of your application. For example, if the DNs for your organization's certificates include the `email` or `userid` fields, the regular expression can be modified to use these values as the authenticated principal's name.

How Spring Security certificate authentication works

Let's review the various actors involved in the review and evaluation of the client certificates and translation into a Spring-Security-authenticated session, with the help of the following diagram:

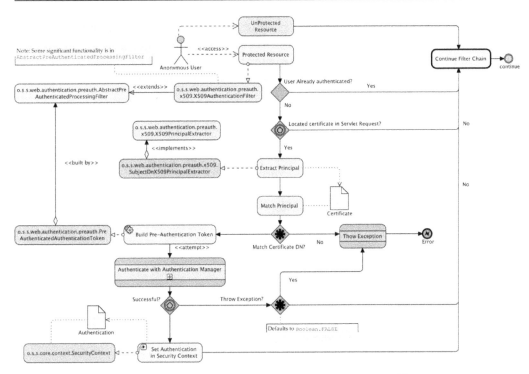

Figure 8.3 – Spring Security certificate authentication workflow

We can see that o.s.s.web.authentication.preauth.x509. X509AuthenticationFilter is responsible for examining the request of an unauthenticated user for the presentation of client certificates. If it sees that the request includes a valid client certificate, it will extract the principal using o.s.s.web.authentication.preauth.x509. SubjectDnX509PrincipalExtractor, using a regular expression matching the certificate owner's DN, as previously described.

> **Important note**
>
> Be aware that although the preceding diagram indicates that examination of the certificate occurs for unauthenticated users, a check can also be performed when the presented certificate identifies a different user than the one that was previously authenticated. This would result in a new authentication request using the newly provided credentials. The reason for this should be clear—any time a user presents a new set of credentials, the application must be aware of this and react in a responsible fashion by ensuring that the user is still able to access it.

Once the certificate has been accepted (or rejected/ignored), as with other authentication mechanisms, an `Authentication` token is built and passed along to `AuthenticationManager` for authentication. We can now review the very brief illustration of the `o.s.s.web.authentication.preauth.PreAuthenticatedAuthenticationProvider` handling of the authentication token:

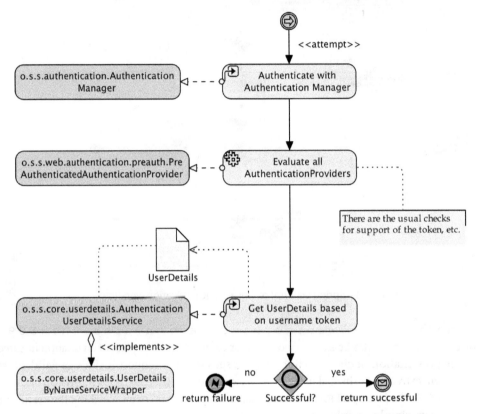

Figure 8.4 – The Spring Security PreAuthenticatedAuthenticationProvider workflow

Though we will not go over them in detail, there are a number of other pre-authenticated mechanisms supported by Spring Security. Some examples include Java EE role mapping (`J2eePreAuthenticatedProcessingFilter`), WebSphere integration (`WebSpherePreAuthenticatedProcessingFilter`), and SiteMinder-style authentication (`RequestHeaderAuthenticationFilter`). If you understand the process flow of client certificate authentication, understanding these other authentication types is significantly easier.

Handling unauthenticated requests with AuthenticationEntryPoint

Since X509AuthenticationFilter will continue processing the request if authentication fails, we'll need to handle situations where the user does not authenticate successfully and has requested a protected resource. The way that Spring Security allows developers to customize this is by plugging in a custom o.s.s.web.AuthenticationEntryPoint implementation. In a default form login scenario, LoginUrlAuthenticationEntryPoint is used to redirect the user to a login page if they have been denied access to a protected resource and are not authenticated.

In contrast, in typical client certificate authentication environments, alternative methods of authentication are simply not supported (remember that Tomcat expects the certificate well before the Spring Security form login takes place anyway). As such, it doesn't make sense to retain the default behavior of redirection to a form login page. Instead, we'll modify the entry point to simply return an HTTP 403 Forbidden message, using o.s.s.web.authentication.Http403ForbiddenEntryPoint. Go ahead and make the following updates in your SecurityConfig.java file, as follows:

```
//src/main/java/com/packtpub/springsecurity/configuration/
SecurityConfig.java
@Bean
public SecurityFilterChain filterChain(HttpSecurity http,
        PersistentTokenRepository persistentTokenRepository,
        Http403ForbiddenEntryPoint forbiddenEntryPoint) throws
Exception {
    http.authorizeRequests( authz -> authz
                .requestMatchers(antMatcher("/webjars/**")).
permitAll()
...
            .exceptionHandling(exceptions -> exceptions
                .authenticationEntryPoint(forbiddenEntryPoint)
                .accessDeniedPage("/errors/403"))
...
    return http.build();
}
@Bean
public Http403ForbiddenEntryPoint forbiddenEntryPoint(){
    return new Http403ForbiddenEntryPoint();
}
```

Now, if a user tries to access a protected resource and is unable to provide a valid certificate, they will be presented with the following page, instead of being redirected to the login page:

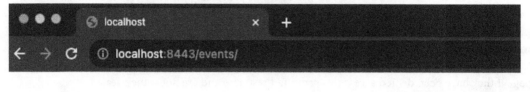

Access to localhost was denied

You don't have authorisation to view this page.

HTTP ERROR 403

Reload

Figure 8.5 – Spring Security forbidden error

> **Important note**
>
> We have removed the username of admin1@example.com, to make sure there is no matching user with the certificate CN.
>
> Your code should look like chapter08.02-calendar.

Other configuration or application flow adjustments that are commonly performed with client certificate authentication are as follows:

- Removal of the form-based login page altogether

- Removal of the logout link (as there's no reason to log out because the browser will always present the user's certificate)

- Removal of the functionality to rename the user account and change the password

- Removal of the user registration functionality (unless you are able to tie it into the issuance of a new certificate)

Supporting dual-mode authentication

It is also possible that some environments may support both certificate-based and form-based authentication. If this is the case in your environment, it is also possible (and trivial) to support it with Spring Security. We can simply leave the default `AuthenticationEntryPoint` interface (redirecting to the form-based login page) intact and allow the user to log in using the standard login form if they do not supply a client certificate.

If you choose to configure your application this way, you'll need to adjust the Tomcat SSL settings (change as appropriate for your application server). Simply change the `clientAuth` directive to `want`, instead of `true`:

```
<Connector port="8443" protocol="HTTP/1.1" SSLEnabled="true"
maxThreads="150" scheme="https" secure="true" sslProtocol="TLS"
keystoreFile="conf/tomcat.keystore" keystorePass="password"
truststoreFile="conf/tomcat.truststore" truststorePass="password"
clientAuth="want"
/>
```

We'll also need to remove the `authenticationEntryPoint()` method that we configured in the previous exercise so that the standard form-based authentication workflow takes over if the user isn't able to supply a valid certificate upon the browser first being queried.

Although this is convenient, there are a few things to keep in mind about dual-mode (form-based and certificate-based) authentication, as follows:

- Most browsers will not re-prompt the user for a certificate if they have failed certificate authentication once, so make sure that your users are aware that they may need to reenter the browser to present their certificate again.

- Recall that a password is not required to authenticate users with certificates; however, if you are still using `UserDetailsService` to support your form-based authenticated users, this may be the same `UserDetailsService` object that you use to give the `PreAuthenticatedAuthenticationProvider` information about your users. This presents a potential security risk, as users who you intend to sign in only with certificates could potentially authenticate using form login credentials.

There are several ways to solve this problem, and they are described in the following list:

- Ensure that the users authenticating with certificates have an appropriately strong password in your user store.

- Consider customizing your user store to clearly identify users who are enabled for form-based login. This can be tracked with an additional field in the table holding user account information, and with minor adjustments to the SQL queries used by the `JpaDaoImpl` object.

- Configure a separate user details store altogether for users who are logging in as certificate-authenticated users, to completely segregate them from users that are allowed to use form-based login.

- Dual-mode authentication can be a powerful addition to your site and can be deployed effectively and securely, provided that you keep in mind the situations under which users will be granted access to it.

Configuring client certificate authentication using Spring beans

Earlier in this chapter, we reviewed the flow of the classes involved in client certificate authentication. As such, it should be straightforward for us to configure the JBCP calendar using explicit beans. By using the explicit configuration, we will have additional configuration options at our disposal. Let's take a look and see how to use explicit configuration:

```java
//src/main/java/com/packtpub/springsecurity/configuration/
SecurityConfig.java
@Bean
public X509AuthenticationFilter x509Filter(){
    return new X509AuthenticationFilter(){{
        setAuthenticationManager(authenticationManager);
    }};
}


@Bean
public PreAuthenticatedAuthenticationProvider
preAuthAuthenticationProvider(final AuthenticationUserDetailsService
authenticationUserDetailsService){
    return new PreAuthenticatedAuthenticationProvider(){{
        setPreAuthenticatedUserDetailsService
(authenticationUserDetailsService);
    }};
}


@Bean
public UserDetailsByNameServiceWrapper
authenticationUserDetailsService(final UserDetailsService
userDetailsService){
    return new UserDetailsByNameServiceWrapper(){{
        setUserDetailsService(userDetailsService);
    }};
}
```

We'll also need to remove the `x509()` method, add `x509Filter` to our filter chain, and add our `AuthenticationProvider` implementation to `AuthenticationManger`:

```
//src/main/java/com/packtpub/springsecurity/configuration/
SecurityConfig.java
@Bean
public AuthenticationManager authManager(HttpSecurity http) throws
Exception {
    AuthenticationManagerBuilder authenticationManagerBuilder =
            http.getSharedObject(AuthenticationManagerBuilder.class);
    http.authenticationProvider(preAuthAuthenticationProvider);
    return authenticationManagerBuilder.build();
}
```

Now, give the application a try. Nothing much has changed from a user perspective, but as developers, we have opened the door to a number of additional configuration options.

> **Important note**
>
> Your code should look like `chapter08.03-calendar`.

Additional capabilities of bean-based configuration

The use of Spring-bean-based configuration provides us with additional capabilities through the exposure of bean properties that aren't exposed via the security namespace style of configuration.

Additional properties available on `X509AuthenticationFilter` are as follows:

Property	Description	Default
`continueFilterChainOn UnsuccessfulAuthentication`	If false, a failed authentication will throw an exception rather than allow the request to continue. This would typically be set in cases where a valid certificate is expected and required to access the secured site. If true, the filter chain will proceed, even if there is a failed authentication.	`true`

Property	Description	Default
checkForPrincipalChanges	If true, the filter will check to see whether the currently authenticated username differs from the username presented in the client certificate. If so, authentication against the new certificate will be performed and the HTTP session will be invalidated (optionally, see the next attribute). If false, once the user is authenticated, they will remain authenticated even if they present different credentials.	false
invalidateSessionOn PrincipalChange	If true, and the principal in the request changes, the user's HTTP session will be invalidated prior to being reauthenticated. If false, the session will remain—note that this may introduce security risks.	true

Table 8.1 – Properties available on X509AuthenticationFilter

The PreAuthenticatedAuthenticationProvider implementation has a couple of interesting properties available to us, which are listed in the following table:

Property	Description	Default
preAuthenticated UserDetailsService	This property is used to build a full UserDetails object from the username extracted from the certificate.	None
throwExceptionWhen TokenRejected	If true, a BadCredentialsException exception will be thrown when the token is not constructed properly (does not contain a username or certificate). It is typically set to true in environments where certificates are used exclusively.	None

Table 8.2 – Properties available on PreAuthenticatedAuthenticationProvider

In addition to these properties, there are a number of other opportunities for implementing interfaces or extending classes involved in certificate authentication to further customize your implementation.

Considerations when implementing client certificate authentication

Client certificate authentication, while highly secure, isn't for everyone and isn't appropriate for every situation.

The pros of client certificate authentication are listed, as follows:

- Certificates establish a framework of mutual trust and verifiability that both parties (client and server) are who they say they are
- Certificate-based authentication, if implemented properly, is much more difficult to spoof or tamper with than other forms of authentication
- If a well-supported browser is used and configured correctly, client certificate authentication can effectively act as a single sign-on solution, enabling transparent login to all certificate-secured applications

The cons of client certificate authentication are listed, as follows:

- The use of certificates typically requires the entire user population to have them. This can lead to both a user training burden and an administrative burden. Most organizations deploying certificate-based authentication on a large scale must have sufficient self-service and helpdesk support for certificate maintenance, expiration tracking, and user assistance.
- The use of certificates is generally an all-or-nothing affair, meaning that mixed-mode authentication and offering support for non-certificated users is not provided due to the complexity of web server configuration or poor application support.
- The use of certificates may not be well supported by all users in your user population, including the ones who use mobile devices.
- The correct configuration of the infrastructure required to support certificate-based authentication may require advanced IT knowledge.

As you can see, there are both benefits and drawbacks to client certificate authentication. When implemented correctly, it can be a very convenient mode of access for your users and has extremely attractive security and non-repudiation properties. You will need to determine your particular situation to see whether or not this type of authentication is appropriate.

Summary

In this chapter, we examined the architecture, flow, and Spring Security support for client certificate-based authentication. We have covered the concepts and overall flow of client certificate (mutual) authentication. We explored the important steps required to configure Apache Tomcat for a self-signed SSL and client certificate scenario.

We also learned about configuring Spring Security to understand certificate-based credentials presented by clients. We covered the architecture of Spring Security classes related to certificate authentication. We also know how to configure a Spring bean-style client certificate environment. We also covered the pros and cons of this type of authentication.

It's quite common for developers unfamiliar with client certificates to be confused by many of the complexities of this type of environment. We hope that this chapter has made this complicated subject a bit easier to understand and implement!

In the next chapter, we will discuss **Open Authorization** (**OAuth 2**) protocol and how you can accomplish single sign-on with **OpenID Connect** (**OIDC**).

Part 3:
Exploring OAuth 2
and SAML 2

This part focuses on OAuth 2, which is a widely adopted method for trusted identity management, empowering users to centrally manage their identities through a single trusted provider. Users benefit from the convenience of securely storing their passwords and personal information with the trusted OAuth 2 provider while retaining the option to disclose personal information as needed. Websites implementing OAuth 2 authentication can trust that users presenting OAuth 2 credentials are authenticated individuals.

In the exploration of SAML 2 support, we delve into the intricacies of integrating **Security Assertion Markup Language (SAML 2.0)** into Spring Security applications. SAML 2.0, an XML-based standard, facilitates the exchange of authentication and authorization data between **Identity Providers (IdP)** and **Service Providers (SP)**, offering seamless integration within Spring Security frameworks.

This part has the following chapters:

- *Chapter 9, Opening up to OAuth 2*
- *Chapter 10, SAML 2 Support*

9

Opening up to OAuth 2

OAuth 2 is a very popular form of trusted identity management that allows users to manage their identity through a single trusted provider. This convenient feature provides users with the security of storing their password and personal information with the trusted OAuth 2 provider, optionally disclosing personal information upon request. Additionally, an OAuth 2-enabled website offers the confidence that the users providing OAuth 2 credentials are who they say they are.

In this chapter, we will cover the following topics:

- Learning how to set up your own OAuth 2 application in less than 5 minutes
- Configuring the JBCP calendar application with a very rapid implementation of OAuth 2
- Learning the conceptual architecture of OAuth 2 and how it provides your site with trustworthy user access
- Implementing OAuth 2-based user registration
- Experimenting with OAuth 2 attribute exchange for user profile functionality
- Configuring OAuth 2 support in Spring Security
- Executing the OAuth 2 provider connection workflow
- Integrating OpenID Connect providers with Spring Security

This chapter's code in action link is here: https://packt.link/ejucD.

The Promising World of OAuth 2

As an application developer, you may have heard the term OAuth 2 thrown around a lot. OAuth 2 has been widely adopted by web service and software companies around the world and is integral to the way these companies interact and share information. But what exactly is it? In a nutshell, OAuth 2 is a protocol that allows distinct parties to share information and resources in a secure and reliable manner.

> **What about OAuth 1.0?**
>
> Built with the same motivation, OAuth 1.0 was designed and ratified in 2007. However, it was criticized for being overly complex and also had issues with imprecise specifications, which led to insecure implementation. All of these issues contributed to the poor adoption of OAuth 1.0, and eventually led to the design and creation of OAuth 2.
>
> OAuth 2 is the successor to OAuth 1.0.
>
> It is also important to note that OAuth 2 is not backward compatible with OAuth 1.0, and so OAuth 2 applications cannot integrate with OAuth 1.0 service providers.

This type of login—through a trusted third party—has been in existence for a long time, in many different forms (for example, *Google Identity Provider* or *Microsoft Entra ID* became two of the more notable central login services on the web for some time).

Here are the key concepts and components of OAuth 2.0:

Roles:

- **Resource Owner (RO)**: The entity that can grant access to a protected resource. Typically, this is the end-user.

- **Client**: The application requesting access to the protected resource on behalf of the resource owner.

- **Authorization Server (AS)**: The server that authenticates the resource owner and issues access tokens after getting proper authorization.

- **Resource Server (RS)**: The server hosting the protected resources that are being accessed.

Authorization Grant:

- OAuth 2.0 defines several authorization grant types, such as Authorization Code, Implicit, Resource Owner Password Credentials, and Client Credentials. The grant type determines the flow and the way the client obtains the access token.

Access Token:

- The access token is a credential representing the authorization granted to the client. It is used to access protected resources on behalf of the resource owner.

Scope:

- Scopes define the extent of the access that a client is requesting. It specifies what actions the client intends to perform on the resource server.

Authorization Endpoint and Token Endpoint:

- The `Authorization Endpoint` facilitates communication with the resource owner to secure an authorization grant, while the Token Endpoint facilitates the exchange of this grant for an access token.

Redirect URI:

- After the resource owner grants permission, the `Authorization Server` redirects the user back to the client application using a redirect URI.

> **Important note**
>
> You can refer to the OAuth 2.0 specification at `https://tools.ietf.org/html/rfc6749`.

The following diagram illustrates the high-level relationship between a site integrating OAuth 2 during the login process and the Facebook OAuth 2 provider, for example:

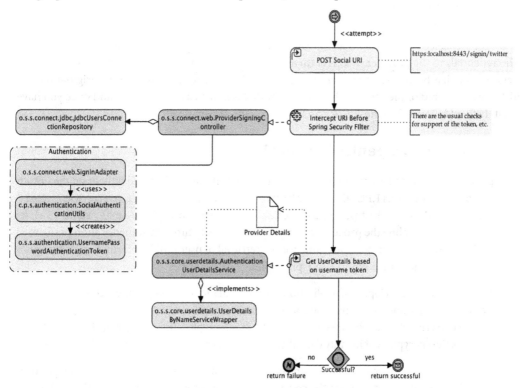

Figure 9.1 – OAuth 2 during the login process and the Facebook OAuth 2 provider

We can see that submitting a form post will initiate a request to the OAuth provider, resulting in the provider displaying an authorization dialog asking the user to allow `jbcpcalendar` to gain permission to specific information from your OAuth provider account. This request contains a `uri` parameter called `code`. Once granted, the user is redirected back to `jbcpcalendar`, and the `code` parameter is included in the

`uri` parameter. Then, the request is redirected to the OAuth provider again, to authorize `jbcpcalendar`. The OAuth provider then responds with an `access_token` that can be used to access the user's OAuth information that `jbcpcalendar` was granted access to.

> **Don't trust OAuth 2 unequivocally!**
>
> Here, you can see a fundamental assumption that can fool users of the system. It is possible for us to sign up for an OAuth 2 provider account, which would make it appear as though we were James Gosling, even though we obviously are not. Do not make the false assumption that just because a user has a convincing-sounding OAuth 2 (or OAuth 2 delegate provider), they are who they say they are without requiring additional forms of identification. Thinking about it another way, if someone came to your door just claiming he was James Gosling, would you let him in without verifying his ID?

The OAuth 2-enabled application then redirects the user to the OAuth 2 provider, and the user presents their credentials to the provider, which is then responsible for making an access decision. Once the access decision has been made by the provider, the provider redirects the user to the originating site, which is now assured of the user's authenticity. OAuth 2 is much easier to understand once you have tried it. Let's add OAuth 2 to the `JBCP calendar` login screen now!

Why do we need OpenID Connect?

OpenID Connect is an authentication protocol designed for interoperability and built upon the OAuth 2.0 framework specifications `RFC 6749` (`https://datatracker.ietf.org/doc/html/rfc6749`) and `RFC 6750` (`https://datatracker.ietf.org/doc/html/rfc6750`). Its primary goal is to streamline the process of verifying user identity through authentication performed by an `Authorization Server`, allowing for the retrieval of user profile information in a manner that is both interoperable and akin to REST principles.

This protocol empowers developers of applications and websites to initiate sign-in processes and receive credible assertions regarding users, ensuring consistency across various platforms, including web-based, mobile, and JavaScript clients. The specification suite is adaptable, supporting a variety of optional features such as the encryption of identity data, discovery of `OpenID Providers`, and session logout.

For developers, OpenID Connect offers a secure and verifiable means of answering the crucial question: "Who is the individual presently using the connected browser or mobile app?" Notably, it alleviates the burden of handling passwords—typically associated with data breaches—by removing the need to set, store, and manage them.

How OpenID Connect Works

OpenID Connect facilitates the establishment of an Internet identity ecosystem by offering seamless integration, robust support, security, and privacy-preserving configuration. It emphasizes interoperability, extends support to a broad array of clients and devices, and allows any entity to function as an **OpenID Provider (OP)**.

Here are the key concepts and components of `OpenID Connect`:

- **RP**, short **for Relying Party**, refers to an application or website that delegates its user authentication function to an **Identity Provider (IDP)**.

- **OP** or **IDP**: An OP is an entity that has implemented the OpenID Connect and OAuth 2.0 protocols. Sometimes, OPs are denoted by the role they fulfill, such as a **Security Token Service (STS)**, **IDP**, or an **Suthorization Server(AS)**.

- **Identity Token**: Serving as the outcome of an authentication process, an identity token includes, at a minimum, a user identifier (referred to as the *sub* or subject claim) and details about when and how the user authenticated. Additional identity data may also be included.

- **Client**: A client is software that requests tokens, either for user authentication or resource access (**RP**). Clients need to be registered with the OP and can take various forms, such as web applications, native mobile and desktop applications, etc.

- **User**: A user is an individual leveraging a registered client to access resources.

After highlighting the OpenID Connect principles as a protocol built on top of the OAuth 2 protocol, we will be learning how to set up OAuth 2 in our `JBCP Calendar` application using some popular providers.

Signing up for an OAuth 2 application

To get the full value out of the exercise in this section (and be able to test logging in), you will need to create an application with a service provider. Currently, Spring Social supports Twitter, Facebook, Google, LinkedIn, and GitHub, and the list is growing.

To get the full value out of the exercises in this chapter, we recommend you have accounts with at least Google. We have set up accounts for the `jbcpcalendar` application, which we will be using for the remainder of this chapter.

Include the following additional dependencies in your `build.gradle` file if you are leveraging the **OAuth 2** feature:

```
//build.gradle
dependencies {
...
    // OAuth2 Configuration:
```

```
    implementation 'org.springframework.boot:spring-boot-starter-
oauth2-client'
...
}
```

> **Important note**
> You should start with the source code in chapter09.00-calendar.

In addition to the Spring Security OAuth 2 dependencies, we will now explore the **OAuth 2** setup in the JBCP Calendar application.

Enabling OAuth 2.0 Login with Spring Security

We can see a common theme among the external authentication providers examined over the next several chapters. Spring Security provides comprehensive OAuth 2 support. This section discusses how to integrate OAuth 2 into your servlet-based application.

The OAuth 2.0 Login feature lets an application have users log in to the application by using their existing account with an OAuth 2.0 Provider (such as GitHub) or OpenID Connect 1.0 Provider (such as Google).

> **Important note**
> OAuth 2.0 Login is implemented by using the Authorization Code Grant, as specified in the OAuth 2.0 Authorization Framework, which you can find here: https://tools.ietf.org/html/rfc6749#section-4.1 and OpenID Connect Core 1.0, found here: https://openid.net/specs/openid-connect-core-1_0.html#CodeFlowAuth.

Initial Setup

This section shows how to configure the OAuth 2.0 Login sample by using Google as the Authentication Provider and covers the following topics:

- Follow the instructions on the OpenID Connect page here: https://developers.google.com/identity/openid-connect/openid-connect, starting in the *Setting up OAuth 2.0* section.

- After completing the Obtain OAuth 2.0 credentials instructions, you should have a new OAuth Client with credentials consisting of a **Client ID** and a **Client Secret**.

This setup is important to configure our application as an **OAuth 2** client.

Setting the Redirect URI

The redirect URI is the path in the application that the end-user's user-agent is redirected back to after they have authenticated with Google and have granted access to the `OAuth Client` (created in the previous step) on the **Consent** page.

In this subsection, ensure that the Authorized redirect URIs field is set to `https://localhost:8443/login/oauth2/code/google`.

> **Important note**
>
> The default redirect URI template is `{baseUrl}/login/oauth2/code/{registrationId}`. The `registrationId` is a unique identifier for the `ClientRegistration`.
>
> If the `OAuth Client` runs behind a proxy server, you should check the proxy server configuration (follow this link: `https://docs.spring.io/spring-security/reference/features/exploits/http.html#http-proxy-server`) to ensure the application is correctly configured. Also, see the supported URI template variables here: `https://docs.spring.io/spring-security/reference/servlet/oauth2/client/authorization-grants.html#oauth2Client-auth-code-redirect-uri` for `redirect-uri`.

Once the redirect URI is established, we will proceed to set up the `application.yml` configuration.

Configure application.yml

Now that you have a new `OAuth Client` with Google, you need to configure the application to use the `OAuth Client` for the authentication flow. To do so, go to `application.yml` and set the following configuration:

```yaml
spring:
  security:
    oauth2:
      client:
        registration:
          google:
            client-id: google-client-id
            client-secret: google-client-secret
```

We have configured here the following client properties:

- `spring.security.oauth2.client.registration` is the base property prefix for OAuth Client properties.

- Following the base property prefix is the ID for the `ClientRegistration`, such as Google.

After configuring the OAuth 2 client properties, we need to register a `SecurityFilterChain` bean.

Register a SecurityFilterChain Bean

The following example shows how to register a `SecurityFilterChain` bean with `@EnableWebSecurity` and enable `OAuth 2.0 login` through `httpSecurity.oauth2Login()`:

```java
//src/main/java/com/packtpub/springsecurity/configuration/
SecurityConfig.java
@Configuration
@EnableWebSecurity
public class SecurityConfig {

    @Bean
    public SecurityFilterChain filterChain(HttpSecurity http,
PersistentTokenRepository persistentTokenRepository) throws Exception
{
        http.authorizeRequests( authz -> authz
                    .requestMatchers(antMatcher("/webjars/**")).
permitAll()
                    .requestMatchers(antMatcher("/css/**")).
permitAll()
                    .requestMatchers(antMatcher("/favicon.ico")).
permitAll()
                    // H2 console:
                    .requestMatchers(antMatcher("/admin/h2/**")).
access("isFullyAuthenticated()")
                    .requestMatchers(antMatcher("/")).permitAll()
                    .requestMatchers(antMatcher("/login/*")).
permitAll()
                    .requestMatchers(antMatcher("/logout")).
permitAll()
                    .requestMatchers(antMatcher("/signup/*")).
permitAll()
                    .requestMatchers(antMatcher("/errors/**")).
permitAll()
                    .requestMatchers(antMatcher("/events/")).
hasRole("ADMIN")
                    .requestMatchers(antMatcher("/**")).
hasAnyAuthority("OIDC_USER", "OAUTH2_USER", "ROLE_USER"))
                .exceptionHandling(exceptions -> exceptions
                    .accessDeniedPage("/errors/403"))
                .formLogin(form -> form
                    .loginPage("/login/form")
                    .loginProcessingUrl("/login")
                    .failureUrl("/login/form?error")
```

```
                  .usernameParameter("username")
                  .passwordParameter("password")
                  .defaultSuccessUrl("/default", true)
                  .permitAll())
             .logout(form -> form
                  .logoutUrl("/logout")
                  .logoutSuccessUrl("/login/form?logout")
                  .permitAll())
             // CSRF is enabled by default, with Java Config
             .csrf(AbstractHttpConfigurer::disable);
        // OAuth2 Config
        http
             .oauth2Login(withDefaults());

        // For H2 Console
        http.headers(headers -> headers.
 frameOptions(FrameOptionsConfig::disable));

        return http.build();
    }


 ... Omitted for brevity
 }
```

The next step after configuring the `SecurityFilterChain` bean is to update the `SpringSecurityUserContext` class.

Update the SpringSecurityUserContext class

The `getCurrentUser` in the `SpringSecurityUserContext` needs to reference the new authenticated user of type `DefaultOidcUser`.

The following example shows how to adapt the current implementation to reference the `DefaultOidcUser` user type:

```
//src/main/java/com/packtpub/springsecurity/service/
SpringSecurityUserContext.java
@Component
public class SpringSecurityUserContext implements UserContext {

    private static final Logger logger = LoggerFactory
            .getLogger(SpringSecurityUserContext.class);


    private final CalendarService calendarService;
```

```java
    public SpringSecurityUserContext(final CalendarService
calendarService) {
        this.calendarService = calendarService;
    }

    @Override
    public CalendarUser getCurrentUser() {
        SecurityContext context = SecurityContextHolder.getContext();
        Authentication authentication = context.getAuthentication();
        if (authentication == null) {
            return null;
        }
        String email;
        if(authentication.getPrincipal() instanceof DefaultOidcUser
oidcUser ) {
            email = oidcUser.getEmail();
        } else if (authentication.getPrincipal() instanceof
DefaultOAuth2User oauth2User) {
            email = oauth2User.getAttribute("email");
        } else {
            User user = (User) authentication.getPrincipal();
            email = user.getUsername();
        }

        if (email == null) {
            return null;
        }
        CalendarUser result = calendarService.findUserByEmail(email);
        if (result == null) {
            throw new IllegalStateException(
                    "Spring Security is not in synch with CalendarUsers.
Could not find user with email " + email);
        }


        logger.info("CalendarUser: {}", result);
        return result;
    }

}
```

After completing the previous steps, we will test the application.

Boot up the Application

1. Launch the sample application and go to `https://localhost:8443/oauth2/authorization/google`. You will be redirected to the default login page, which displays a link for Google.

2. Click on the Google link, and you are then redirected to Google for authentication.

3. The `OAuth Client` retrieves your email address and basic profile information from the `UserInfo Endpoint` (find out more here: `https://openid.net/specs/openid-connect-core-1_0.html#UserInfo`) and establishes an authenticated session.

4. At this point, you should be able to complete a full login using the Google OAuth 2 provider. The redirects that occur are as follows. First, we initiate the OAuth 2 provider login as shown in the following screenshot:

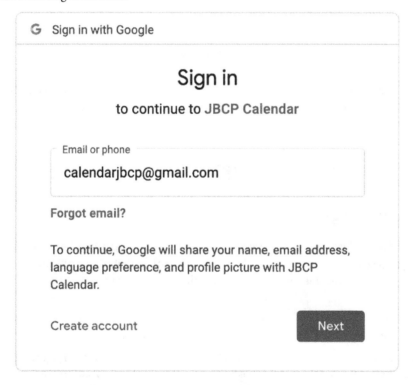

Figure 9.2 – OAuth 2 Social login with Google

5. After filling in the login details, the user is redirected to the `JBCP Calendar` application and automatically logged in using the provider display name:

Chapter 09.01: Below you can find some highlights about myCalendar. Each sample will have a slightly different summary depending on what has been done.

Opening up to OAuth 2

- The functionality hasn't changed since the last checkpoint, but feel free to give it a try.
- Select a link that requires authentication and observe the new login page.
- Try typing an invalid username/password and viewing the error message
- Try logging in as an admin and viewing all events. Notice we are able to view all the events
- Try logging in as a regular user and viewing all events (notice we get an access denied)

- All Events - shows all events for all users, but only allows administrators to access the page.
- My Events - shows all events that user1@example.com is the owner or attendee. We will discuss in Chapter 3 how to obtain the current user's events.
- Create Event - will allow creating a new Event with user1@example.com as the owner. We will discuss in Chapter 3 how to make the current user the owner.
- Logout - we haven't discussed it yet, but you can logout using j_spring_security_logout. Later in in this chapter we will discuss how to customize logout and provide a logout link.
- H2 Database Console - Allows you to interact with the database using a web console. To use it:
  - Click the link above.
  - Ensure that Generic H2 (Embedded) is selected
  - Ensure that org.h2.Driver is the Driver Class
  - Enter **jdbc:h2:mem:dataSource** as the JDBC URL
  - Ensure that the username is sa
  - Ensure the password is left empty
  - Click Connect

Figure 9.3 – Welcome page after successful authentication

At this point, the user exists in the application and is authenticated but not authorized on all the web pages. The page `Create Event` can only be accessed by authenticated users.

> **Important note**
> Your code should now look like that in `chapter09.01-calendar`.

Customize the Login Page

By default, the OAuth 2.0 login page is auto-generated by the `DefaultLoginPageGeneratingFilter`. The default login page shows each configured `OAuth Client` with its `ClientRegistration.clientName` as a link, which is capable of initiating the `Authorization Request` (or OAuth 2.0 Login).

> **Important note**
> For `DefaultLoginPageGeneratingFilter` to show links for configured `OAuth Clients`, the registered `ClientRegistrationRepository` needs to also implement `Iterable<ClientRegistration>`. See `InMemoryClientRegistrationRepository` for reference.

The link's destination for each OAuth Client defaults to the following:

```
OAuth2AuthorizationRequestRedirectFilter.DEFAULT_AUTHORIZATION_
REQUEST_BASE_URI + "/{registrationId}"
```

The following code shows an example of adapting the login.html form:

```
//src/main/resources/templates/login.html
<div class="mb-3">
    <legend>Login With Google</legend>
    <div class="mb-3">
        <a class="btn btn-danger"
            role="button" th:href="@{/oauth2/authorization/
google}">Login with Google</a>
    </div>
</div>
```

Now, you can use the login social button to authenticate your users with Google as an Identity Provider.

We need also to make sure the user is redirected to the jbcpcalendar application and automatically logged in. The following example shows how to adapt a SecurityConfig.java to have a proper redirection after successful authentication:

```
//src/main/java/com/packtpub/springsecurity/configuration/
SecurityConfig.java
@Configuration
@EnableWebSecurity
public class SecurityConfig {

    @Bean
    public SecurityFilterChain filterChain(HttpSecurity http,
GrantedAuthoritiesMapper grantedAuthoritiesMapper) throws Exception {
... omitted for brevity
        // OAuth2 Login
        http
                .oauth2Login(oauth2 -> oauth2
                        .loginPage("/login/form")
                        .defaultSuccessUrl("/default", true));
        return http.build();
    }
}
```

At this point, you should be able to complete a full login using Google's OAuth 2 provider. The redirects that occur are as follows. First, we initiate the OAuth 2 provider login.

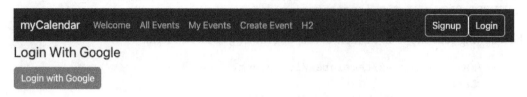

Figure 9.4 – Login screen after successful authentication

We are then redirected to the **provider authorization** page, requesting the user to grant permission to the JBCP Calendar application.

> **Important note**
> Your code should now look like that in chapter09.02-calendar.

Additional OAuth 2 providers

We have successfully integrated a single OAuth 2 provider using one of the popular OAuth 2 providers. There are several other providers available; we are going to add a few more providers so our users have more than one option. Spring Security currently supports the **Google**, **GitHub**, **Facebook**, and **Okta** providers natively. Including additional providers will require configuring custom provider properties.

CommonOAuth2Provider pre-defines a set of default client properties for a number of well-known providers that Spring Security supports natively, as mentioned previously.

For example, the authorization-uri, token-uri, and user-info-uri do not change often for a provider. Therefore, it makes sense to provide default values, to reduce the required configuration.

As demonstrated previously when we configured a Google client, only the client-id and client-secret properties are required.

In order to add GitHub providers to the JBCP calendar application:

1. Register your application in GitHub by following the steps given here: https://docs.github.com/en/apps/oauth-apps/building-oauth-apps/creating-an-oauth-app. At the end, save the client-id and client-secret.

   The **authorization callback URL** should be https://localhost:8443/login/oauth2/code/github.

2. Additional application properties need to be set, and each configured provider will automatically be registered with the `client-id` and `client-secret` keys from the provider application, as follows:

```
//src/main/resources/application.yml
spring:
  security:
    oauth2:
      client:
        registration:
          google:
            client-id: google-client-id
            client-secret: google-client-secret
          github:
            client-id: github-client-id
            client-secret: github -client-secret
```

3. We can now add the new login options to our `login.html` file to include the new provider GitHub:

```
//src/main/resources/templates/login.html

<div class="mb-3">
    <legend>Login With Google</legend>
    <div class="mb-3">
        <a class="btn btn-danger"
            role="button" th:href="@{/oauth2/authorization/
google}">Login with Google</a>
    </div>
    <legend>Login With Github</legend>
    <div class="mb-3">
        <a class="btn btn-dark"
            role="button" th:href="@{/oauth2/authorization/
github}">Login with Github</a>
    </div>
</div>
```

4. Now we have the required details to connect to the additional providers for the JBCP calendar, and we can restart the JBCP calendar application and test logging in with the other OAuth 2 providers.

> **Important note**
> Your code should now look like that in `chapter09.03-calendar`.

When logging in now, we should be presented with additional provider options, as shown in the following screenshot:

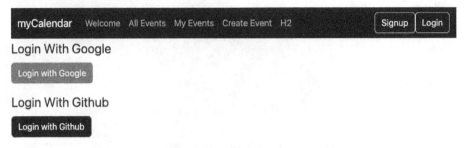

Figure 9.5 – Social login options with Google and GitHub

At this point, you should be able to complete a full login using the Google OAuth 2 provider. The redirects that occur are as follows. First, we initiate the OAuth 2 provider login as shown in the following screenshot:

Figure 9.6 – OAuth 2 social login with GitHub

5. We are then redirected to the provider authorization page, requesting the user to grant permission to the `jbcpcalendar` application as shown in the following screenshot:

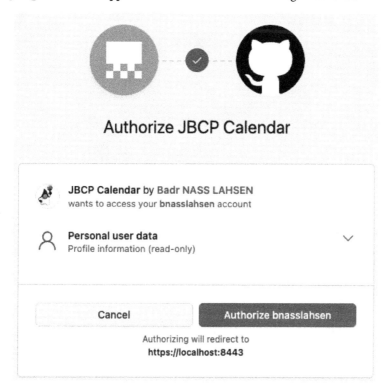

Figure 9.7 – OAuth 2 GitHub consent screen

After authorizing the `jbcpcalendar` application, the user is redirected to the `jbcpcalendar` application and automatically logged in using the provider display name.

Configuring Custom Provider Properties

There are some OAuth 2.0 Providers that support multi-tenancy, which results in different protocol endpoints for each tenant (or sub-domain).

For example, an `OAuth Client` registered with `OKTA` is assigned to a specific sub-domain and has its own protocol endpoints.

To get started, you'll require an **Okta developer account**. You can set one up by visiting `https://developer.okta.com/signup`.

For these cases, `Spring Boot 2.x` provides the following base property for configuring custom provider properties: `spring.security.oauth2.client.provider.[providerId]`.

The following code snippet shows an example:

```
security:
  oauth2:
    client:
      registration:
        okta:
          client-id: okta-client-id
          client-secret: okta-client-secret
          scope: openid,profile,email
      provider:
        okta:
          issuer-uri: https://your-subdomain.okta.com
          authorization-uri: https://your-subdomain.okta.com/oauth2/
v1/authorize
          token-uri: https://your-subdomain.okta.com/oauth2/v1/token
          user-info-uri: https://your-subdomain.okta.com/oauth2/v1/
userinfo
          user-name-attribute: sub
          jwk-set-uri: https://your-subdomain.okta.com/oauth2/v1/keys
```

The base property (`spring.security.oauth2.client.provider.okta`) allows for the custom configuration of protocol endpoint locations.

We can now add the new login options to our `login.html` file, to include with the new OKTA provider:

```
//src/main/resources/templates/login.html
... omitted for brevity
<div class="mb-3">
    <legend>Login With Google</legend>
    <div class="mb-3">
        <a class="btn btn-danger"
           role="button" th:href="@{/oauth2/authorization/
google}">Login with Google</a>
    </div>
    <legend>Login With Github</legend>
    <div class="mb-3">
        <a class="btn btn-dark"
           role="button" th:href="@{/oauth2/authorization/
github}">Login with Github</a>
    </div>
    <legend>Login With OKTA</legend>
    <div class="mb-3">
        <a class="btn btn-success"
           role="button" th:href="@{/oauth2/authorization/okta}">Login
with OKTA</a>
```

```
    </div>
</div>
```

Now we have the required details to connect to the additional providers for the JBCP calendar, and we can restart the JBCP calendar application and test logging in with the custom OAuth 2 OKTA providers.

> **Important note**
>
> Your code should now look like that in chapter09.04-calendar.

When logging in now, we should be presented with additional provider options, as shown in the following screenshot:

Figure 9.8 – Social login options with Google, GitHub, and OKTA

Enabling Proof Key for Code Exchange (PKCE) support

PKCE stands for **Proof Key for Code Exchange**. It is a security feature used in OAuth 2.0 authorization flows to mitigate certain types of attacks, particularly those targeting the authorization code flow.

The traditional OAuth 2.0 authorization code flow, a client application redirects a user to an Authorization Server, the user authenticates and provides consent, and the Authorization Server issues an authorization code to the client. The client then exchanges this code for an access token.

PKCE is designed to prevent authorization code interception attacks. In these attacks, a malicious actor intercepts the authorization code as it's being returned to the client, and then uses it to obtain an access token. PKCE adds an additional layer of security to this process.

The following sequence diagram describes how PKCE works:

OAuth 2.0 Authorization Code Grant with PKCE Flow

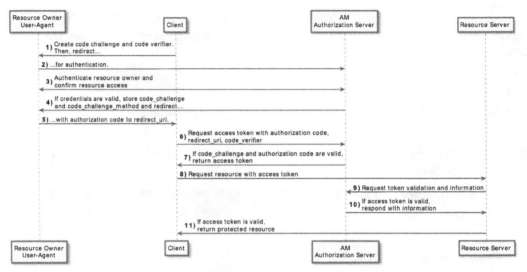

Figure 9.9 – Social login options with Google, GitHub, and OKTA

Public clients receive support through the utilization of PKCE. For further information on PKCE, refer to this link: `https://datatracker.ietf.org/doc/html/rfc7636`. PKCE is automatically employed when the client operates in an untrusted environment (e.g., native or web browser-based applications), rendering it unable to uphold the secrecy of its credentials when the following conditions are true:

- `client-secret` is omitted (or empty)

- `client-authentication-method` is set to **none** (`ClientAuthenticationMethod.NONE`)

If the OAuth 2.0 Provider supports PKCE for confidential clients (find out more about confidential clients: `https://datatracker.ietf.org/doc/html/rfc6749#section-2.1`), you may (optionally) configure it using `DefaultServerOAuth2AuthorizationRequestResolver.setAuthorizationRequestCustomizer(OAuth2AuthorizationRequestCustomizers.withPkce()`.

The following example shows how to adapt a `SecurityConfig.java` to use PKCE by registering your own `OAuth2AuthorizationRequestResolver`:

```
//src/main/java/com/packtpub/springsecurity/configuration/ Security-
Config.java
@Configuration
@EnableWebSecurity
```

```
public class SecurityConfig {

    @Bean
    public SecurityFilterChain filterChain(HttpSecurity http, OAu-
th2AuthorizationRequestResolver pkceResolver) throws Exception {
... omitted for brevity
        // OAuth2 Login
        http
                .oauth2Login(oauth2 -> oauth2
                    .loginPage("/login/form")
                    .authorizationEndpoint(authorization -> au-
thorization.authorizationRequestResolver(pkceResolver))
                    .defaultSuccessUrl("/default", true)
                .userInfoEndpoint(userInfo -> userInfo
                    .userAuthoritiesMapper(grantedAuthoritiesMap-
per)));
        return http.build();
    }

    @Bean
    public OAuth2AuthorizationRequestResolver pkceResolver(ClientReg-
istrationRepository clientRegistrationRepository) {
        DefaultOAuth2AuthorizationRequestResolver resolver = new De-
faultOAuth2AuthorizationRequestResolver(clientRegistrationRepository,
OAuth2AuthorizationRequestRedirectFilter.DEFAULT_AUTHORIZATION_RE-
QUEST_BASE_URI);
        resolver.setAuthorizationRequestCustomizer(OAuth2Authorization-
RequestCustomizers.withPkce());
        return resolver;
    }
}
```

> **Important note**
> Your code should now look like that in `chapter09.05-calendar`.

OpenID Connect 1.0 Logout

OpenID Connect Session Management 1.0 provides the capability to log out the end user through the **Provider** with the use of the **Client**. One available approach is the `RP-Initiated Logout`, detailed at `https://openid.net/specs/openid-connect-rpinitiated-1_0.html`.

In cases where the `OpenID Provider` supports both **Session Management** and **Discovery** (refer to `https://openid.net/specs/openid-connect-discovery-1_0.html` for more information), the client can acquire the `end_session_endpoint` URL from the *OpenID*

Provider's Discovery Metadata. You can achieve this by configuring the `ClientRegistration` with the `issuer-uri`, as outlined in `https://openid.net/specs/openid-connect-session-1_0.html#OPMetadata`:

```yaml
spring:
  security:
    oauth2:
      client:
        registration:
          okta:
            client-id: okta-client-id
            client-secret: okta-client-secret
            ...
        provider:
          okta:
            issuer-uri: https://dev-1234.oktapreview.com
```

Also, you can configure `OidcClientInitiatedLogoutSuccessHandler`, which implements RP-Initiated Logout, as follows:

```java
//src/main/java/com/packtpub/springsecurity/configuration/
SecurityConfig.java
@Configuration
@EnableWebSecurity
public class SecurityConfig {

    private ClientRegistrationRepository clientRegistrationRepository;

    public SecurityConfig(ClientRegistrationRepository
clientRegistrationRepository) {
        this.clientRegistrationRepository =
clientRegistrationRepository;
    }
    @Bean
    public SecurityFilterChain filterChain(HttpSecurity http,
OAuth2AuthorizationRequestResolver pkceResolver) throws Exception {
        http.authorizeRequests(authz -> authz

... omitted for brevity
        // OAuth2 Login
        http
                .oauth2Login(oauth2 -> oauth2
                    .loginPage("/login/form")
                        .authorizationEndpoint(authorization ->
authorization.authorizationRequestResolver(pkceResolver))
```

```
                    .defaultSuccessUrl("/default", true))
                .logout(logout -> logout
                    .logoutSuccessHandler(oidcLogoutSuccessHandler()));
        return http.build();
    }


    private LogoutSuccessHandler oidcLogoutSuccessHandler() {
        OidcClientInitiatedLogoutSuccessHandler
oidcLogoutSuccessHandler =
            new OidcClientInitiatedLogoutSuccessHandler(this.
clientRegistrationRepository);

        // Sets the location that the End-User's User Agent will be
redirected to
        // after the logout has been performed at the Provider
        oidcLogoutSuccessHandler.setPostLogoutRedirectUri("{baseUrl}");

        return oidcLogoutSuccessHandler;
    }
}
```

> **Important note**
> Your code should now look like that in `chapter09.06-calendar`.

Automatic User Registration

Many applications need to hold data about their users locally, even if authentication is delegated to an external provider. This can be done in two steps:

1. Choose a backend for your database, and set up some repositories (using Spring Data, say) for a custom `User` object that suits your needs and can be populated, fully or partially, from external authentication. For our `JBCP Calendar` application, we will adapt `CalendarUser` to add the provider information as follows:

    ```
    @Entity
    @Table(name = "calendar_users")
    public class CalendarUser implements Principal, Serializable {

    ... getter / setter omitted for brevity

        @Id
        @SequenceGenerator(name = "EntityTwoSequence", initialValue
    = 1000)
        @GeneratedValue(generator = "EntityTwoSequence")
    ```

```
    private Integer id;

    private String firstName;

    private String lastName;

    private String email;

    private String provider;

    private String externalId;

    @ManyToMany(fetch = FetchType.EAGER)
    @JoinTable(name = "user_role",
            joinColumns = @JoinColumn(name = "user_id"),
            inverseJoinColumns = @JoinColumn(name = "role_id"))
    private Set<Role> roles;
    /**
}
```

2. Implement and expose `OAuth2UserService` to call the **Authorization Server** as well as your database. Your implementation can be delegated to the default implementation, which will do the heavy lifting of calling the **Authorization Server**. Your implementation should return something that extends your custom `CalendarUser` object and implements `OAuth2User`.

```
@Component
public class CalendarOAuth2UserService implements
OAuth2UserService {

    private final CalendarService calendarService;
    public CalendarOAuth2UserService(CalendarService
calendarService) {
        this.calendarService = calendarService;
    }

    @Override
    public OAuth2User loadUser(OAuth2UserRequest userRequest)
throws OAuth2AuthenticationException {
        DefaultOAuth2UserService delegate = new
DefaultOAuth2UserService();
        OAuth2User user = delegate.loadUser(userRequest);
        String email = user.getAttribute("email");
        CalendarUser calendarUser = calendarService.
findUserByEmail(email);

        if (calendarUser ==null) {
```

```
            calendarUser = new CalendarUser();
            calendarUser.setEmail(email);
            calendarUser.setProvider(userRequest.
    getClientRegistration().getRegistrationId());
            if ("github".equals(userRequest.
    getClientRegistration().getRegistrationId())) {
                calendarUser.setExternalId(user.getAttribute("id").
    toString());
                calendarUser.setFirstName( user.
    getAttribute("name"));
                calendarUser.setLastName(user.
    getAttribute("name"));
            }
        calendarService.createUser(calendarUser);
        }
        return user;
    }
}
```

3. Implement and expose `OidcUserService` to call the `Authorization Server` as well as your database. Your implementation should return something that extends your custom User object and implements `OidcUser`.

```
@Component
public class CalendarOidcUserService extends OidcUserService {

    private final CalendarService calendarService;

    public CalendarOidcUserService(CalendarService
calendarService) {
        this.calendarService = calendarService;
    }


    @Override
    public OidcUser loadUser(OidcUserRequest userRequest) throws
OAuth2AuthenticationException {
        OidcUser user = super.loadUser(userRequest);
        String email = user.getEmail();
        CalendarUser calendarUser = calendarService.
findUserByEmail(email);
        if (calendarUser == null) {
            calendarUser = new CalendarUser();
            calendarUser.setEmail(email);
            calendarUser.setProvider(userRequest.
getClientRegistration().getRegistrationId());
            calendarUser.setExternalId(user.getAttribute("sub"));
```

```
            calendarUser.setFirstName(user.getGivenName());
            calendarUser.setLastName(user.getFamilyName());
        calendarService.createUser(calendarUser);
        }
        return user;
    }
}
```

> **Hint**
>
> Include a new attribute within the User object to establish a connection with a distinct identifier from the external provider (distinct from the user's name, yet uniquely associated with the account on the external platform).

One issue that would need to be resolved if supporting multiple providers is username conflicts between the various provider details returned.

If you log in to the JBCP calendar application with each of the listed providers—which then query the data that was stored in H2 database—you will find the data could be similar, if not exactly the same, based on the user's account details.

In the CALENDAR_USERS table, we have two possible issues;

1. First, we use the email attribute of the UserDetails object as the user ID to find the JBCP calendar users. But the user ID, can be different than an email for some other providers.
2. Second, it is still possible that the user identifier for two different providers will be the same.

We are not going to dive into the various ways to detect and correct this possible issue, but it is worth noting for future reference.

> **Important note**
>
> Your code should now look like that in chapter09.07-calendar.

Mapping User Authorities

The GrantedAuthoritiesMapper receives a collection of granted authorities, including a unique authority of the OAuth2UserAuthority type with the corresponding string identifier OAUTH2_USER (or OidcUserAuthority with the string identifier OIDC_USER).

We will provide a custom implementation of GrantedAuthoritiesMapper and configure it, as follows:

```
//src/main/java/com/packtpub/springsecurity/core/authority/
CalendarUserAuthoritiesMapper.java
```

```
@Component
public class CalendarUserAuthoritiesMapper implements
GrantedAuthoritiesMapper {

    private CalendarUserRepository userRepository;

    public CalendarUserAuthoritiesMapper(CalendarUserRepository
userRepository) {
        this.userRepository = userRepository;
    }


    @Override
    public Collection<? extends GrantedAuthority>
mapAuthorities(Collection<? extends GrantedAuthority> authorities) {
        Set<GrantedAuthority> mappedAuthorities = new HashSet<>();
        authorities.forEach(authority -> {
            String email = null;
            if (authority instanceof OidcUserAuthority
oidcUserAuthority) {
                OidcIdToken idToken = oidcUserAuthority.getIdToken();
                mappedAuthorities.add(oidcUserAuthority);
                email = idToken.getEmail();
            }
            else if (OAuth2UserAuthority.class.isInstance(authority)) {
                OAuth2UserAuthority oauth2UserAuthority =
(OAuth2UserAuthority) authority;
                mappedAuthorities.add(oauth2UserAuthority);
                Map<String, Object> userAttributes = oauth2UserAuthority.
getAttributes();
                email = (String) userAttributes.get("email");
            }
            if (email != null) {
                CalendarUser calendarUser = userRepository.
findByEmail(email);
                List<String> roles = calendarUser.getRoles().stream().
map(Role::getName).toList();
                List<GrantedAuthority> grantedAuthorityList =
AuthorityUtils.createAuthorityList(roles.toArray(new String[0]));
                mappedAuthorities.addAll(grantedAuthorityList);
            }
        });
        return mappedAuthorities;
    }
}
```

The following example shows how to adapt a `SecurityConfig.java` to use the `GrantedAu-thoritiesMapper`:

```
//src/main/java/com/packtpub/springsecurity/configuration/
SecurityConfig.java
@Configuration
@EnableWebSecurity
public class SecurityConfig {

    @Bean
public SecurityFilterChain filterChain(HttpSecurity http,
PersistentTokenRepository persistentTokenRepository,
        GrantedAuthoritiesMapper grantedAuthoritiesMapper) throws
Exception {
... omitted for brevity
        // OAuth2 Login
        http
                .oauth2Login(oauth2 -> oauth2
                    .loginPage("/login/form")
                    .authorizationEndpoint(authorization ->
authorization.authorizationRequestResolver(pkceResolver))
                    .defaultSuccessUrl("/default", true)
                .userInfoEndpoint(userInfo -> userInfo
                    .userAuthoritiesMapper(grantedAuthoritiesMapper)))
                .logout(logout -> logout
                    .logoutSuccessHandler(oidcLogoutSuccessHandler()));
        return http.build();
    }
}
```

With this implementation, you need to make sure the OIDC user roles already exist in the database.

For our JBCP calendar application, let's define a user with the admin role.

For example:

```
//src/main/resources/data.sql
insert into calendar_users(id,email,first_name,last_name) values
(1,'calendarjbcp@gmail.com','Admin','1');
insert into user_role(user_id,role_id) values (1, 1);
```

At this stage, the user `calendarjbcp@gmail.com` has the admin role and can have access to the **All Events** page after successful authentication.

> **Important note**
>
> Your code should now look like that in `chapter09.08-calendar`.

Is OAuth 2 secure?

As support for OAuth 2 relies on the trustworthiness of the OAuth 2 provider and the verifiability of the provider's response, security and authenticity are critical in order for the application to have confidence in the user's OAuth 2-based login.

Fortunately, the designers of the OAuth 2 specification were very aware of this concern, and implemented a series of verification steps to prevent response forgery, replay attacks, and other types of tampering, which are explained as follows:

- **Response forgery** is prevented due to a combination of a shared secret key (created by the OAuth 2-enabled site prior to the initial request) and a one-way hashed message signature on the response itself. A malicious user tampering with the data in any of the response fields without having access to the shared secret key—and signature algorithm—would generate an invalid response.

- **Replay attacks** are prevented due to the inclusion of a nonce, or a one-time use, random key, which should be recorded by the OAuth 2-enabled site so that it cannot ever be reused. In this way, even a user attempting to reissue the response URL would be foiled because the receiving site would determine that the nonce had been previously used, and would invalidate the request.

- The most likely form of attack that could result in a compromised user interaction would be a **man-in-the-middle attack**, where a malicious user could intercept the user's interaction between their computer and the OAuth 2 provider. A hypothetical attacker in this situation could be in a position to record the conversation between the user's browser and the OAuth 2 provider, and record the secret key used when the request was initiated. The attacker, in this case, would need a very high level of sophistication and reasonably a complete implementation of the OAuth 2 signature specification—in short, this is not likely to occur with any regularity.

Summary

In this chapter, we reviewed OAuth 2, a relatively recent technology for user authentication and credentials management. OAuth 2 has a very wide reach on the web and has made great strides in usability and acceptance within the past year or two. Most public-facing sites on the modern web should plan on having some form of OAuth 2 support, and the JBCP calendar application is no exception!

We learned about the following topics: the OAuth 2 authentication mechanism and its high-level architecture and key terminology. We also learned about OAuth 2 login and automatic user registration with the JBCP calendar application. We also covered automatic login with OAuth 2 and the security of OAuth 2's login responses.

We covered one of the simplest single sign-on mechanisms to implement with Spring Security. One of the downsides is that it does not support a standard mechanism for a single logout. In the next chapter, we will explore SAML, another standard, single sign-on protocol that also supports single logout.

Is OAuth 2 Secure?

10

SAML 2 Support

SAML is predominantly employed as a web-based authentication mechanism, relying on the browser agent to facilitate the authentication process. In broad terms, the authentication flow of SAML can be outlined as follows.

Spring Security provides comprehensive SAML 2 support. This section discusses how to integrate SAML 2 into your Servlet-based application.

Starting from 2009, support for relying parties has been available as part of an extension project. In 2019, efforts were initiated to integrate this support into the core of Spring Security. This mirrors a similar process initiated in 2017 for incorporating Spring Security's OAuth 2.0 support.

This chapter will explore the following subjects:

- Fundamental aspects of the SAML protocol
- Establishing your SAML 2 Login using Spring Security
- Acquiring the SAML 2 Authenticated Principal
- Parsing and generating SAML 2.0 metadata
- Tailoring authorities using Spring Security SAML
- Executing Single Logout

This chapter's code in action link is here: `https://packt.link/7qRvM`.

What is SAML?

Security Assertion Markup Language (SAML) stands as a widely embraced open standard based on XML, specifically crafted for the secure exchange of **authentication and authorization (AA)** information among federated organizations. It serves to streamline **Single Sign-On (SSO)** capabilities for browser-based access.

Established in 2005 as an OASIS standard and consistently upheld by the **Organization for the Advancement of Structured Information Standards (OASIS)**, SAML 2.0 amalgamates elements from SAML 1.1, the **Liberty Alliance Identity Federation Framework (ID-FF)** 1.2, and Shibboleth 1.3.

Within the SAML 2.0 specification, three crucial entities assume distinct roles: the principal, the service provider, and the identity provider.

As an illustration, consider Sally accessing her investment account on ucanbeamillionaire.com. To log her in and let her access her account, the site employs SAML for authentication.

SAML 2.0 is widely adopted and used in various scenarios, such as enterprise applications, cloud services, and web-based authentication systems, to establish a secure and interoperable framework for identity and access management.

Key components and concepts of SAML 2.0 include:

- A **Service Provider (SP)** functions as the entity delivering a service, often in the form of an application.

- An **Identity Provider (IdP)** serves as the entity furnishing identities, encompassing the capability to authenticate a user. Typically, the IdP also houses the user profile, which includes additional information such as the first name, last name, job code, phone number, address, and more. The extent of user data required by SPs may vary, ranging from a basic profile (username, email) to a more comprehensive set (job code, department, address, location, manager, etc.), depending on the application.

- A **SAML Request**, which is also known as an authentication request, is initiated by the SP to formally request authentication.

- The IdP generates a **SAML Response**, which includes the actual assertion of the authenticated user. Additionally, the SAML Response may incorporate extra information, such as user profile details and group/role information, based on the capabilities supported by the SP.

- **SP-initiated** sign-in denotes the SAML sign-in flow instigated by the SP. This typically occurs when an end user attempts to access a resource or sign in directly on the SP side, such as when the browser endeavors to access a protected resource on the SP's platform.

- **IdP-initiated** sign-in characterizes the SAML sign-in flow instigated by the IdP. In this scenario, instead of the SAML flow being prompted by a redirection from the SP, the IdP initiates a SAML Response redirected to the SP to validate the user's identity.

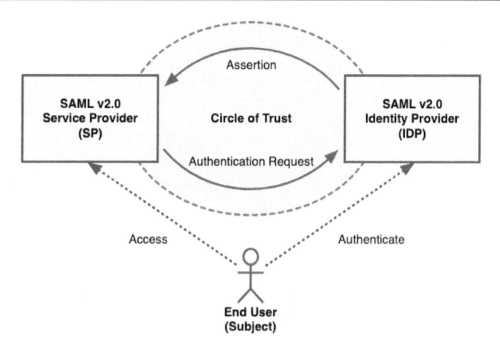

Figure 10.1 – Exploring the SAML protocol

Here are a few key points to consider:

- Direct interaction between the SP and the IdP never occurs. All interactions are facilitated through a browser, which serves as the intermediary for all redirections.
- The SP must be aware of the IdP to redirect to before obtaining information about the user.
- The SP remains unaware of the user's identity until it receives the SAML assertion from the IdP.
- The initiation of this flow is not restricted to the SP; an IdP can also kickstart an authentication flow.
- The SAML authentication flow operates asynchronously. The SP is uncertain if the IdP will complete the entire process. Consequently, the SP does not retain any state related to authentication requests. When the SP receives a response from the IdP, it must include all necessary information.

Following the introduction of the SAML protocol, we'll delve into the functionality of SAML 2.0 Login within the context of Spring Security.

SAML 2.0 Login with Spring Security

The SAML 2.0 Login functionality empowers an application to function as a SAML 2.0 relying party. This enables users to log in to the application using their pre-existing accounts with a SAML 2.0 Asserting Party, such as ADFS, Okta, and other IdPs.

> **Important note**
>
> The implementation of SAML 2.0 Login utilizes the **Web Browser Single Sign-On (SSO) Profile**, as outlined in the SAML 2 Profiles Specification: `https://groups.oasis-open.org/higherlogic/ws/public/document?document_id=35389#page=15`.

To begin our exploration of SAML 2.0 relying party authentication in the context of Spring Security, we observe that Spring Security guides the user to a third party for authentication. This is accomplished through a sequence of redirections:

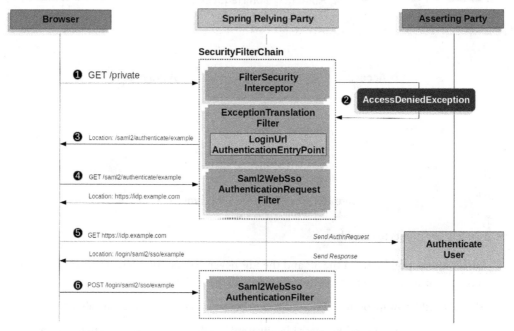

Figure 10.2 – Redirecting to asserting party authentication

Let's delve deeper into this sequence of SAML redirections:

1. Initially, a user submits an unauthenticated request to the `/private` resource without proper authorization.

2. Spring Security's `AuthorizationFilter` signals the denial of the unauthenticated request by throwing an `AccessDeniedException`.

3. Due to the lack of authorization, the `ExceptionTranslationFilter` triggers the start of authentication. The configured `AuthenticationEntryPoint` is an instance of `LoginUrlAuthenticationEntryPoint`, redirecting to the endpoint that generates the `<saml2:AuthnRequest>`, managed by the `Saml2WebSsoAuthenticationRequestFilter`. If multiple asserting parties are configured, it may first redirect to a picker page.

4. Subsequently, the `Saml2WebSsoAuthenticationRequestFilter` generates, signs, serializes, and encodes a `<saml2:AuthnRequest>` using its configured `Saml2Authen-ticationRequestFactory`.

5. The browser then takes the `<saml2:AuthnRequest>` and presents it to the asserting party, initiating the user authentication process. Upon successful authentication, the asserting party returns a `<saml2:Response>` to the browser.

6. The browser proceeds to `POST` the `<saml2:Response>` to the assertion consumer service endpoint.

    The following diagram illustrates the authentication process of a `<saml2:Response>` in Spring Security:

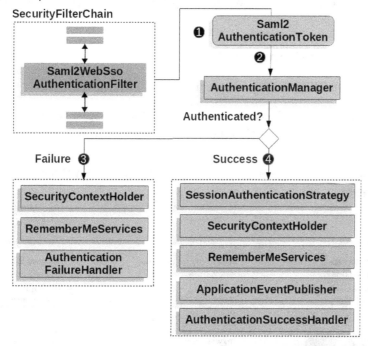

Figure 10.3 – Authenticating a <saml2:Response>

We can summarize the interactions as follow:

7. Upon submission of a `<saml2:Response>` by the browser to the application, the process is handed over to the `Saml2WebSsoAuthenticationFilter`. This filter employs its configured `AuthenticationConverter` to generate a `Saml2AuthenticationToken` by extracting the response from the `HttpServletRequest`. Additionally, the converter resolves the `RelyingPartyRegistration` and provides it to the `Saml2AuthenticationToken`.

8. Subsequently, the filter transfers the token to its configured `AuthenticationManager`, defaulting to the `OpenSamlAuthenticationProvider`.

9. In the event of authentication failure:

- The `SecurityContextHolder` is cleared.

- The `AuthenticationEntryPoint` is invoked to initiate the authentication process anew.

10. If authentication succeeds:

- The `Authentication` is set on the `SecurityContextHolder`.

- The `Saml2WebSsoAuthenticationFilter` invokes `FilterChain#doFilter(request, response)` to proceed with the remaining application logic.

After the introduction of SAML 2.0 Login with Spring Security, we will explore a real-world SAML example using OKTA.

Add a SAML application on OKTA

To begin, you'll need an OKTA developer account.

1. First, visit the OKTA developer website: `https://developer.okta.com/signup`. You will be presented with the following options to create an account:

Figure 10.4 – OKTA developer portal

2. Choose **Access the Okta Developer Edition Service**, then create your developer account.

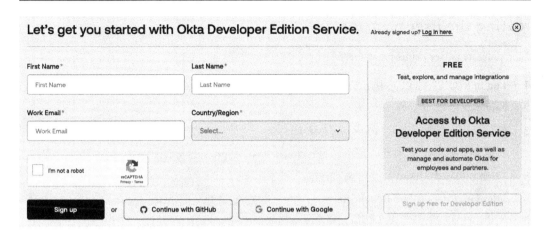

Figure 10.5 – OKTA developer account creation

3. The second step is to log in with your account, then go to **Applications | Create App Integration**.

4. Choose **SAML 2.0** and click **Next**.

5. Give a name to your application like JBCP Calendar SAML and click **Next**.

6. Use the following configuration:

    - Single sign on URL: https://localhost:8443/login/saml2/sso/okta

    - For the recipient URL and Destination URL: (the default)

    - For the audience URI: https://localhost:8443/saml2/service-provider-metadata/okta

7. After that click **Next**. Select these options: **I'm an Okta customer adding an internal app** and **This is an internal app that we have created**.

8. Select **Finish**.

9. OKTA will create your application.

10. Go to the **SAML Signing Certificates** and go to **SHA-2 | Actions | View IdP Metadata**. You can *right-click* and copy this menu item's link or open its URL.

11. Copy the resulting link to your clipboard. It should look something like the following: https://dev-xxxxx.okta.com/app/<random-characters>/sso/saml/metadata.

12. Go to your application's **Assignment** tab and assign access to the **Everyone** group.

Creating the user principal in OKTA

Let's first create a user principal in OKTA.

1. Log in to OKTA and go to the OKTA administrator console.
2. Navigate to the **Users** page.
3. Once logged in, navigate to the **Admin** section.
4. Select **Directory** from the menu. Then choose the **People** sub-menu.
5. Click the **Add Person** button or similar.
6. Fill in user details: Provide the necessary information for the new user, such as first name, last name, email address, and any other required fields. You may also set a username and assign a role or group to the user.

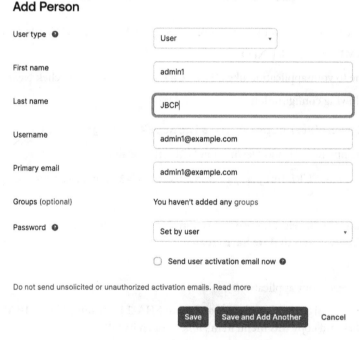

Figure 10.6 – Adding users with OKTA

Additional required dependencies

We include the following additional dependencies in your `build.gradle` file if you are leveraging the OAuth 2 feature:

```
//build.gradle
dependencies {
```

```
...
    constraints {
        implementation "org.opensaml:opensaml-core:4.2.0"
        implementation "org.opensaml:opensaml-saml-api:4.2.0"
        implementation "org.opensaml:opensaml-saml-impl:4.2.0"
    }
    implementation 'org.springframework.security:spring-security-
saml2-service-provider'
...
```

Specifying IdP Metadata

In a Spring Boot application, configure the metadata for an IdP by creating a setup resembling the following:

```
spring:
  security:
    saml2:
      relyingparty:
        registration:
          okta:
            assertingparty:
              metadata-uri: https://dev-xxxxx.okta.com/app/ <random-
characters>/sso/saml/metadata
```

> **Important note**
> Your code should now look like that in `chapter10.01-calendar`.

Retrieving the SAML 2 Authenticated Principal

Once the relying party is appropriately configured for a specific asserting party, it is prepared to receive assertions. Following the validation of an assertion by the relying party, the outcome is a `Saml2Authentication` containing a `Saml2AuthenticatedPrincipal`. Consequently, you can access the principal, as demonstrated in the `SpringSecurityUserContext`:

```
//src/main/java/com/packtpub/springsecurity/service/
SpringSecurityUserContext.java
@Component
public class SpringSecurityUserContext implements UserContext {

    private static final Logger logger = LoggerFactory
            .getLogger(SpringSecurityUserContext.class);
```

```java
    private final CalendarService calendarService;

    public SpringSecurityUserContext(final CalendarService
calendarService) {
        if (calendarService == null) {
            throw new IllegalArgumentException("calendarService cannot
be null");
        }
        this.calendarService = calendarService;
    }


    @Override
    public CalendarUser getCurrentUser() {
        SecurityContext context = SecurityContextHolder.getContext();
        Authentication authentication = context.getAuthentication();
        if (authentication == null) {
            return null;
        }

        if(authentication.getPrincipal() instanceof
DefaultSaml2AuthenticatedPrincipal saml2AuthenticatedPrincipal ) {
            String email = saml2AuthenticatedPrincipal.getName();
            CalendarUser result = calendarService.
findUserByEmail(email);
            if (result == null) {
                throw new IllegalStateException(
                    "Spring Security is not in synch with
CalendarUsers. Could not find user with email " + email);
            }

            logger.info("CalendarUser: {}", result);
            return result;
        }
        return null;
    }


}
```

Important note

Your code should now look like that in `chapter10.02-calendar`.

Spring Security can parse asserting party metadata to produce an `AssertingPartyDetails` instance as well as publish relying party metadata from a `RelyingPartyRegistration` instance.

Parsing SAML 2 metadata

By utilizing `RelyingPartyRegistrations`, it becomes possible to parse metadata from an asserting party. If you're utilizing OpenSAML vendor support, the resultant `AssertingPartyDetails` will be in the form of `OpenSamlAssertingPartyDetails`. Consequently, you can access the underlying OpenSAML XMLObject by following these steps:

```
OpenSamlAssertingPartyDetails details =
(OpenSamlAssertingPartyDetails)
        registration.getAssertingPartyDetails();
EntityDescriptor openSamlEntityDescriptor = details.
getEntityDescriptor();
```

Generating SAML 2 Metadata

You can expose a metadata endpoint with the `saml2Metadata` DSL method, as you can see below:

```
http
        // ...
        .saml2Login(withDefaults())
        .saml2Metadata(withDefaults());
```

Utilize the metadata endpoint for registering your relying party with the asserting party. This typically involves identifying the appropriate form field to provide the metadata endpoint.

The default metadata endpoint is `/saml2/metadata`. It also responds to `/saml2/metadata/{registrationId}` and `/saml2/service-provider-metadata/{registrationId}`.

You can adapt this by calling the `metadataUrl` method in the DSL:

```
.saml2Metadata((saml2) -> saml2.metadataUrl("/saml/metadata"))
```

Adapting RelyingPartyRegistration lookup

To configure your own `Saml2MetadataResponseResolver`, you should use `RelyingPartyRegistration` as described below:

```
@Bean
Saml2MetadataResponseResolver
metadataResponseResolver(RelyingPartyRegistrationRepository
registrations) {
    RequestMatcherMetadataResponseResolver metadata = new
RequestMatcherMetadataResponseResolver(
        (id) -> registrations.findByRegistrationId("relying-party"),
new OpenSamlMetadataResolver());
```

```
        metadata.setMetadataFilename("metadata.xml");
        return metadata;
}
```

Now that we have explored the SAML 2.0 Login functionality with Spring Security, we'll move on to use custom SAML Spring Boot Auto Configuration.

Overriding SAML Spring Boot Auto Configuration

Spring Boot generates two @Bean objects for a relying party.

The first is a SecurityFilterChain that configures the application as a relying party. When including spring-security-saml2-service-provider, the SecurityFilterChain looks like:

You will notice that every authenticated user has a ROLE_USER role by default.

```
//src/main/java/com/packtpub/springsecurity/service/ SecurityConfig.
java
@Configuration
@EnableWebSecurity
public class SecurityConfig {
    @Bean
    public SecurityFilterChain filterChain(HttpSecurity http) throws
Exception {
        http.authorizeHttpRequests( authz -> authz
                    .requestMatchers("/webjars/**").permitAll()
                    .requestMatchers("/css/**").permitAll()
                    .requestMatchers("/favicon.ico").permitAll()
                    // H2 console:
                    .requestMatchers("/admin/h2/**").
fullyAuthenticated()
                    .requestMatchers("/").permitAll()
                    .requestMatchers("/login/*").permitAll()
                    .requestMatchers("/logout").permitAll()
                    .requestMatchers("/signup/*").permitAll()
                    .requestMatchers("/errors/**").permitAll()
                    .requestMatchers("/events/").hasRole("ADMIN")
                    .requestMatchers("/**").hasRole("USER"))

                .exceptionHandling(exceptions -> exceptions
                    .accessDeniedPage("/errors/403"))

                .logout(form -> form
                    .logoutUrl("/logout")
```

```
                    .logoutSuccessUrl("/")
                    .permitAll())
            // CSRF is enabled by default, with Java Config
            .csrf(AbstractHttpConfigurer::disable);

        // @formatter:off
        http
              .saml2Login(withDefaults());
        // For H2 Console
        http.headers(headers -> headers.
  frameOptions(FrameOptionsConfig::disable));
        return http.build();
    }
  ...ommited for breviy


}
```

To test the application, open a web browser and navigate to: `https://localhost:8443`.

Next, proceed to the `/events` page. You should get an **Access Denied** error.

> **Important note**
> Your code should now look like that in `chapter10.03-calendar`.

Creating a custom RelyingPartyRegistrationRepository

Spring Boot creates a `RelyingPartyRegistrationRepository`, which represents the asserting party and relying party metadata. This includes things such as the location of the SSO endpoint the relying party should use when requesting authentication from the asserting party.

You can override the default by publishing your own `RelyingPartyRegistrationRepository` bean. You can also remove the existing `spring.security.saml2.relyingparty.registration` configuration properties programmatically.

For example, you can look up the asserting party's configuration by hitting its metadata endpoint:

```
//src/main/java/com/packtpub/springsecurity/service/ SecurityConfig.
java

@Value("${metadata.location}")
private String assertingPartyMetadataLocation;

@Bean
public RelyingPartyRegistrationRepository relyingPartyRegistrations()
```

```
{
    RelyingPartyRegistration registration = RelyingPartyRegistrations
        .fromMetadataLocation(assertingPartyMetadataLocation)
        .registrationId("okta")
        .build();
    return new
InMemoryRelyingPartyRegistrationRepository(registration);
}
```

Alternatively, you can directly wire up the repository by using the DSL, which also overrides the auto-configured `SecurityFilterChain`:

```
//src/main/java/com/packtpub/springsecurity/service/ SecurityConfig.
java

@Configuration
@EnableWebSecurity
public class SecurityConfig {
    @Bean
    public SecurityFilterChain filterChain(HttpSecurity http) throws
Exception {
... omitted for brevity

        http
            .saml2Login(saml2 -> saml2
                .
relyingPartyRegistrationRepository(relyingPartyRegistrations())
            );
        return http.build();
    }
}
```

> **Important note**
>
> The `registrationId` is a user-defined value chosen to distinguish between different registrations.
>
> Your code should now look like that in `chapter10.04-calendar`.

Creating custom authorities with Spring Security SAML

Upon logging in, you may observe that the displayed page indicates a ROLE_USER authority.

Despite granting access to all users initially, you can configure your **SAML** application to transmit a user's groups as an attribute. Additionally, you have the option to include other attributes such as name and email.

1. Start by choosing to edit your OKTA application SAML settings section.

2. Complete the Group Attribute Statements section.

    *   Name: `groups`

    *   Filter: `Matches regex` and use `.*` for the value

    *   Name format: `Unspecified`

3. You can add additional attributes. As an example:

Name	Name format	Value
`email`	`Unspecified`	`user.email`
`firstName`	`Unspecified`	`user.firstName`
`lastName`	`Unspecified`	`user.lastName`

Figure 10.7 – Additional custom user attributes in OKTA

4. Go to the **Groups** menu and add a group called `ROLE_ADMIN`.

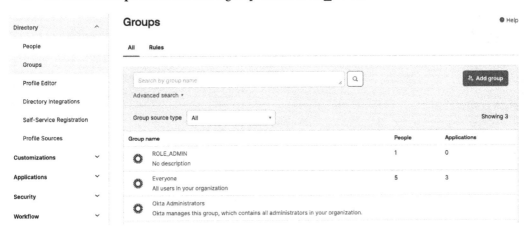

Figure 10.8 – Defining a custom group in OKTA

5.   Then assign to this group the user admin1@example.com.

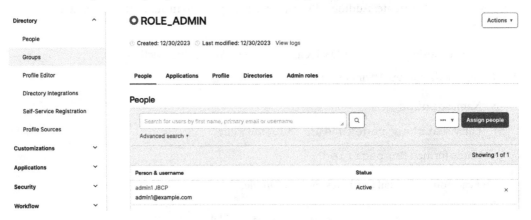

Figure 10.9 – Assigning users to groups in OKTA

6.   Adapt the SecurityConfig.java class to override the default configuration. Then, use a converter to map the values in the groups attribute to Spring Security authorities.

```
//src/main/java/com/packtpub/springsecurity/service/ SecurityConfig.
java
@Configuration
@EnableWebSecurity
public class SecurityConfig {

    @Bean
    public SecurityFilterChain filterChain(HttpSecurity http) throws
Exception {
        http.authorizeRequests( authz -> authz
        ... omitted for brevity
                .requestMatchers(antMatcher("/errors/**")).
permitAll()
                .requestMatchers(antMatcher("/events/")).
hasRole("ADMIN")
                .requestMatchers(antMatcher("/**")).
hasAuthority("Everyone"))

            .exceptionHandling(exceptions -> exceptions
                .accessDeniedPage("/errors/403"))
            .logout(form -> form
                .logoutUrl("/logout")
                .logoutSuccessUrl("/")
                .permitAll())
            .csrf(AbstractHttpConfigurer::disable);
```

```
        OpenSaml4AuthenticationProvider authenticationProvider = new
OpenSaml4AuthenticationProvider();
        authenticationProvider.
setResponseAuthenticationConverter(groupsConverter());

        // @formatter:off
        http
                .saml2Login(saml2 -> saml2
                        .authenticationManager(new
ProviderManager(authenticationProvider)))
                .saml2Logout(withDefaults());
        // For H2 Console
        http.headers(headers -> headers.
frameOptions(FrameOptionsConfig::disable));

        return http.build();
    }

    private Converter<OpenSaml4AuthenticationProvider.ResponseToken,
Saml2Authentication> groupsConverter() {
        Converter<ResponseToken, Saml2Authentication> delegate =
            OpenSaml4AuthenticationProvider.
createDefaultResponseAuthenticationConverter();
        return (responseToken) -> {
            Saml2Authentication authentication = delegate.
convert(responseToken);
            Saml2AuthenticatedPrincipal principal =
(Saml2AuthenticatedPrincipal) authentication.getPrincipal();
            List<String> groups = principal.getAttribute("groups");
            Set<GrantedAuthority> authorities = new HashSet<>();
            if (groups != null) {
                groups.stream().map(SimpleGrantedAuthority::new).
forEach(authorities::add);
            } else {
                authorities.addAll(authentication.getAuthorities());
            }
            return new Saml2Authentication(principal, authentication.
getSaml2Response(), authorities);
        };
    }
}
```

7. Now, you should see your user's groups as authorities. That comes from the OKTA SAML context related to the authenticated user.

You will notice that with these changes, you will now have access to the **All Events** page when you log in with the user `admin1@example.com`.

> **Important note**
> Your code should now look like that in `chapter10.05`.

Performing Single Logout

Spring Security's SAML support includes a logout feature that requires some configuration.

You can use `OpenSSL` to create a private key and certificate. Ensure you provide a value for at least one of the questions during the process, and the setup should be successful.

```
openssl req -newkey rsa:2048 -nodes -keyout rp-private.key -x509 -days
365 -out rp-certificate.crt
```

Copy the generated files to your app's `src/main/resources/credentials` directory.

Configure in `application.yml`, the generated key, the certificates location and the IdP's logout configuration similar to the following:

```
spring:
  security:
    saml2:
      relyingparty:
        registration:
          okta:
            signing:
              credentials:
                - private-key-location: classpath:credentials/rp-
private.key
                  certificate-location: classpath:credentials/rp-
certificate.crt
            assertingparty:
              metadata-uri: https://dev-xxxxx.okta.com/app/ <random-
characters>/sso/saml/metadata
            singlelogout:
              binding: POST
              url: "{baseUrl}/logout/saml2/slo"
```

On the OKTA configuration page:

1. Open the OKTA Admin Console.

2. Choose **Applications | Applications**.

3. Go to the **SAML Settings** section:

    I.   Click **Show Advanced Settings**.

    II.   Choose **Allow application to initiate Single Logout** for **Enable Single Logout**.

    III. Set **Single Logout URL**. For example, `https://localhost:8443/logout/saml2/slo`.

    IV.   Set **SP Issuer** to a URL. For example, `https://localhost:8443/saml2/service-provider-metadata/okta`.

    V.   Click the **Browse** button for **Signature Certificate**, locate the `local.crt` file you created in the previous steps, and click **Upload Certificate**.

4. Click **Next**.

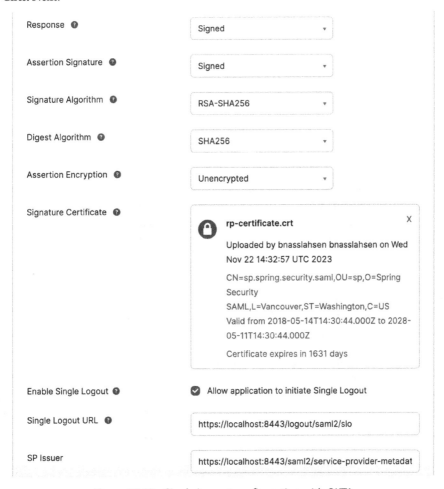

Figure 10.10 – Single Logout configuration with OKTA

5. Click **Finish**.

6. Restart the Spring Boot application. You can now log out from OKTA as well.

Important note

As the SAML 2.0 specification permits multiple values for each attribute, you have the option to use either `getAttribute` to retrieve the list of attributes or `getFirstAttribute` to obtain the first value in the list. The `getFirstAttribute` method proves particularly useful when it's known that there is only one value.

Your code should now look like that in `chapter10.06-calendar`.

Summary

This chapter delved into the realm of **SAML**, a robust standard for achieving SSO in modern identity management. Beginning with an introduction to SAML's foundational principles, it progressed to practical implementation, guiding developers through the seamless integration of **SAML 2** Login within the Spring Security framework.

Key highlights included the practical steps involved in adding a **SAML** application on OKTA, a widely used IdP, and the creation of user principals within OKTA for streamlined user management. Essential dependencies for successful **SAML** integration were outlined, emphasizing the crucial tools and libraries for building a resilient authentication system.

You have gained insights into critical configuration steps, such as specifying IdP metadata to ensure a standardized and secure communication channel. The chapter explored the retrieval of the **SAML 2** authenticated principal and parsing **SAML 2** metadata, and producing **SAML 2** metadata, providing a comprehensive understanding of the technical intricacies involved.

Flexibility in **SAML** integration was introduced through a demonstration of customizing the lookup process for `RelyingPartyRegistration`, overriding **SAML** Spring Boot auto-configuration, and creating a custom `RelyingPartyRegistrationRepository` for advanced customization. Practical guidance was provided on customizing authorities within Spring Security SAML, enabling the effective management of user roles.

This chapter concluded by addressing the crucial aspect of Single Logout, showcasing how SAML supports a standardized mechanism for logging out users across various services. In essence, this chapter equips you with the knowledge and practical insights needed to implement SAML-based authentication, fostering a secure and seamless identity management experience in your applications.

In the next chapter, we'll learn more about Spring Security authorization.

Part 4: Enhancing Authorization Mechanisms

This part delves into fine-grained access control, exploring various methods to implement precise authorization that may impact specific sections of an application page. Initially, we examine two approaches for implementing fine-grained authorization. Subsequently, we explore Spring Security's methodology for securing the business tier through method annotation, utilizing interface-based proxies for **Aspect-Oriented Programming (AOP)**. Furthermore, we investigate annotation-based security's capability for role-based filtering on data collections. Lastly, we compare class-based proxies with interface-based proxies.

Within this section, we delve into the intricate topic of **Access Control Lists (ACLs)**, offering a comprehensive overview of their potential for domain object instance-level authorization. Spring Security offers a robust, albeit complex, ACL module that effectively caters to the needs of small to medium-sized implementations.

Moreover, we undertake the task of crafting custom implementations for Spring Security's essential authorization APIs. This hands-on approach enables a deeper understanding of Spring Security's authorization architecture.

This part has the following chapters:

- *Chapter 11, Fine-Grained Access Control*
- *Chapter 12, Access Control Lists*
- *Chapter 13, Custom Authorization*

Fine-Grained Access Control

In this chapter, we will first examine two ways to implement fine-grained authorization—authorization that may affect portions of a page of the application. Next, we will look at Spring Security's approach to securing the business tier through method annotation and the use of interface-based proxies to accomplish **Aspect-Oriented Programming (AOP)**. Then, we will review an interesting capability of annotation-based security that allows for role-based filtering on collections of data. Lastly, we will look at how class-based proxies differ from interface-based proxies.

During this chapter, we'll cover the following topics:

- Configuring and experimenting with different methods of performing in-page authorization checks on content, given the security context of a user request

- Performing configuration and code annotation to make caller preauthorization a key part of our application's business-tier security

- Several alternative approaches to implement method-level security, and reviewing the pros and cons of each type

- Implementing data-based filters on collections and arrays using method-level annotations

- Implementing method-level security on our `Spring MVC` controllers to avoid configuring `requestMatchers()` methods and `<intercept-url>` elements

This chapter's code in action link is here: `https://packt.link/Mxijd`.

Integrating Spring Expression Language (SpEL)

Spring Security leverages **Spring Expression Language (SpEL)** integration in order to easily articulate various authorization requirements. If you recall, we have already looked at the use of SpEL in *Chapter 2, Getting Started with Spring Security*, when we defined our `requestMatchers()` method:

```
.requestMatchers("/events/").hasRole("ADMIN")
```

Spring Security provides an `o.s.s.access.expression.SecurityExpressionRoot` object that provides the methods and objects available for use, in order to make an access control decision. For example, one of the methods available to use is `hasRole` method, which accepts a string. This corresponds to the value of the access attribute (in the preceding code snippet). In fact, there are a number of other expressions available, as shown in the following table:

Expression	Description
`hasRole(String role)` `hasAuthority(String role)`	Returns `true` if the current user has the specified authority.
`hasAnyRole(String... role)` `hasAnyAuthority(String... authority)`	Returns `true` if the current user has any of the specified authorities.
`authentication`	Obtains the current `Authentication` object from the `SecurityContext` interface returned by the `getContext()` method of the `SecurityContextHolder` class.
`permitAll`	This request does not necessitate authorization and serves as a public endpoint. It's essential to clarify that `Authentication` is never retrieved from the session in this scenario.
`denyAll`	Under no circumstances is the request permitted; it's important to emphasize that `Authentication` is never retrieved from the session in this instance.
`isAnonymous()`	Returns `true` if the current principal is anonymous (is not authenticated).
`isRememberMe()`	Returns `true` if the current principal was authenticated using the remember-me feature.
`isAuthenticated()`	Returns `true` if the user is not an anonymous user (that is, they are authenticated).
`isFullyAuthenticated()`	Returns `true` if the user is authenticated through a means other than **remember-me**.
`hasPermission(Object target, Object permission)`	Returns `true` if the user has permission to access the specified object for the given permission.
`hasPermission(String targetId, String targetType, Object permission)`	Returns `true` if the user has permission to access the specified identifier for a given type and permission.

Table 11.1 – Summary of the authorization rules

We have provided some examples of using these SpEL expressions in the following code snippet. Keep in mind that we will go into more detail throughout this and the next chapter:

```
// allow users with ROLE_ADMIN hasRole('ADMIN')
// allow users that do not have the ROLE_ADMIN

!hasRole('ADMIN')

// allow users that have ROLE_ADMIN or ROLE_ROOT and
// did not use the remember me feature to login
isFullyAuthenticated() and hasAnyRole('ADMIN','ROOT')

// allow if Authentication.getName() equals admin authentication.name
== 'admin'
```

Go ahead and start up the JBCP calendar application. Visit `https://localhost:8443` and log in with the user `user1@example.com` and the password `user1`. You will observe that the **My Events** navigation menu item is displayed, and the **All Events** navigation menu item is displayed as well.

> **Important note**
>
> You should start with the code from `chapter11.00-calendar`.

The WebSecurityExpressionRoot class

The `o.s.s.web.access.expression.WebSecurityExpressionRoot` class makes a few additional properties available to us. These properties, along with the standard properties already mentioned, are made available in the access attribute of the `requestMatchers()` method and in the JSP/Thymeleaf `access` attribute of the `<sec:authorize>` tag, as we will discuss shortly:

Expression	Description
request	The current HttpServletRequest method.
hasIpAddress(String... ipAddress)	Returns true if the current IP address matches the ipAddress value. This can be an exact IP address or the IP address/network mask.

Table 11.2 – hasIpAddress usage with WebSecurityExpressionRoot

The MethodSecurityExpressionRoot class

Method SpEL expressions also provide a few additional properties that can be used through the `o.s.s.access.expression.method.MethodSecurityExpressionRoot` class:

Expression	Description
`target`	Refers to this or the current object being secured.
`returnObject`	Refers to the object returned by the annotated method.
`filterObject`	Can be used on a collection or array in conjunction with `@PreFilter` or `@PostFilter`, to only include the elements that match the expression. The `filterObject` object represents the loop variable of the collection or array.
`#<methodArg>`	Any argument to a method can be referenced by prefixing the argument name with #. For example, a method argument named `id` can be referred to using `#id`.

Table 11.3 – MethodSecurityExpressionRoot properties

If the description of these expressions appears a bit brief, don't worry; we'll work through a number of examples later in this chapter.

We hope that you have a decent grasp of the power of Spring Security's SpEL support. To learn more about SpEL, refer to the Spring reference documentation at: `https://docs.spring.io/spring-framework/reference/core/expressions.html`.

Page-level authorization

Page-level authorization refers to the availability of application features based on the context of a particular user's request. Unlike coarse-grained authorization that we explored in *Chapter 2, Getting Started with Spring Security*, fine-grained authorization typically refers to the selective availability of the portions of a page, rather than restricting access to a page entirely. Most real-world applications will spend a considerable amount of time on the details of fine-grained authorization planning.

Spring Security provides us with the following three methods of selective display functionality:

- **Spring Security JSP** tag libraries allow conditional access declarations to be placed within a page declaration itself, using the standard JSP tag library syntax.

- **Thymeleaf Spring Security** tag libraries allow conditional access declarations to be placed within a page declaration itself, using the standard Thymeleaf tag library syntax.

- Checking user authorization in an MVC application's controller layer allows the controller to make an access decision and bind the results of the decision to the model data provided to the view. This approach relies on standard JSTL conditional page rendering and data binding and is slightly more complicated than Spring Security tag libraries; however, it is more in line with the standard web application MVC logical design.

Any of these approaches are perfectly valid when developing fine-grained authorization models for a web application. Let's explore how each approach is implemented through a JBCP calendar use case.

Conditional rendering with the Thymeleaf Spring Security tag library

The most common functionality used in the Thymeleaf Spring Security tag library is to conditionally render portions of the page based on authorization rules. This is done with the `<sec:authorize*>` tag that functions similarly to the `<if>` tag in the core JSTL library, in that the tag's body will render depending on the conditions provided in the tag attributes. We have already seen a very brief demonstration of how the Spring Security tag library can be used to restrict the viewing of content if the user is not logged in.

Conditional rendering based on URL access rules

The Spring Security tag library provides functionality to render content based on the existing URL authorization rules that are already defined in the security configuration file. This is done via the use of the `authorizeHttpRequests()` method.

If there are multiple HTTP elements, the `authorizeHttpRequests()` method uses the currently matched HTTP element's rules.

For example, we could ensure that the `All Events` navigation menu item is displayed only when appropriate, that is, for users who are administrators—recall that the access rules we've previously defined are as follows:

```
.requestMatchers("/events/").hasRole("ADMIN")
```

Update the `header.html` file to utilize this information and conditionally render the link to the `All Events` page:

```
//src/main/resources/templates/fragments/header.html

<!DOCTYPE html>
<html xmlns:sec="http://www.thymeleaf.org/extras/spring-security"
    xmlns:th="http://www.thymeleaf.org">
...
<li sec:authorize-url="/events/">
    <a id="navEventsLink" th:href="@{/events/}">All Events</a></li>
```

This will ensure that the content of the tag is not displayed unless the user has sufficient privileges to access the stated URL. It is possible to further qualify the authorization check using the HTTP method, by including the method attribute before the URL, as follows:

```
<li sec:authorize-url="/events/">
    <a id="navEventsLink" th:href="@{/events/}">All Events</a></li>
```

Using the authorize-url attribute to define authorization checks on blocks of code is convenient because it abstracts the knowledge of the actual authorization checks from your pages and keeps it in your security configuration file.

Be aware that the HTTP method should match the case specified in your security requestMatchers() method, otherwise they may not match as you expect. Also, note that the URL should always be relative to the web application context root (as your URL access rules are).

For many purposes, the use of the <sec> tag's authorize-url attribute will suffice to correctly display link- or action-related content only when the user is allowed to see it. Remember that the tag need not only surround a link; it could even surround a whole form if the user doesn't have permission to submit it.

Conditional rendering using SpEL

An additional, more flexible method of controlling the display of JSP content is available when the <sec> tag is used in conjunction with a SpEL expression. Let's review what we learned in *Chapter 2, Getting Started with Spring Security*. We could hide the My Events link from any unauthenticated users by changing our header.html file, as follows:

```
//src/main/resources/templates/fragments/header.html

<li sec:authorize="isAuthenticated()">
    <a id="navMyEventsLink" th:href="@{/events/my}">My Events</a></li>
```

The SpEL evaluation is performed by the same code behind the scenes as the expressions utilized in the requestMatchers() method access declaration rules (assuming the expressions have been configured). Hence, the same set of built-in functions and properties are accessible from the expressions built using the <sec> tag.

Both of these methods of utilizing the <sec> tag provide powerful, fine-grained control over the display of page content based on security authorization rules.

Go ahead and start up the JBCP calendar application. Visit https://localhost:8443 and log in with the user user1@example.com and the password user1. You will observe that the **My Events** link is displayed, but the **All Events** navigation menu item is hidden (although it will still be present on the page). Log out and log in as the user admin1@example.com with the password admin1. Now both links are visible.

> **Important note**
> You should start with the code from `chapter11.01-calendar`.

Using controller logic to conditionally render content

In this section, we will demonstrate how we can use Java-based code to determine if we should render some content. We can choose to only show the **Create Event** link on the **Welcome** page to users who have a username that contains `user`. This will hide the **Create Event** link on the **Welcome** page from users who are not logged in as administrators.

The welcome controller from the sample code for this chapter has been updated to populate the model with an attribute named `showCreateLink`, derived from the method name, as follows:

```
//src/main/java/com/packtpub/springsecurity/web/controllers/
WelcomeControll er.java
@ModelAttribute("showCreateLink")
public boolean showCreateLink(Authentication authentication) {
    // NOTE We could also get the Authentication from
SecurityContextHolder.getContext().getAuthentication()
    return authentication != null && authentication.getName().
contains("user");
}
```

You may notice that Spring MVC can automatically obtain the `Authentication` object for us. This is because Spring Security maps our current `Authentication` object to the `HttpServletRequest`. `getPrincipal()` method. Since Spring MVC will automatically resolve any object of the `java.security.Principal` type to the value of `HttpServletRequest.getPrincipal()`, specifying `Authentication` as an argument to our controller is an easy way to access the current `Authentication` object. We could also decouple the code from Spring Security by specifying an argument of the `Principal` type instead. However, we chose `Authentication` in this scenario to help demonstrate how everything connects.

If we were working in another framework that did not know how to do this, we could obtain the `Authentication` object using the `SecurityContextHolder` class, as we did in *Chapter 3, Custom Authentication*. Also note that if we were not using Spring MVC, we could just set the `HttpServletRequest` attribute directly rather than populating it on the model. The attribute that we populated on the request would then be available to our JSP, just as it is when using a `ModelAndView` object with Spring MVC.

Next, we will need to use the `HttpServletRequest` attribute in our `index.html` file to determine if we should display the `Create Event` link. Update `index.html`, as follows:

```
//src/main/resources/templates/fragments/header.html
<li th:if="${showCreateLink}" class="nav-item"><a class="nav-link"
id="navCreateEventLink"
                    th:href="@{/events/form}">Create Event</a>
```

Now, start the application, log in using `admin1@example.com` as the username and `admin1` as the password, and visit the **All Events** page. You should no longer see the **Create Events** navigation menu item (although it will still be present on the page).

> **Important note**
>
> You should start with the code from `chapter11.02-calendar`.

The WebInvocationPrivilegeEvaluator class

There may be times when an application will not be written using JSPs and will need to be able to determine access based upon a URL, as we did with `<... sec:authorize- url="/events/">`. This can be done by using the `o.s.s.web.access.WebInvocationPrivilegeEvaluator` interface, which is the same interface that backs the JSP tag library.

In the following code snippet, we demonstrate the use of `WebInvocationPrivilegeEvaluator` by populating our model with an attribute named `showAdminLink`. We are able to obtain `WebInvocationPrivilegeEvaluator` using the `@Autowired` annotation:

```
//src/main/java/com/packtpub/springsecurity/web/controllers/
WelcomeControll er.java
<li th:if="${showAdminLink}" class="nav-item"><a class="nav-link"
id="navH2Link"
                        target="_blank"
                        th:href="@{/admin/h2}">H2</a></li>
```

If the framework you are using is not being managed by Spring, `@Autowire` will not be able to provide you with `WebInvocationPrivilegeEvaluator`. Instead, you can use Spring's `org.springframework.web.context.WebApplicationContextUtils` interface to obtain an instance of `WebInvocationPrivilegeEvaluator`, as follows:

```
ApplicationContext context = WebApplicationContextUtils
        .getRequiredWebApplicationContext(servletContext);
WebInvocationPrivilegeEvaluator privEvaluator = context.
getBean(WebInvocationPrivilegeEvaluator.class);
```

To try it out, go ahead and update `index.html` to use the `showAdminLink` request attribute, as follows:

```
//src/main/resources/templates/index.html
<li th:if="${showAdminLink}">
    <a class="link-warning" id="h2Link" target="_blank" th:href="@
{admin/h2/}">H2
...
</li>
```

Restart the application and view the **Welcome** page before you have logged in. The **H2** link should not be visible. Log in as `admin1@example.com/admin1`, and you should see it.

> **Important note**
> You should start with the code from `chapter11.03-calendar`.

What is the best way to configure in-page authorization?

In many cases, the use of the `authorize-url` attribute of the tag can appropriately isolate the code from changes in authorization rules. You should use the `authorize-url` attribute of the tag in the following scenarios:

- The tag is preventing display functionality that can be clearly identified by a single URL

- The contents of the tag can be unambiguously isolated to a single URL

  Unfortunately, in a typical application, the likelihood that you will be able to use the `authorize-url` attribute of the tag frequently is somewhat low. The reality is that applications are usually much more complex than this and require more involved logic when deciding to render portions of a page.

It's tempting to use the Thymeleaf Spring Security tag library to declare bits of rendered pages as off-limits based on security criteria in other methods. However, there are a number of reasons why (in many cases) this isn't a great idea, as follows:

- Complex conditions beyond role membership are not supported by the tag library. For example, if our application incorporated customized attributes on the `UserDetails` implementation, IP filters, geolocation, and so on, none of these would be supported by the standard `<sec>` tag.

- These could, however, conceivably be supported by the custom tags or using SpEL expressions. Even in this case, the page is more likely to be directly tied to business logic rather than what is typically encouraged.

- The `<sec>` tag must be referenced on every page that it's used in. This leads to potential inconsistencies between the rulesets that are intended to be common, but may be spread across different physical pages. A good object-oriented system design would suggest that conditional rule evaluations be located in only one place, and logically referred to from where they should be applied.

- It is possible (and we illustrate this using our common header page) to encapsulate and reuse portions of pages to reduce the occurrence of this type of problem, but it is virtually impossible to eliminate in a complex application.

- There is no way to validate the correctness of rules stated at compile time. Whereas compile-time constants can be used in typical Java-based, object- oriented systems, the tag library requires (in typical use) hardcoded role names where a simple typo might go undetected for some time.

- To be fair, such typos could be caught easily by comprehensive functional tests on the running application, but they are far easier to test using a standard Java component unit testing techniques.

- We can see that, although the *template-based approach* for conditional content rendering is convenient, there are some significant downsides.

All of these issues can be solved by the use of code in controllers that can be used to push data into the application view model. Additionally, performing advanced authorization determinations in code allows for the benefits of reuse, compile-time checks, and proper logical separation of the model, view, and controller.

Method-level security

Our primary focus up to this point in the book has been on securing the web-facing portion of the JBCP calendar application; however, in real-world planning of secured systems, equal attention should be paid to securing the service methods that allow users access to the most critical part of any system—its data.

Why we secure in layers?

Let's take a minute to see why it is important to secure our methods, even though we have already secured our URLs.

1. Start the JBCP calendar application up. Log in using user1@example.com as the username and user1 as the password.

2. Visit the **All Events** page https://localhost:8443/events/. You will see the custom **Access Denied** page.

3. Now, add backdoor to the end of the URL in the browser so that the URL is now https://localhost:8443/events/backdoor. You will now see a response with the same data as the **All Events** page. This data should only be visible to an administrator, but we have bypassed it by finding a URL that was not configured properly.

4. We can also view the details of an event that we do not own and are not invited to. Change backdoor with 102 so that the URL is now https://localhost:8443/events/102: You will now see an **Vacation Event** that is not listed on your **My Events** page. This should not be visible to us because we are not an administrator, and this is not our event.

As you can see, our URL rules are not quite strong enough to entirely secure our application. These exploits do not even need to take advantage of more complex problems, such as differences in how containers handle URL normalization. In short, there are often ways to bypass URL-based security. Let's see how adding a security layer to our business tier can help with our new security vulnerability.

Securing the business tier

Spring Security has the ability to add a layer of authorization (or authorization-based data pruning) to the invocation of any Spring-managed bean in your application. While many developers focus on web-tier security, business-tier security is arguably just as important, as a malicious user may be able to penetrate the security of your web tier or access services exposed through a non-UI frontend, such as a web service.

Let's examine the following logical diagram to see why we're interested in applying a secondary layer of security:

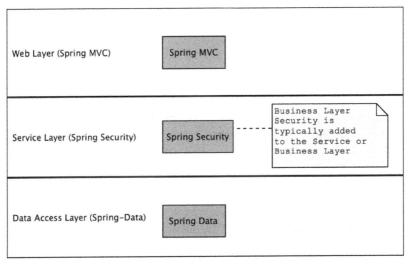

Figure 11.1 – Logical application Layers

Spring Security has the following two main techniques for securing methods:

- **Preauthorization:** This technique ensures that certain constraints are satisfied prior to the execution of a method that is being allowed, for example, if a user has a particular GrantedAuthority, such as ROLE_ADMIN. Failure to satisfy the declared constraints means that the method call will fail.

- **Postauthorization:** This technique ensures that the calling principal still satisfies declared constraints after the method returns. This is rarely used but can provide an extra layer of security around some complex, interconnected business tier methods.

The preauthorization and postauthorization techniques provide formalized support for what are generally called preconditions and postconditions in a classic, object-oriented design. Preconditions and postconditions allow a developer to declare through runtime checks that certain constraints around a method's execution must always hold true. In the case of security preauthorization and postauthorization, the business tier developer makes a conscious decision about the security profile of particular methods by encoding expected runtime conditions as part of an interface or class API declaration. As you may imagine, this requires a great deal of forethought to avoid unintended consequences!

Adding the @PreAuthorize method annotation

Our first design decision will be to augment method security at the business tier by ensuring that a user must be logged in as an ADMIN user before he/she is allowed to access the getEvents() method. This is done with a simple annotation added to the method in the service interface definition, as follows:

```
import org.springframework.security.access.prepost.PreAuthorize;
...
public interface CalendarService {
...
    @PreAuthorize("hasRole('ADMIN')")
    List<Event> getEvents();
}
```

This is all that is required to ensure that anyone invoking our getEvents() method is an administrator. Spring Security will use a runtime AOP pointcut to execute BeforeAdvice on the method, and throw o.s.s.access.AccessDeniedException if the security constraints aren't met.

Instructing Spring Security to use method annotations

We'll also need to make a one-time change to SecurityConfig.java, where we've got the rest of our Spring Security configuration. Simply add the following annotation to the class declaration:

```
//src/main/java/com/packtpub/springsecurity/configuration/SecurityC
onfig.java
@Configuration
@EnableWebSecurity
@EnableMethodSecurity
public class SecurityConfig {
```

In the next section, let's test the method security validation.

Validating method security

Don't believe it was that easy? Log in with user1@example.com as the username and user1 as the password, and try accessing https://localhost:8443/events/backdoor. You should see the **Access Denied** page now.

> **Important note**
> You should start with the code from chapter11.04-calendar.

If you look at the *Tomcat console*, you'll see a very long stack trace, starting with the following output:

```
org.springframework.security.access.AccessDeniedException: Access
Denied
 at org.springframework.security.authorization.method.
AuthorizationManagerBeforeMethodInterceptor.attemptAuthorization
 at org.springframework.security.authorization.method.
AuthorizationManagerBeforeMethodInterceptor.invoke
 at org.springframework.aop.framework.ReflectiveMethodInvocation.
proceed
 at org.springframework.aop.framework.
CglibAopProxy$CglibMethodInvocation.proceed
 at org.springframework.aop.framework.
CglibAopProxy$DynamicAdvisedInterceptor.intercept
 at com.packtpub.springsecurity.service.
DefaultCalendarService$$SpringCGLIB$$0.getEvents
 at com.packtpub.springsecurity.web.controllers.EventsController.
events
```

Based on the **Access Denied** page, and the stack trace clearly pointing to the getEvents method invocation, we can see that the user was appropriately denied access to the business method because it lacked the GrantedAuthority of ROLE_ADMIN. If you run the same with the username admin1@example.com and the password admin1, you will discover that access will be granted.

Isn't it amazing that with a simple declaration in our interface, we're able to ensure that the method in question is secure? But how does AOP work?

Interface-based proxies

In the given example from the previous section, Spring Security used an interface-based proxy to secure our getEvents method. Let's take a look at the simplified pseudocode of what happened to understand how this works:

```
DefaultCalendarService originalService = context.getBean
(CalendarService.class)
CalendarService secureService = new CalendarService() {
//... other methods just delegate to originalService ...
    public List<Event> getEvents() {
        if(!permitted(originalService.getEvents)) {
            throw AccessDeniedException()
        }
        return originalCalendarService.getEvents()
    }
};
```

You can see that Spring creates the original `CalendarService` just as it normally does. However, it instructs our code to use another implementation of `CalendarService` that performs a security check before returning the result of the original method. The secure implementation can be created with no prior knowledge of our interface because Spring uses Java's `java.lang.reflect.Proxy` APIs to dynamically create new implementations of the interface.

Note that the object returned is no longer an instance of `DefaultCalendarService`, since it is a new implementation of `CalendarService`, that is, it is an anonymous implementation of `CalendarService`. This means that we must program against an interface in order to use the secure implementation, otherwise, a `ClassCastException` exception will occur.

To learn more about Spring AOP, refer to the Spring reference documentation at `https://docs.spring.io/spring-framework/reference/core/aop/proxying.html#aop-understanding-aop-proxies`.

In addition to the `@PreAuthorize` annotation, there are several other ways of declaring security preauthorization requirements on methods. We can examine these different ways of securing methods and then evaluate their pros and cons in different circumstances.

JSR-250 compliant standardized rules

JSR-250 Common Annotations for the Java platform defines a series of annotations, some that are security-related, which are intended to be portable across JSR-250 compliant runtime environments. The Spring Framework became compliant with JSR-250 as part of the Spring 2.x release, including the Spring Security framework.

Gradle dependencies

There are several optional dependencies that may be required, depending on what features you decide to use for example to enable the support of JSR 250 `@RolesAllowed` annotation. Many of these dependencies are commented as Spring Boot includes them already in the starter parent.

You will find that our `build.gradle` file already includes the above dependency (transitively):

```
//build.gradle
// Required for JSR-250 based security:
// JSR-250 Annotations
implementation 'jakarta.annotation:jakarta.annotation-api:2.1.1'
```

While JSR-250 annotations are not as expressive as Spring native annotations, they have the benefit that the declarations they provide are compatible across implementing Jakarta EE application servers. Depending on your application's needs and requirements for portability, you may decide that the trade-off of reduced specificity is worth the portability of the code.

To implement the rule we specified in the first example, we make a few changes by performing the following steps:

1. First, we need to update our `SecurityConfig` file to use the JSR-250 annotations:

    ```
    //src/main/java/com/packtpub/springsecurity/configuration/
    SecurityConfig.java

    @Configuration
    @EnableWebSecurity
    @EnableMethodSecurity(jsr250Enabled = true)
    public class SecurityConfig {
    ```

2. Lastly, the `@PreAuthorize` annotation needs to change to the `@RolesAllowed` annotation. As we might anticipate, the `@RolesAllowed` annotation does not support SpEL expressions, so we edit `CalendarService` as follows:

    ```
    @RolesAllowed("ADMIN")
    List<Event> getEvents();
    ```

3. Restart the application, log in as `user1@example.com/user1`, and try to access `https://localhost:8443/events/backdoor`. You should see **the Access Denied** page again.

> **Important note**
>
> You should start with the code from `chapter11.05-calendar`.

Note that it's also possible to provide a list of allowed `GrantedAuthority` names using the standard Java 5 String array annotation syntax:

```
@RolesAllowed({"ADMIN", "USER"})
List<Event> getEvents();
```

There are also two additional annotations specified by JSR-250, namely `@PermitAll` and `@DenyAll`, which function as you might expect, permitting and denying all requests to the method in question.

> **Important note**
>
> Be aware that method-level security annotations can be applied at the class level as well! Method-level annotations, if supplied, will always override annotations specified at the class level. This can be helpful if your business needs to dictate the specification of security policies for an entire class.
>
> Take care to use this functionality in conjunction with good comments and coding standards, so that developers are very clear about the security characteristics of the class and its methods.

Method security using Spring's @Secured annotation

Spring itself provides a simpler annotation style that is similar to the JSR-250 @RolesAllowed annotation. The @Secured annotation is functionally and syntactically the same as @RolesAllowed. The only notable differences are that it does not require the external dependency, cannot be processed by other frameworks, and the processing of these annotations must be explicitly enabled with another attribute on the @EnableMethodSecurity annotation:

```
//src/main/java/com/packtpub/springsecurity/configuration/
SecurityConfig.java
@Configuration
@EnableWebSecurity
@EnableMethodSecurity(jsr250Enabled = true)
public class SecurityConfig {}
```

As @Secured functions in the same way as the JSR standard @RolesAllowed annotation, there's no real compelling reason to use it in new code, but you may run across it in older Spring code.

Method security rules incorporating method parameters

Logically, writing rules that refer to method parameters in their constraints seem sensible for certain types of operations. For example, it might make sense for us to restrict the findForUser(int userId) method to meet the following constraints:

- The userId argument must be equal to the current user's ID

- The user must be an administrator (in this case, it is valid for the user to see any event)

While it's easy to see how we could alter the rule to restrict the method invocation only to administrators, it's not clear how we would determine if the user were attempting to change their own password.

Fortunately, SpEL binding, used by the Spring Security method annotations, supports more sophisticated expressions, including expressions that incorporate method parameters. You will also want to ensure that you have enabled pre- and post-annotations in the SecurityConfig file, as follows:

```
//src/main/java/com/packtpub/springsecurity/configuration/
SecurityConfig.java
@Configuration
@EnableWebSecurity
@EnableMethodSecurity
public class SecurityConfig { }
//Lastly, we can update our CalendarService interface as follows:
@PreAuthorize("hasRole('ROLE_ADMIN') or principal.id == #userId")
List<Event> findForUser(int userId);
```

You can see here that we've augmented the SpEL directive we used in the first exercise with a check against the ID of the principal and against the `userId` method parameter (`#userId`, the method parameter name, is prefixed with a # symbol). The fact that this powerful feature of method parameter binding is available should get your creative juices flowing and allow you to secure method invocations with a very precise set of logical rules.

> **Important note**
>
> Our principal is currently an instance of `CalendarUser` due to the custom authentication setup from *Chapter 3, Custom Authentication*. This means that the principal has all of the properties that our `CalendarUser` application has on it. If we had not done this customization, only the properties of the `UserDetails` object would be available.

SpEL variables are referenced with the hash (#) prefix. One important note is that in order for method argument names to be available at runtime, debugging symbol table information must be retained after compilation. Common methods to retain the debugging symbol table information are listed as follows:

1. If you are using the `javac` compiler, you will need to include the `-g` flag when building your classes.

2. When using the `<javac>` task in Ant, add the attribute `debug="true"`.

3. In Gradle, ensure to add `--debug` when running the main method, or the `bootRun` task.

4. In Maven, ensure the `maven.compiler.debug=true` property (the default is `true`).

5. Consult your compiler, **build tool**, or IDE documentation for assistance on configuring this same setting in your environment.

6. Start up your application and try logging in with `user1@example.com` as the username and `user1` as the password.

7. On the **Welcome** page, request the **My Events**. (email=admin1@example.com) link to see an **Access Denied** page.

8. Try again with **My Events** (email=user1@example.com) to see it work.

Note that the displayed user on the **My Events** page matches the currently logged-in user. Now, try the same steps and log in as `admin1@example.com/admin1`. You will be able to see both pages since you are logged in as a user with the `ROLE_ADMIN` permission.

> **Important note**
> You should start with the code from `chapter11.06-calendar`.

Method security rules incorporating returned values

Just as we were able to leverage the parameters to the method, we can also leverage the returned value of the method call. Let's update the getEvent method to meet the following constraints on the returned value:

- The attendee's ID must be the current user's ID

- The owner's ID must be the current user's ID

- The user must be an administrator (in this case, it is valid for the user to see any event)

Add the following code to the CalendarService interface:

```
@PostAuthorize("hasRole('ROLE_ADMIN') or " +
        "principal.id == returnObject.owner.id or " +
        "principal.id == returnObject.attendee.id")
Event getEvent(int eventId);
```

Now, try logging in with the username user1@example.com and the password user1. Next, try accessing the **Vacation Event** using the link on the **Welcome** page. You should now see the **Access Denied** page.

If you log in using the username user2@example.com and the password user2, the **Vacation Event** will display as expected since user2@example.com is the attendee at the **Vacation Event**.

> **Important note**
> You should start with the code from chapter11.07-calendar.

Securing method data using role-based filtering

The two final Spring Security-dependent annotations are @PreFilter and @PostFilter, which are used to apply security-based filtering rules to collections or arrays (with @PostFilter only). This type of functionality is referred to as security trimming or security pruning and involves using the security credentials of principal at runtime to selectively remove members from a set of objects. As you might expect, this filtering is performed using SpEL expression notation within the annotation declaration.

We'll work through an example with JBCP calendar, as we want to filter the getEvents method to only return the events that this user is allowed to see. In order to do this, we remove any existing security annotations and add the @PostFilter annotation to our CalendarService interface, as follows:

```
@PostAuthorize("hasRole('ROLE_ADMIN') or " +
        "principal.id == returnObject.owner.id or " +
```

```
            "principal.id == returnObject.attendee.id")
    Event getEvent(int eventId);
```

> **Important note**
>
> You should start with the code from `chapter11.08-calendar`.

Remove the `requestMatchers()` method, restricting access to `/events/` URL so that we can test our annotation. Start up the application and view the **All Events** page when logged in with the username `user1@example.com` and password `user1`. You will observe that only the events that are associated with our user are displayed.

With `filterObject` acting as the loop variable that refers to the current event, Spring Security will iterate over the `List<Event>` returned by our service and modify it to only contain the `Event` objects that match our SpEL expression.

In general, the `@PostFilter` method behaves in the following way. For brevity, we refer to the collection as the method return value, but be aware that `@PostFilter` works with either collection or array method return types.

The `filterObject` object is rebound to the SpEL context for each element in the collection. This means that if your method is returning a collection with 100 elements, the SpEL expression will be evaluated for each.

The SpEL expression must return a Boolean value. If the expression evaluates to true, the object will remain in the collection, while if the expression evaluates to false, the object will be removed.

In most cases, you'll find that collection post filtering saves you from the complexity of writing boilerplate code that you would likely be writing anyway. Take care that you understand how `@PostFilter` works conceptually; unlike `@PreAuthorize`, `@PostFilter` specifies method behavior and not a precondition. Some object-oriented purists may argue that `@PostFilter` isn't appropriate for inclusion as a method annotation, and such filtering should instead be handled through code in a method implementation.

> **Important note**
>
> Be aware that the actual collection returned from your method will be modified! In some cases, this isn't desirable behavior, so you should ensure that your method returns a collection that can be safely modified. This is especially important if the returned collection is an ORM-bound one, as post-filter modifications could inadvertently be persisted to the ORM data store!

Spring Security also offers functionality to prefilter method parameters that are collections; let's try implementing that now.

Prefiltering collections with @PreFilter

The `@PreFilter` annotation can be applied to a method to filter collection elements that are passed into a method based on the current security context. Functionally, once it has a reference to a collection, this annotation behaves exactly the same as the `@PostFilter` annotation, with a couple of exceptions, as follows:

- The `@PreFilter` annotation supports only collection arguments and does not support array arguments.

- The `@PreFilter` annotation takes an additional, optional `filterTarget` attribute which is used to specifically identify the method parameter and filter it when the annotated method has more than one argument.

- As with `@PostFilter`, keep in mind that the original collection passed to the method is permanently modified. This may not be desirable behavior, so ensure that callers know that the collection's security may be trimmed after the method is invoked!

Imagine if we had a `save` method that accepted a collection of event objects, and we wanted to only allow the saving of events that were owned by the currently logged-in user. We could do this as follows:

```
@PreFilter("principal.id == filterObject.owner.id")
void save(Set<Event> events);
```

Much like our `@PostFilter` method, this annotation causes Spring Security to iterate over each event with the loop variable `filterObject`. It then compares the current user's ID against the event owner's ID. If they match, the event is retained. If they do not match, the result is discarded.

Comparing method authorization types

The following quick reference chart may assist you in selecting a type of method authorization check to use:

Method authorization type	Specified as	JSR standard	Allows SpEL expressions
`@PreAuthorize`, `@PostAuthorize`	Annotation	No	Yes
`@RolesAllowed`, `@PermitAll`, `@DenyAll`	Annotation	Yes	No
`@Secure`	Annotation	No	No
`protect-pointcut`	XML	No	No

Table 11.4 – Method authorization types

Most Java consumers of Spring Security will probably opt to use the JSR-250 annotations for maximum compatibility and reuse their business classes (and relevant constraints) across an IT organization. Where needed, these basic declarations can be replaced with the annotations that tie the code to the Spring Security implementation itself.

If you are using Spring Security in an environment that doesn't support annotations, unfortunately, your choices are somewhat limited to method security enforcement. Even in this situation, the use of AOP provides a reasonably rich environment in which we can develop basic security declarations.

Practical considerations for annotation-based security

One thing to consider is that when returning a collection of real-world applications, there is likely to be some sort of paging. This means that our `@PreFilter` and `@PostFilter` annotations cannot be used as the sole means of selecting which objects to return. Instead, we need to ensure that our queries only select the data that the user is allowed to access.

This means that the security annotations become redundant checks. However, it is important to remember our lesson at the beginning of this chapter; we want to secure layers in case one layer is able to be bypassed.

Summary

In this chapter, we have covered most of the remaining areas in standard Spring Security implementations that deal with authorization. We've learned enough to take a thorough pass through the JBCP calendar application and verify that proper authorization checks are in place in all tiers of the application, to ensure that malicious users cannot manipulate or access data to which they do not have access.

We developed two techniques for micro-authorization, namely filtering out in-page content based on authorization or other security criteria using the Thymeleaf Spring Security tag library and Spring MVC controller data binding. We also explored several methods of securing business functions and data in the business tier of our application and supporting a rich, declarative security model that was tightly integrated with the code. We also learned how to secure our Spring MVC controllers and the differences between interface and class proxy objects.

At this point, we've wrapped up coverage of much of the important Spring Security functionality that you're likely to encounter in most standard, secure web application development scenarios.

In the next chapter, we will discuss the ACL (domain object model) module of Spring Security. This will allow us to explicitly declare authorization, rather than relying on existing data.

12

Access Control Lists

In this chapter, we will address the complex topic of **Access Control Lists** (**ACLs**), which can provide a rich model of domain object instance-level authorization. Spring Security ships with a robust, but complicated, ACL module that can serve the needs of small to medium-sized implementations reasonably well.

In this chapter, we'll cover the following topics:

- Understanding the conceptual model of an ACL

- Reviewing the terminology and application of ACL concepts in the Spring Security ACL module

- Building and reviewing the database schema required to support Spring ACL

- Configuring the **Jim Bob CP Calendar** (**JBCP**) calendar to use ACL-secured business methods via annotations and Spring beans

- Performing advanced configuration, including customized ACL permissions, ACL-enabled **JavaServer Page** (**JSP**) tag checks and method security, mutable ACLs, and smart caching

- Examining architectural considerations and planning scenarios for ACL deployment

This chapter's code in action link is here: `https://packt.link/hRby2`.

The conceptual module of an ACL

The final piece of the non-web tier security puzzle is security at the business object level, applied at or below the business tier. Security at this level is implemented using a technique known as ACL, or ACLs. To sum up the objective of ACLs in a single sentence, ACLs allow the specification of a set of group permissions based on the unique combination of a group, business object, and logical operation.

For example, an ACL declaration for the JBCP calendar might declare that a given user must write access to his or her own event. This can be shown as follows:

Username	Group	Object	Permissions
josh		event_01	read, write
	ROLE_USER	event_123	read
	ANONYMOUS	Any event	none

Table 12.1 – Example of user ACL declaration

You can see that this ACL is eminently readable by a human—josh has read and write access to his own event (event_01); other registered users can read the events of josh, but anonymous users cannot.

This type of rule matrix is, in a nutshell, what ACL attempts to synthesize about a secured system and its business data into a combination of code, access checking, and metadata. Most true ACL-enabled systems have extremely complex ACL lists and may conceivably have millions of entries across the entire system. Although this sounds frighteningly complex, proper up-front reasoning and implementation with a capable security library can make ACL management quite feasible.

If you use a Microsoft Windows or Unix/Linux-based computer, you experience the magic of ACLs every single day. Most modern computer **Operating Systems** (**OSs**) use ACL directives as part of their file storage systems, allowing permission granting based on a combination of a user or group, file, or directory, and permission. In Microsoft Windows, you can view some of the ACL capabilities of a file by right-clicking on a file and examining its security properties (**Properties | Security**), as shown in the following screenshot:

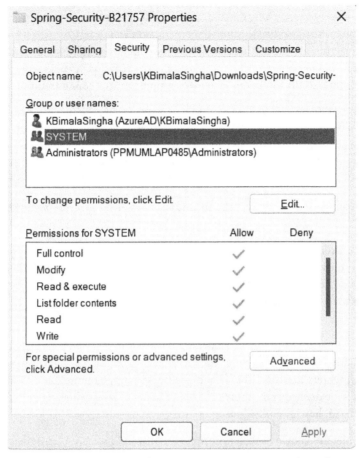

Figure 12.1 – An example of ACL capabilities with Microsoft Windows

You will be able to see that the combinations of inputs to the ACL are visible and intuitive as you navigate through the various groups or users and permissions.

In this section, we've explored the conceptual module of ACL. In the following section, we will proceed to delve into the workings of ACLs in Spring Security.

ACLs in Spring Security

Spring Security supports ACL-driven authorization checks against access to individual domain objects by individual users of the secured system. As in the OS filesystem example, it is possible to use the Spring Security ACL components to build logical tree structures of both business objects and groups or principals. The intersection of permissions (inherited or explicit) on both the requestor and the requestee is used to determine allowed access.

It's quite common for users approaching the ACL capability of Spring Security to be overwhelmed by its complexity, combined with a relative dearth of documentation and examples. This is compounded by the fact that setting up the ACL infrastructure can be quite complicated, with many interdependencies and reliance on bean-based configuration mechanisms, which are quite unlike much of the rest of Spring Security (as you'll see in a moment when we set up the initial configuration).

The Spring Security ACL module was written to be a reasonable baseline, but users intending to build extensively on the functionality will likely run into a series of frustrating limitations and design choices, which have gone (for the most part) uncorrected as they were first introduced in the early days of Spring Security. Don't let these limitations discourage you! The ACL module is a powerful way to embed rich access controls in your application, and further scrutinize and secure user actions and data.

Before we dig into configuring Spring Security ACL support, we need to review some key terminology and concepts.

The main unit of secured actor identity in the Spring ACL system is the **Security IDentifier** (**SID**). The SID is a logical construct that can be used to abstract the identity of either an individual principal or a group (`GrantedAuthority`). The SID object defined by the ACL data model you construct is used as the basis for explicit and derived access control rules when determining the allowed level of access for a particular principal.

If SIDs are used to define actors in the ACL system, the opposite half of the security equation is the definition of the secured objects themselves. The identification of individual secured objects is called (unsurprisingly) an **Object Identity**. The default Spring ACL implementation of an object identity requires ACL rules to be defined at the individual object instance level, which means, if desired, every object in the system can have an individual access rule.

Individual access rules are known as **Access control Entries** (**ACEs**). An ACE is the combination of the following factors:

- The SID for the actor to which the rule applies
- The object identity to which the rule applies
- The permission that should be applied to the given `SID` and the stated object identity
- Whether or not the stated permission should be allowed or denied for the given `SID` and object identity

The purpose of the Spring ACL system is to evaluate each secured method invocation and determine whether the object or objects being acted on in the method should be allowed as per the applicable ACEs. Applicable ACEs are evaluated at runtime, based on the caller and the objects in play.

Spring Security ACL is flexible in its implementation. Although most of this chapter details the out-of-the-box functionality of the Spring Security ACL module, keep in mind, however, that many of the rules indicated represent default implementations, which in many cases can be overridden based on more complex requirements.

Spring Security uses helpful value objects to represent the data associated with each of these conceptual entities. These are listed in the following table:

ACL conceptual object	Java object
SID	`o.s.s.acls.model.Sid`
Object identity	`o.s.s.acls.model.ObjectIdentity`
ACL	`o.s.s.acls.model.Acl`
ACE	`o.s.s.acls.model.AccessControlEntry`

Table 12.2 – Spring Security ACL java objects

Let's work through the process of enabling Spring Security ACL components for a simple demonstration in the JBCP calendar application.

Basic configuration of Spring Security ACL support

Although we hinted previously that configuring ACL support in Spring Security requires bean-based configuration (which it does), you can use ACL support while retaining the simpler security XML namespace configuration if you choose. In the remaining examples in this chapter, we will be focusing on Java-based configuration.

Gradle dependencies

As with most of the chapters, we will need to add some dependencies in order to use the functionality in this chapter. A list of the dependencies we have added with comments about when they are needed can be checked as follows:

```
//build.gradle
//Spring ACL
implementation "org.springframework.security:spring-security-acl"
```

Once you've updated your project dependencies, we can investigate the implementation of fine-grained permission access controls in our JBCP calendar application.

Defining a simple target scenario

Our simple target scenario is to grant `user2@example.com` read access to only the birthday party event.

All other users will not have any access to any events. You will observe that this differs from our other examples since `user2@example.com` is not otherwise associated with the birthday party event.

Although there are several ways to set up ACL checking, our preference is to follow the annotation-based approach that we used in this chapter's method-level annotations. This nicely abstracts the use of ACLs away from the actual interface declarations and allows for replacement (if you want) of the role declarations with something other than ACLs at a later date (should you so choose).

We'll add an annotation to the `CalendarService.getEvents` method, which filters each event based upon the current user's permission to the event:

```
//src/main/java/com/packtpub/springsecurity/service/CalendarService.
java
@PostFilter("hasPermission(filterObject, 'read')")
List<Event> getEvents();
```

> **Important note**
> You should start with the code from `chapter12.00-calendar`.

Adding ACL tables to the H2 database

The first thing we'll need to do is add the required tables and data to support persistent ACL entries in our in-memory H2 database. To do this, we'll add a new SQL **Data Definition Language (DDL)** file and the corresponding data to our embedded database declaration in `schema.sql`. We will break down each of these files later in the chapter.

We have included the following `schema.sql` file with this chapter's source code, which is based upon the schema files included in the Spring Security reference's *Appendix, Additional Reference Material*:

```
// src/main/resources/schema.sql
--- ACLs ---------------------------------------------
-- ACL Schema --
create table acl_sid (
  id bigint generated by default as identity(start with 23) not null
primary key,
  principal boolean not null,
  sid varchar_ignorecase(100) not null,
  constraint uk_acl_sid unique(sid,principal) );
```

```
create table acl_class (
  id bigint generated by default as identity(start with 100) not null
primary key,
  class varchar_ignorecase(500) not null,
  constraint uk_acl_class unique(class) );

create table acl_object_identity (
  id bigint generated by default as identity(start with 33) not null
primary key,
  object_id_class bigint not null,
  object_id_identity bigint not null,
  parent_object bigint,
  owner_sid bigint not null,
  entries_inheriting boolean not null,
  constraint uk_acl_objid unique(object_id_class,object_id_identity),
  constraint fk_acl_obj_parent foreign key(parent_object)references
acl_object_identity(id),
  constraint fk_acl_obj_class foreign key(object_id_class)references
acl_class(id),
  constraint fk_acl_obj_owner foreign key(owner_sid)references acl_
sid(id) );

create table acl_entry (
  id bigint generated by default as identity(start with 100) not null
primary key,
  acl_object_identity bigint not null,
  ace_order int not null,
  sid bigint not null,
  mask integer not null,
  granting boolean not null,
  audit_success boolean not null,
  audit_failure boolean not null,
  constraint uk_acl_entry unique(acl_object_identity,ace_order),
  constraint fk_acl_entry_obj_id foreign key(acl_object_identity)
  references acl_object_identity(id),
  constraint fk_acl_entry_sid foreign key(sid) references acl_sid(id)
);
-- the end --
```

The preceding code will result in the following database schema:

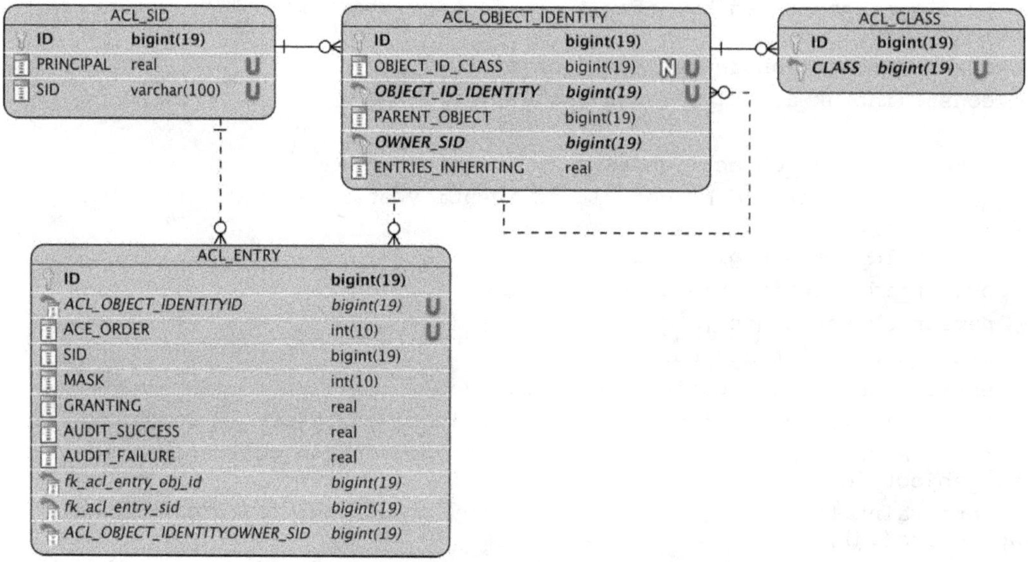

Figure 12.2 – ACL database schema

You can see how the concepts of `SIDs`, `OBJECT_IDENTITY`, and `ACEs` map directly to the database schema. Conceptually, this is convenient, as we can map our mental model of the ACL system and how it is enforced directly to the database.

If you've cross-referenced this with the H2 database schema supplied with the Spring Security documentation, you'll note that we've made a few tweaks that commonly bite users. These are as follows:

- Change the `ACL_CLASS.CLASS` column to `500` characters, from the default value of `100`. Some long, fully qualified class names don't fit in `100` characters.

- Name the foreign keys with something meaningful so that failures are more easily diagnosed.

If you are using another database, such as Oracle, you'll have to translate the DDL into DDL and data types specific to your database.

Once we configure the remainder of the ACL system, we'll return to the database to set up some basic ACEs to prove the ACL functionality in its most primitive form.

Configuring SecurityExpressionHandler

We'll need to configure `@EnableMethodSecurity` to enable annotations (where we'll annotate based on the expected ACL privilege) and reference a custom access decision manager.

We will also need to provide an `o.s.s.access.expression.SecurityExpressionHandler` implementation that is aware of how to evaluate permissions. Update your `SecurityConfig.java` configuration, as follows:

```
//src/main/java/com/packtpub/springsecurity/configuration/
SecurityConfig.java
@Configuration
@EnableWebSecurity
@EnableMethodSecurity(securedEnabled = true)
public class SecurityConfig {

    @Bean
    public SecurityFilterChain filterChain(HttpSecurity http) throws
Exception {
        http.authorizeHttpRequests( authz -> authz
// NOTE: "/events/" is now protected by ACL:
//.requestMatchers(antMatcher("/events/")).hasRole("ADMIN")
...
        return http.build();
    }
}
```

This is a bean reference to the `DefaultMethodSecurityExpressionHandler` object that we have defined in `AclConfig.java` file for you, as follows:

```
//src/main/java/com/packtpub/springsecurity/configuration/AclConfig.
java
    @Bean
    public DefaultMethodSecurityExpressionHandler expressionHandler(){
        DefaultMethodSecurityExpressionHandler dmseh = new
DefaultMethodSecurityExpressionHandler();
        dmseh.setPermissionEvaluator(permissionEvaluator());
        dmseh.setPermissionCacheOptimizer(permissionCacheOptimizer());
        return dmseh;
    }
```

With even a relatively straightforward ACL configuration, as we have in our scenario, there are a number of required dependencies to set up. As we mentioned previously, the Spring Security ACL module comes out of the box with a number of components that you can assemble to provide a decent set of ACL capabilities.

The AclPermissionCacheOptimizer object

The `DefaultMethodSecurityExpressionHandler` object has two dependencies. The `AclPermissionCacheOptimizer` object is used to prime the cache with all of the ACLs for a collection of objects in a single JDBC select statement. The relatively simple configuration included with this chapter can be checked, as follows:

```
//src/main/java/com/packtpub/springsecurity/configuration/AclConfig.
java @Bean

@Bean
public AclPermissionCacheOptimizer
permissionCacheOptimizer(MutableAclService aclService){
    return new AclPermissionCacheOptimizer(aclService);
}
```

Optimizing AclPermission Cache

The `DefaultMethodSecurityExpressionHandler` object then delegates to a `PermissionEvalulator` instance. For the purposes of this chapter, we are using ACLs so that the bean we will use `AclPermissionEvaluator`, which will read the ACLs that we define in our database. You can view the provided configuration for `permissionEvaluator`, as follows:

```
//src/main/java/com/packtpub/springsecurity/configuration/AclConfig.j
ava

@Bean
public AclPermissionEvaluator permissionEvaluator(MutableAclService
aclService){
    return new AclPermissionEvaluator(aclService);
}
```

The JdbcMutableAclService object

At this point, we have seen a reference to th with the `aclService` ID twice. The `aclService` ID resolves to an implementation of `o.s.s.acls.model.AclService` that is responsible (through delegation) for translating information about the object being secured by ACLs into expected ACEs:

```
//src/main/java/com/packtpub/springsecurity/configuration/AclConfig.j
ava

@Bean
public MutableAclService aclService(LookupStrategy lookupStrategy,
SpringCacheBasedAclCache aclCache){
    return new JdbcMutableAclService(dataSource,
                lookupStrategy,
                aclCache);
}
```

We'll use o.s.s.acls.jdbc.JdbcMutableAclService, which is the default implementation of o.s.s.acls.model.AclService. This implementation comes out of the box and is ready to use the schema that we defined in the last step of this exercise. The JdbcMutableAclService object will additionally use recursive SQL and post-processing to understand the object and SID hierarchies and ensure that representations of these hierarchies are passed back to AclPermissionEvaluator.

The BasicLookupStrategy class

The JdbcMutableAclService class uses the same JDBC dataSource instance that we've defined with the embedded database declaration, and it also delegates to an implementation of o.s.s.acls.jdbc.LookupStrategy, which is solely responsible for actually making database queries and resolving requests for ACLs. The only LookupStrategy implementation supplied with Spring Security is o.s.s.acls.jdbc.BasicLookupStrategy, and is defined as follows:

```
//src/main/java/com/packtpub/springsecurity/configuration/AclConfig.j
ava
@Bean
  public LookupStrategy lookupStrategy(AclCache aclCache,
                AclAuthorizationStrategy aclAuthorizationStrategy,
ConsoleAuditLogger consoleAuditLogger){
      return new BasicLookupStrategy(
              dataSource,
              aclCache,
              aclAuthorizationStrategy,
              consoleAuditLogger);
  }
```

Now, BasicLookupStrategy is a relatively complex beast. Remember that its purpose is to translate a list of the ObjectIdentity declarations to be protected into the actual, applicable ACE list from the database.

As ObjectIdentity declarations can be recursive, this proves to be quite a challenging problem, and a system that is likely to experience heavy use should consider the SQL's, which is generated for performance, impact on the database.

Querying with the lowest common denominator

Be aware that BasicLookupStrategy is intended to be compatible with all databases by strictly sticking with **American National Standards Institute (ANSI)** SQL syntax, notably left [outer] joins. Some older databases do not support this join syntax, so, ensure that you verify that the syntax and structure of SQL are compatible with your database!

There are also most certainly more efficient database-dependent methods of performing hierarchical queries using non-standard SQL, for example, Oracle's CONNECT BY statement and the **common table expression** (CTE) capability of many other databases, including PostgreSQL and Microsoft SQL Server.

Much as you learned in the example in *Chapter 4, JDBC-based Authentication*, using a custom schema for the JdbcDaoImpl implementation of the UserDetailsService properties are exposed to allow for configuration of the SQL utilized by BasicLookupStrategy. Consult the Javadoc and the source code itself to see how they are used so that they can be correctly applied to your custom schema.

We can see that LookupStrategy requires a reference to the same JDBC dataSource instance that AclService utilizes. The other three references bring us almost to the end of the dependency chain.

AclCache interface

The o.s.s.acls.model.AclCache interface declares an interface for a caching ObjectIdentity to ACL mappings, to prevent redundant (and expensive) database lookups. Spring Security supports any implementation of JCache (JSR-107).

For example, to enable third-party support for Ehcache, which is an open-source, memory- and disk-based caching library that is widely used in many open-source and commercial Java products, you need to add the following Gradle dependency:

```
//build.gradle
//Enabling Ehcache support
implementation "org.ehcache:ehcache"
```

We will set up ConcurrentMapCache in our example by updating the configuration in AclConfig. java:

```
//src/main/java/com/packtpub/springsecurity/configuration/AclConfig.
java

@Bean
public AclCache aclCache( Cache concurrentMapCache,
        PermissionGrantingStrategy permissionGrantingStrategy,
AclAuthorizationStrategy aclAuthorizationStrategy){
    return new SpringCacheBasedAclCache(concurrentMapCache,
            permissionGrantingStrategy,
            aclAuthorizationStrategy);
}

@Bean
public PermissionGrantingStrategy permissionGrantingStrategy(){
    return new
```

```
DefaultPermissionGrantingStrategy(consoleAuditLogger());
}

@Bean
public ConcurrentMapCache concurrentMapCache(){
    return new ConcurrentMapCache("aclCache");
}

@Bean
public CacheManager cacheManager(){
    return new ConcurrentMapCacheManager();
}
```

The ConsoleAuditLogger class

The next simple dependency hanging off of `o.s.s.acls.jdbc.BasicLookupStrategy` is an implementation of the `o.s.s.acls.domain.AuditLogger` interface, which is used by the `BasicLookupStrategy` class to audit ACL and ACE lookups. Similar to the `AclCache` interface, only one implementation is supplied with Spring Security, and it simply logs to the console. We'll configure it with another one-line bean declaration:

```
//src/main/java/com/packtpub/springsecurity/configuration/AclConfig.j
ava
@Bean
public ConsoleAuditLogger consoleAuditLogger(){
    return new ConsoleAuditLogger();
}
```

The AclAuthorizationStrategyImpl interface

The final dependency to resolve is an implementation of the `o.s.s.acls.domain.AclAuthorizationStrategy` interface, which actually has no immediate responsibility at all during the load of the ACL from the database. Instead, the implementation of this interface is responsible for determining whether a runtime change to an ACL or ACE is allowed, based on the type of change.

We'll explain this more later when we cover mutable ACLs, as the logical flow is both somewhat complicated and not pertinent to getting our initial configuration complete. The final configuration requirements are as follows:

```
//src/main/java/com/packtpub/springsecurity/configuration/AclConfig.j
ava
@Bean
public AclAuthorizationStrategy aclAuthorizationStrategy() {
    return new AclAuthorizationStrategyImpl(
```

```
                    new SimpleGrantedAuthority("ROLE_ADMINISTRATOR")
        );
    }
```

You might wonder what the reference to the bean with the `adminAuthority` ID is for—`AclAuthorizationStrategyImpl` provides the ability to specify `GrantedAuthority`, which is required to allow specific operations at runtime on mutable ACLs. We'll cover these later in this chapter.

We're finally done with the basic configuration of an out-of-the-box Spring Security ACL implementation. The next and final step requires that we insert a simple ACL and ACE into the H2 database and test it out!

Creating a simple ACL entry

Recall that our very simple scenario is to only allow `user2@example.com` access to the birthday party event and ensure that no other events are accessible.

You may find it helpful to refer back several pages (*Figure 12.2*) to the database schema diagram to follow which data we are inserting and why.

We have already included a file named `data.sql` in the sample application.

All of the SQL explained in this section will be from the file—you may feel free to experiment and add more test cases based on the sample SQL we've provided. In fact, we encourage you experiment with sample data!

Let's take a look at the following steps for creating a simple ACL entry:

1. First, we'll need to populate the `ACL_CLASS` table with any or all of the domain object classes, which may have ACL rules—in the case of our example, this is simply our `Event` class:

```
//src/main/resources/data.sql
insert into acl_class (id, class) values (10, 'com.packtpub.
springsecurity.domain.Event');
```

    We chose to use primary keys that are between 10 and 19 for the `ACL_CLASS` table, 20 and 29 for the `ACL_SID` table, and so on.

    This will help make it easier to understand which data associates with which table. Note that our `Event` table starts with a primary key of `100`. These conveniences are done for example purposes and are not suggested for production purposes.

2. Next, the `ACL_SID` table is seeded with SIDs that will be associated with the ACEs. Remember that SIDs can either be roles or users—we'll populate the roles and `user2@example.com` here.

3. While the SID object for roles is straightforward, the SID object for a user is not quite as clear cut. For our purposes, the username is used for the SID. To learn more about how the SIDs are resolved for roles and users, refer to `o.s.s.acls.domain.SidRetrievalStrategyImpl`. If the defaults do not meet your needs, a custom `o.s.s.acls.model.SidRetrievalStrategy` default can be injected into `AclPermissionCacheOptimizer` and `AclPermissionEvaluator`. We will not need this sort of customization for our example, but it is good to know that it is available if necessary:

```
//src/main/resources/data.sql
-- User specific:
insert into acl_sid (id, principal, sid) values (20, true, 'user2@
example.com');
-- Role specific:
insert into acl_sid (id, principal, sid) values (21, false, 'ROLE_
USER');
insert into acl_sid (id, principal, sid) values (22, false, 'ROLE_
ADMIN');
```

The table where things start getting complicated is the `ACL_OBJECT_IDENTITY` table, which is used to declare individual domain object instances, their parent (if any), and owning `SID`.

For example, this table represents the `Event` objects that we are securing. We'll insert a row with the following properties:

- A domain object of the `Event` type that is a foreign key, `10`, to our `ACL_CLASS` table via the `OBJECT_ID_CLASS` column.

- A domain object primary key of `100` (the `OBJECT_ID_IDENTITY` column). This is a foreign key (although not enforced with a database constraint) to our `Event` object.

- The `SID` owner of `user2@example.com`, which is a foreign key, `20`, to `ACL_SID` via the `OWNER_SID` column.

The SQL to represent our events with IDs of `100` (birthday event), `101`, and `102` is as follows:

```
//src/main/resources/data.sql
-- object identity
-- Event entry for user2 SID
insert into acl_object_identity (id,object_id_identity,object_id_
class,parent_object,owner_sid,entries_inheriting)
values (30,100, 10, null, 20, false);


-- Event entry for ROLE_USER SID
insert into acl_object_identity (id,object_id_identity,object_id_
class,parent_object,owner_sid,entries_inheriting)
values (31,101, 10, null, 21, false);
```

```
-- Event entry for ROLE_ADMIN SID
insert into acl_object_identity (id,object_id_identity,object_id_
class,parent_object,owner_sid,entries_inheriting)
values (32,102, 10, null, 21, false);
```

Keep in mind that the owning SID could also represent a role—both types of rules function equally well as far as the ACL system is concerned.

Finally, we'll add an ACE related to this object instance, which declares that user2@example.com is allowed read access to the birthday event:

```
//src/main/resources/data.sql
-- ACEntry list ---------------------------------
-- mask == R
-- Entry for Event entry for user2 SID
insert into acl_entry (acl_object_identity, ace_order, sid, mask,
granting, audit_success, audit_failure)
values(30, 1, 20, 1, true, true, true);
```

The MASK column here represents a bitmask, which is used to grant permission assigned to the stated SID on the object in question. We'll explain the details of this later in this chapter—unfortunately, it doesn't tend to be as useful as it may sound.

Now, we can start the application and run through our sample scenario. Try logging in with user2@example.com/user2 and accessing the **All Events** page. You will see that only the birthday event is listed.

When logged in with admin1@example.com/admin1 and viewing the **All Events** page, no events will be displayed.

However, if we navigated directly to an event, it would not be protected. Can you figure out how to secure direct access to an event based on what you learned in this chapter?

If you have not figured it out yet, you can secure direct access to an event by making the following update to CalendarService.java, as follows:

```
//src/main/java/com/packtpub/springsecurity/service/CalendarService.
java
@PostAuthorize("hasPermission(filterObject, 'read') ")
Event getEvent(int eventId);
```

We now have a basic working setup of ACL-based security (albeit, a very simple scenario).

Let's move on to some more explanations of concepts we saw during this walkthrough, and then review a couple of considerations in a typical Spring ACL implementation that you should consider before using it.

> **Important note**
>
> You should start with the code from `chapter12.01-calendar`.

It is worth noting that we have not created new ACL entries when we create events. Thus, in the current state, if you create an event, you will receive an error similar to the following:

```
Exception during execution of Spring Security application! Unable
to find ACL information for object identity 'org.springframework.
security.acls.domain.ObjectIdentityImpl[Type: com.packtpub.
springsecurity.domain.Event; Identifier: 1]'
```

After examining the process of creating a straightforward ACL entry, let's now delve into a comprehensive understanding of how permissions operate, focusing on advanced ACL configurations.

Advanced ACL topics

Some high-level topics that we skimmed over during the configuration of our ACL environment had to do with ACE permissions and the use of the `GrantedAuthority` indicators to assist the ACL environment in determining whether certain types of runtime changes to ACLs were allowed. Now that we have a working environment, we'll review these more advanced topics.

How permissions work

Permissions are no more than single logical identifiers represented by bits in an integer. An ACE grants permissions to `SID`s based on the bitmask, which comprises the logical and of all permissions applicable to that ACE.

The default permission implementation, `o.s.s.acls.domain.BasePermission`, defines a series of integer values representing common ACL authorization verbs. These integer values correspond to single bits set in an integer, so a value of `BasePermission`, `WRITE`, with an integer value of 1 has a bitwise value of 2^1 or 2.

These are illustrated in the following diagram:

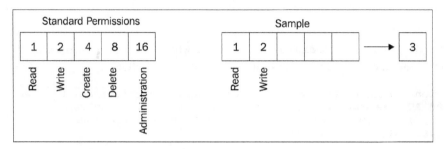

Figure 12.3 – Default and custom permissions bitmask

We can see that the Sample permission bitmask would have an integer value of 3, due to the application of both the Read and Write permissions to the permission value.

All of the standard integer single permission values shown in the preceding diagram are defined in the BasePermission object as static constants.

The logical constants that are included in BasePermission are just a sensible baseline of commonly used permissions in ACE and have no semantic meaning within the Spring Security framework. It's quite common for very complex ACL implementations to invent their own custom permissions, augmenting best practice examples with domain- or business-dependent ones.

One issue that often confuses users is how the bitmasks are used in practice, given that many databases either do not support bitwise logic or do not support it in a scalable way. Spring ACL intends to solve this problem by putting more of the load of calculating appropriate permissions related to bitmasks on the application rather than on the database.

It's important to review the resolution process, where we see how AclPermissionEvaluator resolves permissions declared on the method itself (in our example, with the @PostFilter annotation) to real ACL permissions.

The following diagram illustrates the process that Spring ACL performs to evaluate the declared permission against the relevant ACEs for the requesting principal:

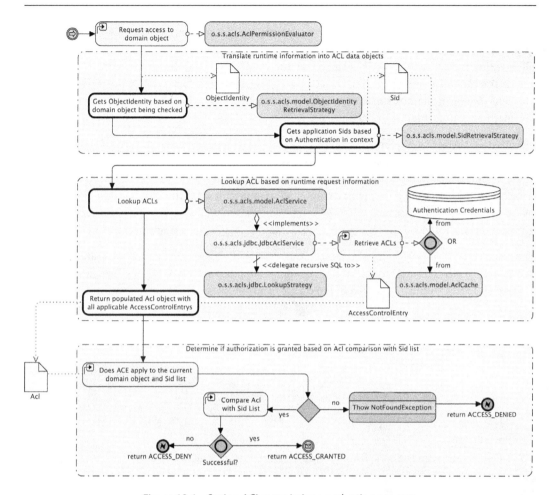

Figure 12.4 – Spring ACL permissions evaluation process

We see that `AclPermissionEvaluator` relies on classes implementing two interfaces, `o.s.s.acls.model.ObjectIdentityRetrievalStrategy` and `o.s.s.acls.model.SidRetrievalStrategy`, to retrieve `ObjectIdentity` and SIDs appropriate for the authorization check. The important thing to note about these strategies is how the default implementation classes actually determine the `ObjectIdentity` and SIDs objects to return, based on the context of the authorization check.

The `ObjectIdentity` object has two properties, `type` and `identifier`, that are derived from the object being checked at runtime and used to declare ACE entries. The default `ObjectIdentityRetrievalStrategy` interface uses the fully-qualified class name to populate the `type` property. The `identifier` property is populated with the result of a method with the `Serializable getId()` signature, invoked on the actual object instance.

As your object isn't required to implement an interface to be compatible with ACL checks, the requirement to implement a method with a specific signature can be surprising for developers implementing Spring Security ACL. Plan ahead and ensure that your domain objects contain this method! You may also implement your own `ObjectIdentityRetrievalStrategy` class (or implement the out-of-the-box subclass) to call a method of your choice. The name and type signature of the method is, unfortunately, not configurable.

Unfortunately, the actual implementation of `AclImpl` directly compares the permission specified in our **Spring Expression Language (SpEL)** expression specified in our `@PostFilter` annotation, and the permission stored on the ACE in the database, without using bitwise logic. You will need to take care when declaring a user with a combination of permissions, as either `AclEntryVoter` must be configured with all combinations of permissions or the ACEs need to ignore the fact that the permission field is intended to store multiple values and instead store a single permission per ACE.

If you want to verify this with our simple scenario, change the READ permission we granted to the `user2@example.com` SID to the bitmask combination of `Read` and `Write`, which translates to a value of 3. This would be updated in the `data.sql` file, as follows:

```
//src/main/resources/data.sql
-- READ / WRITE Entry for Event entry for user2 SID
insert into acl_entry
(acl_object_identity, ace_order, sid, mask, granting, audit success,
audit_failure)
values(30, 1, 20, 3, true, true, true);
```

> **Important note**
>
> You should start with the code from `chapter12.02-calendar`.

The custom ACL permission declaration

As stated in the earlier discussion on permission declarations, permissions are nothing but logical names for integer bit values. As such, it's possible to extend the `o.s.s.acls.domain.BasePermission` class and declare your own permissions. We'll cover a very straightforward scenario here, where we create a new ACL permission called ADMIN_READ.

This is a permission that will be granted only to administrative users and will be assigned to protect resources that only administrators can read. Although a contrived example for the JBCP calendar application, this type of use of custom permissions occurs quite often in situations dealing with **Personally Identifiable Information (PII)** (for example, social security number, and so on—recall that we covered PII in *Chapter 1, Anatomy of an Unsafe Application*).

Let's get started with making the changes required to support this by performing the following steps:

1. The first step is to extend the `BasePermission` class with our own `com.packtpub.springsecurity.acls.domain.CustomPermission` class, as follows:

    ```
    package com.packtpub.springsecurity.acls.domain;
    public class CustomPermission extends BasePermission {

        public static final Permission ADMIN_READ = new
    CustomPermission(1 << 5, 'M'); // 32

        public CustomPermission(int mask, char code) {
            super(mask, code);
        }

    }
    ```

2. Next, we will need to configure the `o.s.s.acls.domain.PermissionFactory` default implementation, `o.s.s.acls.domain.DefaultPermissionFactory`, to register our custom permission logical value. The role of `PermissionFactory` is to resolve permission bitmasks into logical permission values (which can be referenced by the constant value, or by name, such as `ADMIN_READ`, in other areas of the application). The `PermissionFactory` instance requires that any custom permission is registered with it for proper lookup. We have included the following configuration that registers our `CustomPermission` class, as follows:

    ```
    //src/main/java/com/packtpub/springsecurity/configuration/
    AclConfig.java
    @Bean
    public DefaultPermissionFactory permissionFactory(){
        return new DefaultPermissionFactory(CustomPermission.class);
    }
    ```

3. Next, we will need to override the default `PermissionFactory` instance for our `BasicLookupStrategy` and `AclPermissionEvaluator` interfaces with the customized `DefaultPermissionFactory` interface:

    ```
    //src/main/java/com/packtpub/springsecurity/configuration/
    AclConf ig.java
    @Bean
    public AclPermissionEvaluator
    permissionEvaluator(MutableAclService aclService){
            AclPermissionEvaluator pe = new
    AclPermissionEvaluator(aclService);
            pe.setPermissionFactory(permissionFactory());
            return pe;
    }
    ```

```
@Bean
public LookupStrategy lookupStrategy(AclCache aclCache,
        AclAuthorizationStrategy aclAuthorizationStrategy,
ConsoleAuditLogger consoleAuditLogger) {
        BasicLookupStrategy lookupStrategy = new
BasicLookupStrategy(
                dataSource,
                aclCache,
                aclAuthorizationStrategy,
                consoleAuditLogger);
        lookupStrategy.
setPermissionFactory(permissionFactory());
        return lookupStrategy;
}
```

4. We also need to add the SQL query to utilize the new permission to grant access to the conference call (`acl_object_identity ID of 31`) event to `admin1@example.com`. Make the following updates to `data.sql`:

```
-- custom permission
insert into acl_sid (id, principal, sid) values (23, true,
'admin1@example.com');
insert into acl_entry
(acl_object_identity, ace_order, sid, mask, granting, audit_
success, audit_failure)
values(31, 1, 23, 32, true, true, true);
```

We can see that the new integer bitmask value of 32 has been referenced in the ACE data. This intentionally corresponds to our new `ADMIN_READ` ACL permission, as defined in Java code. The conference call event is referenced by its primary key (stored in the `object_id_identity` column) value of 31, in the `ACL_OBJECT_IDENTITY` table.

5. The last step is to update our `CalendarService's getEvents()` method to utilize our new permission, as follows:

```
@PostFilter("hasPermission(filterObject, 'read') " +
        "or hasPermission(filterObject, 'admin_read')")
List<Event> getEvents()
```

With all of these configurations in place, we can start up the site again and test out the custom ACL permission. Based on the sample data we have configured, here is what should happen when the various available users click on categories:

Username/password	Birthday party event	Conference call event	Other events
`user2@example.com/user2`	Allowed via READ	Denied	Denied
`admin1@example.com/admin1`	Denied	Allowed via ADMIN_READ	Denied
`user1@example.com/user1`	Denied	Denied	Denied

Table 12.3 – Users ACL Java objects

We can see that, even with the use of our simple cases, we've now been able to extend the Spring ACL functionality in a very limited way to illustrate the power of this fine-grained access control system.

> **Important note**
> You should start with the code from `chapter12.03-calendar`.

Enabling ACL permission evaluation

We saw in *Chapter 2, Getting Started with Spring Security*, that the Spring Security JSP tag library offers functionality to expose authentication-related data to the user and to restrict what the user can see based on a variety of rules. So far in this book, we have used the `Thymeleaf Security tag libraries`, which are built on top of Spring Security.

The very same tag library can also interact with an ACL-enabled system right out of the box! From our simple experiments, we have configured a simple ACL authorization scenario around the first two categories in the list on the home page. Let's look at the following steps and learn how to enable ACL permission evaluation in our Thymeleaf pages:

1. First, we will need to remove our `@PostFilter` annotation from the `getEvents()` method in our `CalendarService` interface in order to give our JSP tag library a chance to filter out the events that are not allowed for display. Go ahead and remove `@PostFilter` now, as follows:

```
// src/main/java/com/packtpub/springsecurity/service/
CalendarService.java

List<Event> getEvents();
```

2. Now that we have removed `@PostFilter`, we can utilize the `<sec:authorize-acl>` tag to hide the events that the user doesn't actually have access to. Refer to the table in the preceding section as a refresher on the access rules we've configured up to this point!

3. We'll wrap the display of each event with the `<sec:authorize-acl>` tag, declaring the list of permissions to check on the object to be displayed:

```
//src/main/resources/templates/events/list.html
<th:block th:each="event : ${events}">
    <tr sec:authorize-acl="${event} :: '1,32'" >
        <td th:text="${#calendars.format(event.dateWhen, 'yyyy-
MM-dd HH:mm')}">today</td>
        <td th:text="${event.owner.name}">Chuck Norris</td>
        <td th:text="${event.attendee.name}">Josh Knutson</td>
        <td><a th:href="@{'/events/{id}'(id=${event.id})}"
th:text="${event.summary}">-1</a></td>
    </tr>
</th:block>
```

4. Think for a moment about what we want to occur here—we want the user to see only the items they actually have the READ or ADMIN_READ (our custom permission) access. However, to use the tag library, we need to use the permission mask, which can be referenced from the following table:

Name	Mask
READ	1
WRITE	2
ADMIN_READ	32

Table 12.4 – Permissions mask table

Behind the scenes, the tag implementation utilizes the same `SidRetrievalStrategy` and `ObjectIdentityRetrievalStrategy` interfaces that were discussed earlier in this chapter. So, the computation of access checking follows the same workflow as it does with ACL-enabled voting on method security. As we will see in a moment, the tag implementation will also use the same `PermissionEvaluator`.

We have already enabled our @EnableMethodSecurity annotation with an expressionHandler element that references DefaultMethodSecurity ExpressionHandler. The DefaultMethodSecurityExpressionHandler implementation is aware of our AclPermissionEvaluator interface, but we must also make Spring Security's web tier aware of AclPermissionEvalulator. If you think about it, this symmetry makes sense, since securing methods and HTTP requests are protecting two very different resources. Fortunately, Spring Security's abstractions make this rather simple.

5. Add a DefaultWebSecurityExpressionHandler handler that references the bean with the ID of permissionEvaluator, which we have already defined:

```
//src/main/java/com/packtpub/springsecurity/configuration/
AclConfig.java

@Bean
public DefaultWebSecurityExpressionHandler
webExpressionHandler(AclPermissionEvaluator permissionEvaluator)
{
    DefaultWebSecurityExpressionHandler webExpressionHandler =
new DefaultWebSecurityExpressionHandler();
    webExpressionHandler.
setPermissionEvaluator(permissionEvaluator);
    return webExpressionHandler;
}
```

You can see how these steps are very similar to how we added support for permission handling to our method security. This time, it was a bit simpler, since we were able to reuse the same bean with the ID of PermissionEvaluator, which we already configured.

Start up our application and try accessing the **All Events** page as different users. You will find that the events that are not allowed for a user. They are now hidden using our tag library instead of the @ PostFilter annotation.

We are still aware that accessing an event directly would allow a user to see it. However, this could easily be added by combining what you learned in this chapter with what you learned about the @ PostAuthorize annotation in this chapter.

> **Important note**
> You should start with the code from chapter12.04-calendar.

Mutable ACLs and authorization

Although the JBCP calendar application doesn't implement full user administration functionality, it's likely that your application will have common features, such as new user registration and administrative user maintenance. To this point, a lack of these features—which we have worked around using SQL inserts at application startup—hasn't stopped us from demonstrating many of the features of Spring Security and Spring ACL.

However, the proper handling of runtime changes to declared ACLs, or the addition or removal of users in the system, is critical to maintaining the consistency and security of the ACL-based authorization environment. Spring ACL solves this issue through the concept of the mutable ACL (`o.s.s.acls.model.MutableAcl`).

Extending the standard ACL interface, the `MutableAcl` interface allows for runtime manipulation of ACL fields to change the in-memory representation of a particular ACL. This additional functionality includes the ability to create, update, or delete ACEs, change ACL ownership, and other useful functions.

We might expect, then, that the Spring ACL module would come out of the box with a way to persist runtime ACL changes to the JDBC datastore, and indeed it does. The `o.s.s.acls.jdbc.JdbcMutableAclService` class may be used to create, update, and delete the `MutableAcl` instances in the database, as well as to do general maintenance on the other supporting tables for ACLs (handling `SID`s, `ObjectIdentity`, and domain object class names).

Recall from earlier in the chapter that the `AclAuthorizationStrategyImpl` class allows us to specify the administrative role for actions on mutable ACLs. These are supplied to the constructor as part of the bean configuration. The constructor arguments and their meaning are as follows:

Arg #	What it does
1	It indicates the authority that a principal is required to have to take ownership of an ACL-protected object at runtime
2	It indicates the authority that a principal is required to have to change the auditing of an ACL-protected object at runtime
3	It indicates the authority that a principal is required to have to make any other kind of change (create, update, and delete) to an ACL-protected object at runtime

Table 12.5 – Constructor arguments for AclAuthorizationStrategyImpl

It may be confusing that we only specified a single constructor argument when there are three arguments listed. The `AclAuthorizationStrategyImpl` class can also accept a single `GrantedAuthority`, which will then be used for all three arguments. This is convenient if we want the same `GrantedAuthority` to be used for all of the operations.

The JdbcMutableAclService interface contains a number of methods used to manipulate ACL and ACE data at runtime. While the methods themselves are fairly understandable (createAcl, updateAcl, and deleteAcl), the correct way to configure and use JdbcMutableAclService is often difficult for even advanced Spring Security users.

Let's modify CalendarService to create a new ACL for newly created events.

Currently, if a user creates a new event, it will not be visible to the user in the **All Events** view, since we are using the <sec:authorize-acl> tag to only display event objects that the user has access to. Let's update our DefaultCalendarService interface so that when a user creates a new event, they are granted read access to that event and it will be displayed for them on the **All Events** page.

Let's look at the following steps to add ACLs to newly created events:

1. The first step is to update our constructor to accept MutableAclService and UserContext:

    ```
    //src/main/java/com/packtpub/springsecurity/service/
    DefaultCalendarService.java
    @Repository
    public class DefaultCalendarService implements CalendarService {
        private final EventDao eventDao;
        private final CalendarUserDao userDao;
        private final MutableAclService aclService;
        private final UserContext userContext;


        public DefaultCalendarService(EventDao eventDao,
    CalendarUserDao userDao, MutableAclService aclService,
    UserContext userContext) {
            this.eventDao = eventDao;
            this.userDao = userDao;
            this.aclService = aclService;
            this.userContext = userContext;
        }
    ```

2. Then, we need to update our createEvent method to also create an ACL for the current user. Make the following changes:

    ```
    //src/main/java/com/packtpub/springsecurity/service/
    DefaultCalendarService.java
    @Transactional
    @Override
    public int createEvent(Event event) {
        int result = eventDao.createEvent(event);
        event.setId(result);
        // Add new ACL Entry:
        MutableAcl acl = aclService.createAcl(new
    ```

```
ObjectIdentityImpl(event));
    PrincipalSid sid = new PrincipalSid(userContext.
getCurrentUser().getEmail());
    acl.setOwner(sid);
    acl.insertAce(0, BasePermission.READ, sid, true);
    aclService.updateAcl(acl);
    return result;
}
```

3. The `JdbcMutableAclService` interface uses the current user as the default owner for the created `MutableAcl` interface. We chose to explicitly set the owner again to demonstrate how this can be overridden.

4. We then add a new ACE and save our ACL. That's all there is to it.

5. Start the application and log in with `user1@example.com/user1`.

6. Visit the **All Events** page and see that there are no events currently listed. Then, create a new event and it will be displayed the next time you visit the **All Events** page. If you log in as any other user, the event will not be visible on the **All Events** page.

    However, it will potentially be visible to the user since we have not applied security to other pages. Again, we encourage you to attempt to secure these pages on your own.

> **Important note**
>
> You should start with the code from `chapter12.05-calendar`.

Upon grasping the workings of Spring Security ACL, the subsequent sections will address the factors to be considered in a standard ACL deployment.

Considerations for a typical ACL deployment

Actually, deploying Spring ACL in a true business application tends to be quite involved. We wrap up coverage of Spring ACL with some considerations that arise in most Spring ACL implementation scenarios.

ACL scalability and performance modeling

For small and medium-sized applications, the addition of ACLs is quite manageable, and while it adds overhead to database storage and runtime performance, the impact is not likely to be significant. However, depending on the granularity with which ACLs and ACEs are modeled, the numbers of database rows in a medium- to large-sized application can be truly staggering and can task even the most seasoned database administrator.

Let's assume we were to extend ACLs to cover an extended version of the JBCP calendar application. Let's assume that users can manage accounts, post pictures to events, and administer (add/remove users) from an event. We'll model the data as follows:

- All users have accounts.

- 10% of users are able to administer an event. The average number of events that a user can administer will be two.

- Events will be secured (read-only) per customer, but also need to be accessible (read/write) by administrators.

- 10% of all customers will be allowed to post pictures. The average number of posts per user will be 20.

- Posted pictures will be secured (read/write) per user, as well as administrators. Posted pictures will be read-only for all other users.

Given what we know about the ACL system, we know that the database tables have the following scalability attributes:

Table	Scales with data	Scalability notes
ACL_CLASS	No	One row is required per domain class.
ACL_SID	Yes (users)	One row is required per role (GrantedAuthority). One row is required for each user account (if individual domain objects are secured per user).
ACL_OBJECT_IDENTITY	Yes (domain class instances per class)	One row is required per instance of a secured domain object.
ACL_ENTRY	Yes (domain object instances individual ACE entries)	One row is required per ACE; may require multiple rows for a single domain object.

Table 12.6 – Database table's scalability attributes

We can see that ACL_CLASS doesn't really have scalability concerns (most systems will have fewer than 1,000 domain classes).

The ACL_SID table will scale linearly based on the number of users in the system. This is probably not a matter of concern because other user-related tables will scale in this fashion as well (user account, and so on).

The two tables of concern are ACL_OBJECT_IDENTITY and ACL_ENTRY. If we model the estimated rows required to model an order for an individual customer, we come up with the following estimates:

Table	ACL data per event	ACL data per picture post
ACL_OBJECT_IDENTITY	One row is required for a single event.	One row is required for a single post.
ACL_ENTRY	Three rows—one row is required for read access by the owner (the user SID) and two rows are required (one for read access, one for write access) for the administrative group SID.	Four rows—one row is required for read access by the user group SID, one row is required for write access by the owner, and two rows are required for the administrative group SID (as with events).

Table 12.7 – Scalability estimates per events or posted pictures

We can then take the usage assumptions from the previous page and calculate the following ACL scalability matrix as follows:

Table/Object	Scale factor	Estimates (Low)	Estimates (High)
Users		10,000	1,000,000
Events	# Users * 0.1 * 2	2,000	200,000
Picture Posts	# Users * 0.1 * 20	20,000	2,000,000
ACL_SID	# Users	10,000	1,000,000
ACL_OBJECT_IDENTITY	# Events + # Picture Posts	220,000	2,200,000
ACL_ENTRY	(# Events * 3) + (# Picture Posts * 4)	86,000	8,600,000

Table 12.8 – ACL scalability matrix

From these projections based on only a subset of the business objects likely to be involved and secured in a typical ACL implementation, you can see that the number of database rows devoted to storing ACL information is likely to grow linearly (or faster) in relation to your actual business data. Especially in large system planning, forecasting the amount of ACL data that you are likely to use is extremely important. It is not uncommon for very complex systems to have hundreds of millions of rows related to ACL storage.

Do not discount custom development costs

Utilizing a Spring ACL-secured environment often requires significant development work above and beyond the configuration steps we've described to this point. Our sample configuration scenario has the following limitations:

- No facility is provided for responding to the manipulation modification of events or modification of permissions.

- Not all of the applications use permissions. For example, the **My Events** page and directly navigating to an event are both not secured.

The application does not effectively use ACL hierarchies. These limitations would significantly impact the functionality if we were to roll out ACL security to the whole site. This is why it is critical that when planning Spring ACL rollout across an application, you must carefully review all the places where the domain data is manipulated and ensure that these locations correctly update ACL, ACE rules, and invalidate caches. Typically, the securing of methods and data takes place at the service or business application layer, and the hooks required to maintain ACLs and ACEs occur at the data access layer.

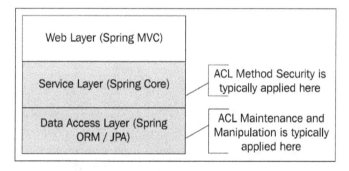

Figure 12.6 – Spring ACL permissions evaluation process

If you are dealing with a reasonably standard application architecture, with proper isolation and encapsulation of functionality, it's likely that there's an easily identified central location for these changes. On the other hand, if you're dealing with an architecture that has devolved (or was never designed well in the first place), then adding ACL functionality and supporting hooks in data manipulation code can prove to be very difficult.

As previously hinted, it's important to keep in mind that the Spring ACL architecture hasn't changed significantly since the days of `Acegi 1.x (Parent project of Spring Security)`.

The following are some of the most important and commonly encountered issues with the Spring ACL architecture:

- The ACL infrastructure requires a numeric primary key. For applications that use a **globally unique identifier (GUID)** or **Universal Unique Identifier (UUID)** primary key (which occurs more frequently due to more efficient support in modern databases), this can be a significant limitation.

- Several inconsistencies exist between the method of configuring Spring ACL and the rest of Spring Security. In general, it is likely that you will run into areas where class delegates or properties are not exposed through **Dependency Injection (DI)**, necessitating an override and rewrite strategy that can be time consuming and expensive to maintain.

- The permission bitmask is implemented as an integer and thus has 32 possible bits. It's somewhat common to expand the default bit assignments to indicate permissions on individual object properties (for example, assigning a bit to read the social security number of an employee). Complex deployments may have well over 32 properties per domain object, in which case the only alternative would be to remodel your domain objects around this limitation.

Depending on your application requirements, it is likely that you will encounter additional issues, especially with regard to the number of classes requiring change when implementing certain types of customizations.

Should I use Spring Security ACL?

Just like the details of applying Spring Security are highly business dependent, so is the application of Spring ACL support. In fact, this tends to be even more true of ACL support due to its tight coupling to business methods and domain objects. We hope that this guide to Spring ACL has explained the important high-level and low-level configurations and concepts required to analyze Spring ACL for use in your application and can assist you in determining and matching its capabilities to real-world use.

Summary

In this chapter, we focused on security based on ACLs and the specific details of how this type of security is implemented by the Spring ACL module.

We reviewed the basic concept of ACLs, and many reasons why they can be very effective solutions to authorization. Also, we learned about the key concepts related to the Spring ACL implementation, including ACEs, SIDs, and object identity. We examined the database schema and logical design required to support a hierarchical ACL system. We configured all the required Spring beans to enable the Spring ACL module and enhanced one of the service interfaces to use annotated method authorization.

We then tied the existing users in our database, and business objects used by the site itself, into a sample set of ACE declarations and supporting data. We reviewed the concepts around Spring ACL permission handling. We expanded our knowledge of the Spring Security Thymeleaf tag library and SpEL (for method security) to utilize ACL checks.

We discussed the mutable ACL concept and reviewed the basic configuration and custom coding required in a mutable ACL environment. We developed a custom ACL permission and configured the application to demonstrate its effectiveness. We configured and analyzed the use of the `Spring` cache manager to reduce the database impact of Spring ACL. We analyzed the impact and design considerations of using the Spring ACL system in a complex business application.

This wraps up our discussion about Spring Security ACLs. In the next chapter, we'll dig a bit further into how Spring Security works.

Custom Authorization

In this chapter, we will write some custom implementations for `Spring Security`'s key authorization APIs. Once we have done this, we will use our understanding of the custom implementations to understand how `Spring Security`'s authorization architecture works.

Throughout this chapter, we will cover the following topics:

- Gaining an understanding of how authorization works
- Writing a custom `SecurityMetaDataSource` backed by a database instead of `requestMatchers()` methods
- Creating custom **Spring Expression Language (SpEL)** expressions
- Implementing a custom `PermissionEvaluator` object that allows our permissions to be encapsulated
- Declaring a custom `AuthorizationManager`

This chapter's code in action link is here: `https://packt.link/e630f`.

Authorizing the Requests

As in the authentication process, `Spring Security` provides an `o.s.s.web.access.intercept.FilterSecurityInterceptor` servlet filter, which is responsible for coming up with a decision as to whether a particular request will be accepted or denied. At the point the filter is invoked, the principal has already been authenticated, so the system knows that a valid user has logged in; remember that we implemented the `List<GrantedAuthority> getAuthorities()` method, which returns a list of authorities for the principal, in *Chapter 3, Custom Authentication*. In general, the authorization process will use the information from this method (defined by the `Authentication` interface) to determine, for a particular request, whether or not the request should be allowed.

This method serves as a means for an `AuthorizationManager` instance to acquire a precise String representation of the `GrantedAuthority`. By providing a representation as a String, most `AuthorizationManager` implementations can easily `read` the `GrantedAuthority`. If a `GrantedAuthority` cannot be accurately represented as a String, it is considered `complex`, and the `getAuthority()` method must return `null`.

A prime example of a complex `GrantedAuthority` would be an implementation storing a list of operations and authority thresholds related to various customer account numbers. Trying to represent this intricate `GrantedAuthority` as a String would pose considerable challenges. Consequently, the `getAuthority()` method should return `null`. This signals to any `AuthorizationManager` that it needs to support the specific `GrantedAuthority` implementation to comprehend its contents.

`Spring Security` features a concrete `GrantedAuthority` implementation named `SimpleGrantedAuthority`. This implementation enables the conversion of any user-specified String into a `GrantedAuthority`. All `AuthenticationProvider` instances integrated into the security architecture utilize `SimpleGrantedAuthority` to populate the `Authentication` object.

By default, role-based authorization rules involve the prefix ROLE_. Therefore, if an authorization rule mandates a security context to possess the role of USER, `Spring Security` will automatically seek a `GrantedAuthority#getAuthority` that returns ROLE_USER.

Remember that authorization is a binary decision—a user either has access to a secured resource or does not. There is no ambiguity when it comes to authorization.

A smart object-oriented design is pervasive within the `Spring Security` framework, and authorization decision management is no exception.

In `Spring Security`, the `o.s.s.access.AccessDecisionManager` interface specifies two simple and logical methods that fit sensibly into the processing decision flow of requests, as follows:

- `supports`: This logical operation actually comprises two methods that allow the `AccessDecisionManager` implementation to report whether or not it supports the current request.

- `decide`: This allows the `AccessDecisionManager` implementation to verify, based on the request context and security configuration, whether or not access should be allowed and the request accepted. The `Decide` method actually has no return value, and instead reports the denial of a request by throwing an exception to indicate rejection.

Specific types of exceptions can further dictate the action to be taken by the application to resolve authorization decisions. The `o.s.s.access.AccessDeniedException` interface is the most common exception thrown in the area of authorization and merits special handling by the filter chain.

The implementation of `AccessDecisionManager` is completely configurable using standard Spring bean binding and references. The default `AccessDecisionManager` implementation provides an access granting mechanism based on `AccessDecisionVoter` and vote aggregation.

A voter is an actor in the authorization sequence whose job is to evaluate any or all of the following things:

- The context of the request for a secured resource (such as a URL requesting an IP address)
- The credentials (if any) presented by the user
- The secured resource being accessed
- The configuration parameters of the system, and the resource itself

After demonstrating the process of authorizing requests, we will delve into the management of invocations.

Handling of Invocations

`Spring Security` offers interceptors that are responsible for governing access to secure objects, be it method invocations or web requests. `AuthorizationManager` instances play a crucial role in making pre-invocation decisions regarding whether the invocation is permitted to proceed. Additionally, these instances contribute to post-invocation decisions, determining whether a particular value may be returned.

The AuthorizationManager class

`AuthorizationManager` takes precedence over both `AccessDecisionManager` and `AccessDecisionVoter`. Applications that customize either an `AccessDecisionManager` or an `AccessDecisionVoter` are advised to transition using `AuthorizationManager`.

`Spring Security`'s request-based, method-based, and message-based authorization components invoke `AuthorizationManager` instances, assigning them the responsibility of making definitive access control decisions.

The `check` method of `AuthorizationManager` receives all the pertinent information necessary to render an authorization decision. Specifically, passing the secure object allows the examination of arguments within the actual invocation of the secure object. For instance, if the secure object is a `MethodInvocation`, querying it for any client argument becomes straightforward. Subsequently, security logic can be implemented in the `AuthorizationManager` to ensure that the principal is authorized to operate on that customer. Implementations are expected to return a positive `AuthorizationDecision` if access is granted, a negative `AuthorizationDecision` if access is denied, and a null `AuthorizationDecision` when abstaining from making a decision.

The verify function invokes `check` and throws an `AccessDeniedException` if a negative `AuthorizationDecision` is reached.

Delegate-based AuthorizationManager Implementations

Although users have the flexibility to implement their own AuthorizationManager to govern all facets of authorization, Spring Security comes with a delegating AuthorizationManager designed to work in tandem with individual AuthorizationManagers.

The RequestMatcherDelegatingAuthorizationManager aligns the request with the most suitable delegate AuthorizationManager. For method security, AuthorizationManagerBeforeMethodInterceptor and AuthorizationManagerAfterMethodInterceptor can be employed.

The relevant classes for AuthorizationManager implementations are outlined for reference in *Figure 13.1*:

Figure 13.1 – Implementations of AuthorizationManager

With this approach, a collection of AuthorizationManager implementations can be consulted for an authorization decision.

In the following subsections, we will take a deeper look at some authorization managers.

AuthorityAuthorizationManager

The predominant AuthorizationManager provided by Spring Security is the AuthorityAuthorizationManager. It is configured with a specific set of authorities to check for in the current Authentication. If the Authentication contains any of the configured authorities, it will yield a positive AuthorizationDecision; otherwise, it will result in a negative AuthorizationDecision.

AuthenticatedAuthorizationManager

Another manager available is the `AuthenticatedAuthorizationManager`. It proves useful in distinguishing between *anonymous*, *fully-authenticated*, and *remember-me* authenticated users. Some websites grant limited access under *remember-me* authentication but necessitate users to confirm their identity by logging in for complete access.

AuthorizationManagers

`AuthorizationManagers` also offer useful static factories for combining individual `AuthorizationManagers` into more intricate expressions.

Custom AuthorizationManagers

Certainly, you have the option to implement a custom `AuthorizationManager`, allowing for the inclusion of virtually any access control logic. It may be tailored to your application, related to business logic, or involve security administration logic. For instance, you can create an implementation capable of querying Open Policy Agent or your own authorization database.

After delving into the management of invocations, we will proceed to examine the customization of `AccessDecisionManager` and `AccessDecisionVoter`.

Modifying AccessDecisionManager and AccessDecisionVoter

Before the introduction of `AuthorizationManager`, `Spring Security` introduced `AccessDecisionManager` and `AccessDecisionVoter`.

In certain scenarios, such as when migrating an older application, it might be preferable to incorporate an `AuthorizationManager` that invokes an `AccessDecisionManager` or `AccessDecisionVoter`.

To invoke an existing `AccessDecisionManager`, you can use:

```
@Component
public class AccessDecisionManagerAuthorizationManagerAdapter
implements AuthorizationManager {
    private final AccessDecisionManager accessDecisionManager;
    private final SecurityMetadataSource securityMetadataSource;

    @Override
    public AuthorizationDecision check(Supplier<Authentication>
authentication, Object object) {
        try {
            Collection<ConfigAttribute> attributes = this.
```

```
securityMetadataSource.getAttributes(object);
        this.accessDecisionManager.decide(authentication.get(),
object, attributes);
        return new AuthorizationDecision(true);
    } catch (AccessDeniedException ex) {
        return new AuthorizationDecision(false);
    }
  }


    @Override
    public void verify(Supplier<Authentication> authentication, Object
object) {
        Collection<ConfigAttribute> attributes = this.
securityMetadataSource.getAttributes(object);
        this.accessDecisionManager.decide(authentication.get(), object,
attributes);
    }
}
```

Subsequently, integrate it into your `SecurityFilterChain`.

Alternatively, if you wish to only invoke an `AccessDecisionVoter`, you can use:

```
@Component
public class AccessDecisionVoterAuthorizationManagerAdapter implements
AuthorizationManager {
    private final AccessDecisionVoter accessDecisionVoter;
    private final SecurityMetadataSource securityMetadataSource;

    @Override
    public AuthorizationDecision check(Supplier<Authentication>
authentication, Object object) {
        Collection<ConfigAttribute> attributes = this.
securityMetadataSource.getAttributes(object);
        int decision = this.accessDecisionVoter.vote(authentication.
get(), object, attributes);
        switch (decision) {
            case ACCESS_GRANTED:
                return new AuthorizationDecision(true);
            case ACCESS_DENIED:
                return new AuthorizationDecision(false);
        }
        return null;
    }
}
```

Afterward, integrate it into your `SecurityFilterChain`.

Legacy Authorization Components

In this section, we will take a closer look at certain authorization components that existed in `Spring Security` but have been deprecated with the introduction of `Spring Security 6`.

The AccessDecisionManager

The `AbstractSecurityInterceptor` invokes the `AccessDecisionManager`, which is tasked with making conclusive access control decisions. The `AccessDecisionManager` interface encompasses three methods:

```
void decide(Authentication authentication, Object secureObject,
        Collection<ConfigAttribute> attrs) throws
AccessDeniedException;

boolean supports(ConfigAttribute attribute);

boolean supports(Class clazz);
```

The `decide` method of the `AccessDecisionManager` receives all the pertinent information required to make an authorization decision. Specifically, passing the secure object allows the inspection of arguments within the actual invocation of the secure object. For instance, if the secure object is a `MethodInvocation`, you can inquire about any `Customer` argument in the `MethodInvocation` and then implement security logic in the `AccessDecisionManager` to verify whether the principal is authorized to operate on that customer. Implementations are expected to throw an `AccessDeniedException` if access is denied.

The `supports(ConfigAttribute)` method is invoked by the `AbstractSecurityInterceptor` during startup to determine whether the `AccessDecisionManager` can handle the provided `ConfigAttribute`. The `supports(Class clazz)` method is called by a security interceptor implementation to ensure that the configured `AccessDecisionManager` supports the type of secure object presented by the security interceptor.

AccessDecisionManager Implementations Based on Voting

Although users have the flexibility to implement their own `AccessDecisionManager` to oversee all aspects of authorization, `Spring Security` provides various `AccessDecisionManager` implementations grounded in a voting mechanism. The relevant classes are explained in the **Voting Decision Manager**.

The `AccessDecisionManager` interface is illustrated in the following figure:

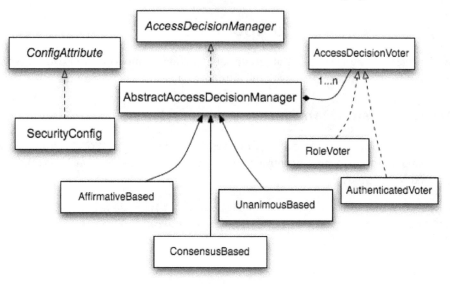

Figure 13.2 – Voting Decision Manager

Through this approach, a sequence of `AccessDecisionVoter` implementations is consulted for an authorization decision. The `AccessDecisionManager` subsequently determines whether or not to throw an `AccessDeniedException` based on its evaluation of the votes.

`Spring Security` provides three concrete `AccessDecisionManager` implementations to aggregate votes. The `ConsensusBased` implementation allows or denies access based on the consensus of non-abstain votes. Configurable properties govern behavior in case of vote equality or if all votes abstain. The `AffirmativeBased` implementation grants access if one or more `ACCESS_GRANTED` votes are received (ignoring deny votes as long as there is at least one grant vote). Similar to `ConsensusBased`, it has a parameter controlling behavior if all voters abstain. The `UnanimousBased` implementation requires unanimous `ACCESS_GRANTED` votes for access, disregarding abstains. It denies access with any `ACCESS_DENIED` vote. Like the others, it has a parameter governing behavior if all voters abstain.

Custom `AccessDecisionManager` instances can be implemented to customize vote tallying. For instance, votes from a specific `AccessDecisionVoter` might carry additional weight, and a deny vote from a particular voter could have a veto effect.

RoleVoter

The `RoleVoter`, the most commonly utilized `AccessDecisionVoter` provided by `Spring Security`, interprets configuration attributes as role names and votes to grant access if the user has been assigned that role.

It casts a vote if any `ConfigAttribute` starts with the `ROLE_` prefix. Access is granted if there is a `GrantedAuthority` that returns a String representation (via the `getAuthority()` method) exactly matching one or more `ConfigAttribute` instances, starting with the `ROLE_` prefix. If there is no precise match for any `ConfigAttribute` starting with `ROLE_`, `RoleVoter` votes to deny access. If no `ConfigAttribute` starts with `ROLE_`, the voter abstains.

AuthenticatedVoter

Another implicit voter is the `AuthenticatedVoter`, useful for distinguishing between *anonymous*, *fully-authenticated*, and *remember-me* authenticated users. Many websites allow limited access under *remember-me* authentication but necessitate user confirmation of identity by logging in for full access.

The processing of the `IS_AUTHENTICATED_ANONYMOUSLY` attribute for granting anonymous access is handled by the `AuthenticatedVoter`, as seen in previous examples.

Custom Voters

Implementing a custom `AccessDecisionVoter` enables the inclusion of virtually any access control logic. It may be tailored to your application, related to business logic, or involve security administration logic. For instance, a blog article on the Spring website outlines using a voter to deny real-time access to users with suspended accounts.

Expression-based request authorization

As you might expect, **SpEL** handling is supplied by a different `Voter` implementation, `o.s.s.web.access.expression.WebExpressionVoter`, which understands how to evaluate the SpEL expressions. The `WebExpressionVoter` class relies on an implementation of the `SecurityExpressionHandler` interface for this purpose. The `SecurityExpressionHandler` interface is responsible both for evaluating the expressions and for supplying the security-specific methods that are referenced in the expressions. The default implementation of this interface exposes methods defined in the `o.s.s.web.access.expression.WebSecurityExpressionRoot` class.

The flow and relationship between these classes are shown in the following diagram:

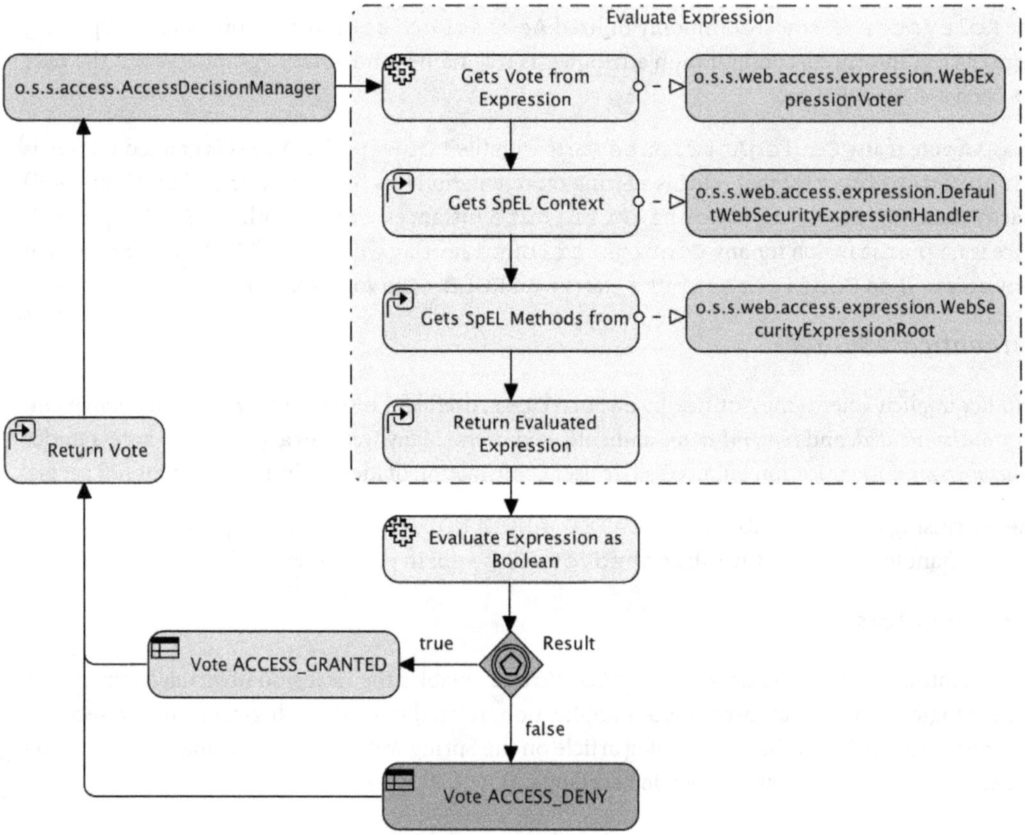

Figure 13.3 – Relationship between WebSecurityExpressionRoot and AccessDecisionManager

Now that we know how request authorization works, let's solidify our understanding by making a few custom implementations of some key interfaces.

The real power of `Spring Security`'s authorization is demonstrated by how adaptable it is to custom requirements. Let's explore a few scenarios that will help reinforce our understanding of the overall architecture.

Dynamically defining access control to URLs

`Spring Security` provides several methods for mapping `ConfigAttribute` objects to a resource. For example, the `requestMatchers()` method ensures it is simple for developers to restrict access to specific HTTP requests in their web applications. Behind the scenes, an implementation of `o.s.s.acess.SecurityMetadataSource` is populated with these mappings and queried to determine what is required in order to be authorized to make any given HTTP request.

While the `requestMatchers()` method is very simple, there may be times when it would be desirable to provide a custom mechanism for determining the URL mappings. An example of this might be if an application needs to be able to dynamically provide access control rules. Let's demonstrate what it would take to move our URL authorization configuration into a database.

Configuring the RequestConfigMappingService

The first step is to be able to obtain the necessary information from the database. This will replace the logic that reads in the `requestMatchers()` methods from our security bean configuration. In order to do this, the chapter's sample code contains `JpaRequestConfigMappingService`, which will obtain a mapping of an `Ant Pattern` and an expression from the database represented as `RequestConfigMapping`. The rather simple implementation is as follows:

```
//src/main/java/com/packtpub/springsecurity/web/access/intercept/
JpaRequestConfigMappingService.java
@Repository
public class JpaRequestConfigMappingService implements
RequestConfigMappingService {

    private final SecurityFilterMetadataRepository
securityFilterMetadataRepository;

    public JpaRequestConfigMappingService(final
SecurityFilterMetadataRepository securityFilterMetadataRepository) {
        this.securityFilterMetadataRepository =
securityFilterMetadataRepository;
    }

    public List<RequestConfigMapping> getRequestConfigMappings() {
        return securityFilterMetadataRepository
            .findAll()
            .stream()
            .sorted(Comparator.
comparingInt(SecurityFilterMetadata::getSortOrder))
            .map(md -> new RequestConfigMapping(
                new AntPathRequestMatcher(md.getAntPattern()),
                new SecurityConfig(md.getExpression()))).
toList();
    }

}
```

It is important to notice that, just as with the `requestMatchers()` methods, order matters. Therefore, we ensure the results are sorted by the `sort_order` column. The service creates an `AntRequestMatcher` and associates it with `SecurityConfig`, an instance of `ConfigAttribute`. This will provide a mapping of the HTTP request to `ConfigAttribute` objects that can be used by `Spring Security` to secure our URLs.

We need to create a domain object to use for **Jakarta Persistence (JPA)** to map to, as follows:

```java
//src/main/java/com/packtpub/springsecurity/domain/
SecurityFilterMetadata.java
@Entity
@Table(name = "security_filter_metadata")
public class SecurityFilterMetadata implements Serializable {

    @Id
    @GeneratedValue(strategy = GenerationType.AUTO)
    private Integer id;

    private String antPattern;
    private String expression;
    private Integer sortOrder;
... setters / getters ...

}
```

Finally, we need to create a Spring Data repository object, as follows:

```java
//src/main/java/com/packtpub/springsecurity/repository/
SecurityFilterMetadataRepository.java
public interface SecurityFilterMetadataRepository extends
JpaRepository<SecurityFilterMetadata, Integer> {}
```

In order for the new service to work, we will need to initialize our database with the schema and the access control mappings, just as with the service implementation. The `security_filter_metadata` table schema can be auto-generated by `spring-data-jpa`.

We can then use the same `requestMatchers()` mappings from our `SecurityConfig.java` file to produce the `data.sql` file:

```sql
//src/main/resources/data.sql
insert into security_filter_metadata(id,ant_pattern,expression,sort_order) values (115, '/','permitAll',15);
insert into security_filter_metadata(id,ant_pattern,expression,sort_order) values (120, '/login/*','permitAll',20);
insert into security_filter_metadata(id,ant_pattern,expression,sort_order) values (130, '/logout','permitAll',30);
insert into security_filter_metadata(id,ant_pattern,expression,sort_
```

```
order) values (140, '/signup/*','permitAll',40);
insert into security_filter_metadata(id,ant_pattern,expression,sort_
order) values (150, '/errors/**','permitAll',50);
insert into security_filter_metadata(id,ant_pattern,expression,sort_
order) values (160, '/admin/**','hasRole("ADMIN")',60);
insert into security_filter_metadata(id,ant_pattern,expression,sort_
order) values (170, '/events/','hasRole("ADMIN")',70);
insert into security_filter_metadata(id,ant_pattern,expression,sort_
order) values (180, '/**','hasRole("USER")',80);
```

Once the `RequestConfigMappingService` is configured, we'll explore the implementation of a custom `SecurityMetadataSource`.

Custom SecurityMetadataSource implementation

In order for `Spring Security` to be aware of our URL mappings, we need to provide a custom `FilterInvocationSecurityMetadataSource` implementation. The `FilterInvocationSecurityMetadataSource` package extends the `SecurityMetadataSource` interface which, given a particular HTTP request, is what provides `Spring Security` with the information necessary for determining whether access should be granted. Let's take a look at how we can utilize our `RequestConfigMappingService` interface to implement a `SecurityMetadataSource` interface:

```java
//src/main/java/com/packtpub/springsecurity/web/access/intercept/
FilterInvocationServiceSecurityMetadataSource.java
@Component
public class FilterInvocationServiceSecurityMetadataSource implements
FilterInvocationSecurityMetadataSource {
    private FilterInvocationSecurityMetadataSource delegate;

    private final RequestConfigMappingService
requestConfigMappingService;

    public FilterInvocationServiceSecurityMetadataSource
(RequestConfigMappingService filterInvocationService) {
        this.requestConfigMappingService = filterInvocationService;
    }

    public Collection<ConfigAttribute> getAllConfigAttributes() {
        return this.delegate.getAllConfigAttributes();
    }

    public Collection<ConfigAttribute> getAttributes(Object object) {
        if (delegate == null)
            getDelegate();
        return this.delegate.getAttributes(object);
```

```
    }

    public boolean supports(Class<?> clazz) {
        return this.delegate.supports(clazz);
    }

    public void getDelegate() {
        List<RequestConfigMapping> requestConfigMappings =
requestConfigMappingService.getRequestConfigMappings();
        LinkedHashMap<RequestMatcher, Collection<ConfigAttribute>>
requestMap = new LinkedHashMap<RequestMatcher,
Collection<ConfigAttribute>>(requestConfigMappings.size());
        for (RequestConfigMapping requestConfigMapping :
requestConfigMappings) {
            RequestMatcher matcher = requestConfigMapping.
getMatcher();
            requestMap.put(matcher, requestConfigMapping.
getAttributes());
        }
        this.delegate = new
ExpressionBasedFilterInvocationSecurityMetadataSource(requestMap, new
DefaultWebSecurityExpressionHandler());
    }
}
```

We are able to use our `RequestConfigMappingService` interface to create a map of `RequestMatcher` objects that map to `ConfigAttribute` objects. We then delegate to an instance of `ExpressionBasedFilterInvocationSecurityMetadataSource` to do all the work. For simplicity, the current implementation would require restarting the application to pick up changes. However, with a few minor changes, we could avoid this inconvenience.

Registering a custom SecurityMetadataSource

Now, all that is left is for us to configure `FilterInvocationServiceSecurityMetadataSource`. The only problem is that `Spring Security` does not support configuring a custom `FilterInvocationServiceSecurityMetadataSource` interface directly. This is not too difficult, so we will register this `SecurityMetadataSource` with our `FilterSecurityInterceptor` in our `SecurityConfig` file:

```
//src/main/java/com/packtpub/springsecurity/configuration/
SecurityConfig.java
@Configuration
@EnableWebSecurity
@EnableMethodSecurity
public class SecurityConfig {
```

```
@Bean
public SecurityFilterChain filterChain(HttpSecurity http,
        FilterInvocationServiceSecurityMetadataSource metadataSource,
        AccessDecisionManager accessDecisionManager) throws Exception {
    http.authorizeRequests().anyRequest().authenticated();
    http.authorizeRequests().
accessDecisionManager(accessDecisionManager);
    http.authorizeRequests()
            .withObjectPostProcessor(new
ObjectPostProcessor<FilterSecurityInterceptor>() {
            public <O extends FilterSecurityInterceptor> O
postProcess(
                    O fsi) {
                fsi.setPublishAuthorizationSuccess(true);
                fsi.setSecurityMetadataSource(metadataSource);
                return fsi;
            }
        });
...omitted for brevity
    return http.build();
}
```

This sets up our custom SecurityMetadataSource interface with the FilterSecurityInterceptor object as the default metadata source.

Now that the database is being used to map our security configuration, we can remove the requestMatchers() method from our SecurityConfig.java file.

You should now be able to start the application and test to ensure that our URLs are secure, as they should be. Our users will not notice a difference, but we know that our URL mappings are persisted in a database now.

> **Important note**
> Your code should now look like this: calendar13.01-calendar.

Creating a custom expression

The o.s.s.access.expression.SecurityExpresssionHandler interface is how Spring Security abstracts how the Spring expressions are created and initialized. Just as with the SecurityMetadataSource interface, there is an implementation for creating expressions for web requests and creating expressions for securing methods. In this section, we will explore how we can easily add new expressions.

Configuring a custom SecurityExpressionRoot

Let's assume that we want to support a custom Web Expression named isLocal that will return true if the host is localhost and false otherwise. This new method could be used to provide additional security for our SQL console by ensuring that it is only accessed from the same machine that the web application is deployed from.

This is an artificial example that does not add any security benefits since the host comes from the headers of the HTTP request. This means a malicious user could inject a header stating that the host is localhost even if they are requesting to an external domain.

All of the expressions that we have seen are available because the SecurityExpressionHandler interface makes them available via an instance of o.s.s.access.expression. SecurityExpressionRoot. If you open this object, you will find the methods and properties we use in Spring expressions (that is, hasRole, hasPermission, and so on), which are common in both web and method security. A subclass provides the methods that are specific to the web and method expressions. For example, o.s.s.web.access.expression.WebSecurityExpressionRoot provides the hasIpAddress method for web requests.

To create a custom web SecurityExpressionhandler, we will first need to create a subclass of WebSecurityExpressionRoot that defines our isLocal method, as follows:

```
//src/main/java/com/packtpub/springsecurity/web/access/expression/
CustomWebSecurityExpressionRoot.java
public class CustomWebSecurityExpressionRoot extends
WebSecurityExpressionRoot {

    public CustomWebSecurityExpressionRoot(Authentication a,
FilterInvocation fi) {
        super(a, fi);
    }

    public boolean isLocal() {
        return "localhost".equals(request.getServerName());
    }

}
```

> **Important note**
>
> It is important to note that getServerName() returns the value that is provided in the Host header value. This means that a malicious user can inject a different value into the header to bypass constraints. However, most application servers and proxies can enforce the value of the Host header. Please read the appropriate documentation before leveraging such an approach to ensure that malicious users do not inject a Host header value to bypass such a constraint.

Configuring a custom SecurityExpressionHandler

In order for our new method to become available, we need to create a custom `SecurityExpressionHandler` interface that utilizes our new root object. This is as simple as extending `WebSecurityExpressionHandler`, as follows:

```
//src/main/java/com/packtpub/springsecurity/web/access/expression/
CustomWebSecurityExpressionHandler.java
@Component
public class CustomWebSecurityExpressionHandler extends
DefaultWebSecurityExpressionHandler {

    private final AuthenticationTrustResolver trustResolver = new
AuthenticationTrustResolverImpl();

    @Override
    protected SecurityExpressionOperations
createSecurityExpressionRoot(Authentication authentication,
FilterInvocation fi) {
        WebSecurityExpressionRoot root = new
CustomWebSecurityExpressionRoot(authentication, fi);
        root.setPermissionEvaluator(getPermissionEvaluator());
        root.setTrustResolver(trustResolver);
        root.setRoleHierarchy(getRoleHierarchy());
        return root;
    }
}
```

We perform the same steps that the superclass does, except that we use `CustomWebSecurityExpressionRoot`, which contains the new method.

The `CustomWebSecurityExpressionRoot` becomes the root of our SpEL expression.

> **Important note**
>
> For further details, refer to the SpEL documentation within the Spring Reference documentation: https://docs.spring.io/spring-framework/reference/core/expressions.html.

Configuring and using CustomWebSecurityExpressionHandler

We now need to configure `CustomWebSecurityExpressionHandler`. Fortunately, this can be done easily using the `Spring Security` namespace configuration support. Add the following configuration to the `SecurityConfig.java` file:

```
// Web Expression Handler:
http.authorizeRequests()
        .expressionHandler(customWebSecurityExpressionHandler);
```

Now, let's update our initialization SQL query to use the new expression. Update the `data.sql` file so that it requires the user to be ROLE_ADMIN and requested from the local machine. You will notice that we are able to write local instead of `isLocal`, since SpEL supports Java Bean conventions:

```
//src/main/resources/data.sql
insert into security_filter_metadata(id,ant_pattern,expression,sort_
order) values (160, '/admin/**','local and hasRole("ADMIN")',60);
```

Restart the application and access the H2 console using `localhost:8080/admin/h2` and `admin1@example.com/admin1` to see the admin console.

If the H2 console is accessed using `127.0.0.1:8080/admin/h2` and `admin1@example.com/admin1`, the **Access Denied** page will be displayed.

> **Important note**
> Your code should now look like this: `calendar13.02-calendar`.

Alternative to a CustomWebSecurityExpressionHandler

Following an examination of the `CustomWebSecurityExpressionHandler` usage, we will investigate alternative approaches by employing a custom `PermissionEvaluator` to enhance the security of our `CalendarService`.

How does method security work?

The access decision mechanism for method security—whether or not a given request is allowed—is conceptually the same as the access decision logic for web request access. `AccessDecisionManager` polls a set of `AccessDecisionVoter` instances, each of which can provide a decision to grant or deny access or abstain from voting. The specific implementation of `AccessDecisionManager` aggregates the voter decisions and arrives at an overall decision to allow for the method invocation.

Web request access decision-making is less complicated, due to the fact that the availability of servlet filters makes the interception (and summary rejection) of securable requests relatively straightforward. As method invocation can happen from anywhere, including areas of code that are not directly configured by Spring Security, the Spring Security designers chose to use a Spring-managed **Aspect-Oriented Programming (AOP)** approach to recognize, evaluate, and secure method invocations.

The following high-level flow illustrates the main players involved in authorization decisions for method invocation:

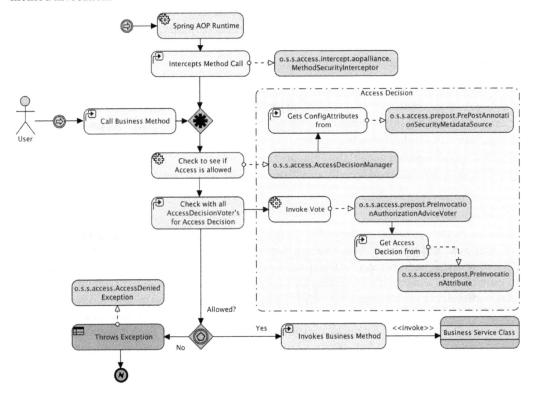

Figure 13.4 – Main classes involved in authorization decisions for method invocation

We can see that Spring Security's o.s.s.access.intercept.aopalliance. MethodSecurityInterceptor is invoked by the standard Spring AOP runtime to intercept method calls of interest. From here, the logic of whether or not to allow a method call is relatively straightforward, as per the previous flow diagram.

At this point, we might wonder about the performance of the method security feature. Obviously, `MethodSecurityInterceptor` can't be invoked for every method call in the application—so how do annotations on methods or classes result in AOP interception?

First of all, AOP proxying isn't invoked for all Spring-managed beans by default. Instead, if `@EnableMethodSecurity` is defined in the Spring Security configuration, a standard Spring AOP `o.s.beans.factory.config.BeanPostProcessor` will be registered that will introspect the AOP configuration to see whether any AOP advisors indicate that proxying (and the interception) is required. This workflow is standard Spring AOP handling (known as **AOP auto-proxying**) and doesn't inherently have any functionality specific to Spring Security. All registered `BeanPostProcessor` instances run upon initialization of the Spring `ApplicationContext`; after all, Spring bean configurations have occurred.

The AOP auto-proxy functionality queries all registered `PointcutAdvisor` instances, to see if there are AOP pointcuts that resolve method invocations that should have AOP advice applied. Spring Security implements the `o.s.s.access.intercept.aopalliance.MethodSecurityMetadataSourceAdvisor` class, which examines any and all configured method security annotations and sets up appropriate AOP interception. Take note that only interfaces or classes with declared method security annotations will be proxied for AOP!

> **Important note**
>
> Be aware that it is strongly encouraged to declare AOP rules (and other security annotations) on interfaces, and not on implementation classes. The use of classes, while available using CGLIB proxying with Spring, may unexpectedly change the behavior of your application, and is generally less semantically correct than security declarations (through AOP) on interfaces. `MethodSecurityMetadataSourceAdvisor` delegates the decision to affect methods with the AOP advice to an `o.s.s.access.method.MethodSecurityMetadataSource` instance. The different forms of method security annotation each have their own `MethodSecurityMetadataSource` implementation, which is used to introspect each method, class in turn, and add AOP advice to be executed at runtime.

The following diagram illustrates how this process occurs:

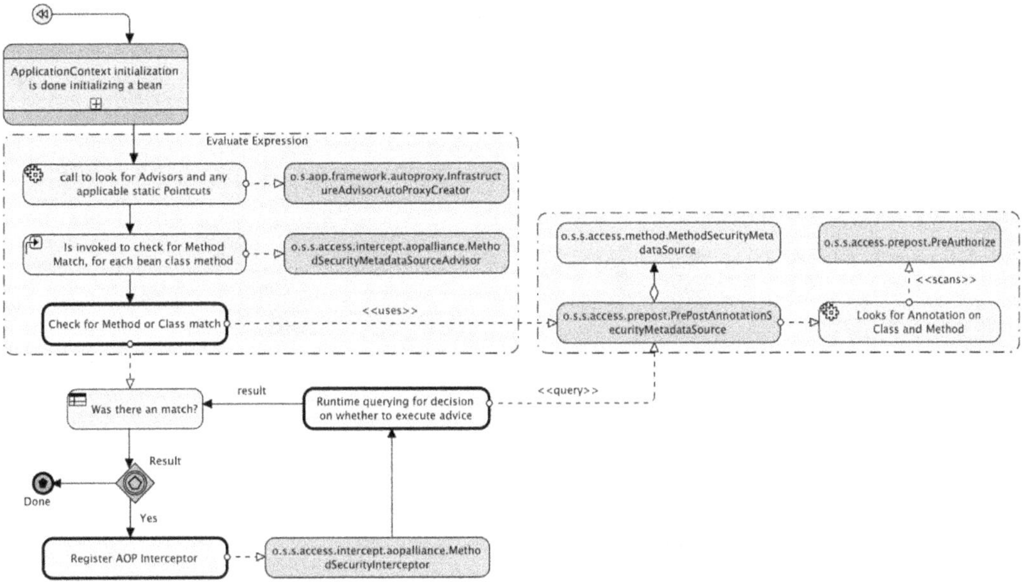

Figure 13.5 – AOP interceptors for method security

Depending on the number of Spring beans configured in your application and the number of secured method annotations you have, adding method security proxying may increase the time required to initialize your `ApplicationContext`. Once your Spring context is initialized, however, there is a negligible performance impact on individual proxied beans.

Now that we have an understanding of how we can use AOP to apply `Spring Security`, let's strengthen our grasp of `Spring Security` authorization by creating a custom `PermissionEvaluator`.

Creating a custom PermissionEvaluator

In the previous chapter, we demonstrated that we could use `Spring Security`'s built-in `PermissionEvaluator` implementation, `AclPermissionEvaluator`, to restrict access to our application. While powerful, this can often be more complicated than necessary. We have also discovered how `SpEL` can formulate complex expressions that are able to secure our application. While simple, one of the downsides of using complex expressions is that the logic is not centralized. Fortunately, we can easily create a custom `PermissionEvaluator` that is able to centralize our authorization logic and still avoid the complexity of using ACLs.

The CalendarPermissionEvaluator Class

A simplified version of our custom `PermissionEvaluator` that does not contain any validation can be seen as follows:

```java
//src/main/java/com/packtpub/springsecurity/access/
CalendarPermissionEvalua tor.java
public final class CalendarPermissionEvaluator implements
PermissionEvaluator {

    private final EventDao eventDao;

    public CalendarPermissionEvaluator(EventDao eventDao) {
        this.eventDao = eventDao;
    }

    @Override
    public boolean hasPermission(Authentication authentication, Object
targetDomainObject, Object permission) {
        if(targetDomainObject instanceof Event) {
            return hasPermission(authentication, (Event)
targetDomainObject, permission);
        }
        return targetDomainObject == null;
    }

    @Override
    public boolean hasPermission(Authentication authentication,
Serializable targetId, String targetType,
                                 Object permission) {
        if(!Event.class.getName().equals(targetType)) {
            throw new IllegalArgumentException("targetType is not
supported. Got "+targetType);
        }
        if(!(targetId instanceof Integer)) {
            throw new IllegalArgumentException("targetId type is not
supported. Got "+targetType);
        }
        Event event = eventDao.getEvent((Integer)targetId);
        return hasPermission(authentication, event, permission);
    }

    private boolean hasPermission(Authentication authentication, Event
event, Object permission) {
        if(event == null) {
            return true;
```

```
        }
        String currentUserEmail = authentication.getName();
        String ownerEmail = extractEmail(event.getOwner());
        if("write".equals(permission)) {
            return currentUserEmail.equals(ownerEmail);
        } else if("read".equals(permission)) {
            String attendeeEmail = extractEmail(event.getAttendee());
            return currentUserEmail.equals(attendeeEmail) ||
currentUserEmail.equals(ownerEmail);
        }
        throw new IllegalArgumentException("permission "+permission+"
is not supported.");
    }

    private String extractEmail(CalendarUser user) {
        if(user == null) {
            return null;
        }
        return user.getEmail();
    }
}
```

The logic is fairly similar to the Spring expressions that we have already used, except that it differentiates read and write access. If the current user's username matches the owner's email of the Event object, then both read and write access is granted. If the current user's email matches the attendee's email, then read access is granted. Otherwise, access is denied.

> **Important note**
>
> It should be noted that a single PermissionEvaluator is used for every domain object. So, in a real-world situation, we must perform instanceof checks first. For example, if we were also securing our CalendarUser objects, these could be passed into this same instance. For a full example of these minor changes, refer to the sample code included in the book.

Configuring CalendarPermissionEvaluator

We can then leverage the CustomAuthorizationConfig.java configuration that is provided with this chapter to provide an ExpressionHandler that uses our CalendarPermissionEvaluator, like so:

```
//src/main/java/com/packtpub/springsecurity/configuration/
CustomAuthorizationConfig.java
@Bean
public DefaultMethodSecurityExpressionHandler
defaultExpressionHandler(EventDao eventDao){
```

```
    DefaultMethodSecurityExpressionHandler deh = new
DefaultMethodSecurityExpressionHandler();
    deh.setPermissionEvaluator(
            new CalendarPermissionEvaluator(eventDao));
    return deh;
}
```

The configuration should look similar to the configuration from *Chapter 12, Access Control Lists*, except that we now use our `CalendarPermissionEvalulator` class instead of `AclPermissionEvaluator`.

Securing our CalendarService

Lastly, we can secure our `CalendarService getEvent(int eventId)` method with a `@PostAuthorize` annotation. You will notice that this step is exactly the same as what we did in *Chapter 1, Anatomy of an Unsafe Application*, and we have only changed the implementation of `PermissionEvaluator`:

```
//src/main/java/com/packtpub/springsecurity/service/CalendarService.
java

@PostAuthorize("hasPermission(returnObject,'read')")
Event getEvent(int eventId);
```

If you have not done so already, restart the application, log in with the username/password admin1@example.com/admin1, and visit the **Conference Call Event** (events/101) using the link on the **Welcome** page. The **Access Denied** page will be displayed.

> **Important note**
>
> Your code should now look like this: `calendar13.03-calendar`.

However, we would like ROLE_ADMIN users to be able to access all events.

Benefits of a custom PermissionEvaluator

With only a single method being protected, it would be trivial to update the annotation to check whether the user has the role of ROLE_ADMIN or has permission. However, if we had protected all of our service methods that use an event, it would have become quite cumbersome. Instead, we could just update our `CalendarPermissionEvaluator`. Make the following changes:

```
    private boolean hasPermission(Authentication authentication, Event
event, Object permission) {
        if(event == null) {
            return true;
        }
// Custom Role verification
```

```
GrantedAuthority adminRole = new SimpleGrantedAuthority("ROLE_ADMIN");
if(authentication.getAuthorities().contains(adminRole)) {
    return true;
... omitted for brevity
}
  }
```

Now, restart the application and repeat the previous exercise. This time, the **Conference Call Event** will display successfully.

You can see that the ability to encapsulate our authorization logic can be extremely beneficial. However, sometimes, it may be useful to extend the expressions themselves.

> **Important note**
> Your code should now look like this: `calendar13.04-calendar`.

Removing the CustomWebSecurityExpressionHandler class

There is a much simpler way of defining custom **Web Expressions**.

In our previous example, you can remove the following classes: `CustomWebSecurity ExpressionHandler` and `CustomWebSecurityExpressionRoot`.

Declare a Spring bean containing the custom `Web Expression`:

```
//src/main/java/com/packtpub/springsecurity/access/expression/
CustomWebExpression.java
@Component
public class CustomWebExpression {

    public boolean isLocalHost(final HttpServletRequest request) {
        return "localhost".equals(request.getServerName());
    }
}
```

In the `CustomAuthorizationConfig` class, add the following bean:

```
//src/main/java/com/packtpub/springsecurity/configuration/
CustomAuthorizationConfig.java

@Bean
public DefaultWebSecurityExpressionHandler
customWebSecurityExpressionHandler (){
    return new DefaultWebSecurityExpressionHandler();
}
```

We can then remove the following declaration to the `CustomWebSecurityExpressionHandler` inside the `SecurityConfig` class:

```
//src/main/java/com/packtpub/springsecurity/configuration/
SecurityConfig.java
// Line of Expression Handler needs to be removed
http.authorizeRequests()
.expressionHandler(customWebSecurityExpressionHandler);
```

Now, let's update our initialization SQL query to adapt the syntax of the new expression. Update the `data.sql` as follows:

```
//src/main/resources/data.sql
insert into security_filter_metadata(id,ant_pattern,expression,sort_
order) values (160, '/admin/**', '@customWebExpression.
isLocalHost(request) and hasRole("ADMIN")',60);
```

Restart the application and test the application access with the user `admin1@example.com/admin1` for both URLs:

- `http://127.0.0.1:8080/admin/h2`: Access should be denied

- `http://localhost:8080/admin/h2`: Access should be allowed

> **Important note**
> Your code should now look like this: `calendar13.05-calendar`.

Declaring a custom AuthorizationManager

`Spring Security 6` has deprecated the usage of the `AccessDecissionManager` and `AccessDecisionVoter`.

The recommended approach is to implement a custom `AuthorizationManager`, as explained in the introduction of this chapter. To achieve this goal, you can follow the next steps.

First, we will create a custom implementation of the `AuthorizationManager` that checks for the allowed permissions based on the `security_filter_metadata` table definition:

```
//src/main/java/com/packtpub/springsecurity/access/
CustomAuthorizationManager.java
@Component
public class CustomAuthorizationManager implements
AuthorizationManager<RequestAuthorizationContext> {

    private final
SecurityExpressionHandler<RequestAuthorizationContext>
```

```
expressionHandler;

    private final RequestConfigMappingService
requestConfigMappingService;

    private static final Logger logger = LoggerFactory.
getLogger(CustomAuthorizationManager.class);

    public
CustomAuthorizationManager(DefaultHttpSecurityExpressionHandler
expressionHandler, RequestConfigMappingService
requestConfigMappingService) {
        this.expressionHandler = expressionHandler;
        this.requestConfigMappingService = requestConfigMappingService;
    }

    @Override
    public AuthorizationDecision check(Supplier<Authentication>
authentication, RequestAuthorizationContext context) {
        List<RequestConfigMapping> requestConfigMappings =
requestConfigMappingService.getRequestConfigMappings();
        LinkedHashMap<RequestMatcher, Collection<ConfigAttribute>>
requestMap = new LinkedHashMap<>(requestConfigMappings.size());

        for (RequestConfigMapping requestConfigMapping :
requestConfigMappings) {
            RequestMatcher matcher = requestConfigMapping.getMatcher();

            if (matcher.matches(context.getRequest())) {
                requestMap.put(matcher, requestConfigMapping.
getAttributes());

                String expressionStr = requestConfigMapping.
getAttributes().iterator().next().getAttribute();
                Expression expression = this.expressionHandler.
getExpressionParser().parseExpression(expressionStr);

                try {
                    EvaluationContext evaluationContext = this.
expressionHandler.createEvaluationContext(authentication, context);
                    boolean granted = ExpressionUtils.
evaluateAsBoolean(expression, evaluationContext);
                    return new ExpressionAuthorizationDecision(granted,
expression);
                } catch (AccessDeniedException ex) {
                    logger.error("Access denied exception: {}",
ex.getMessage());
```

```
                    return new AuthorizationDecision(false);
                }
            }
        }

        return new AuthorizationDecision(false);
    }
}
```

Then, we will inject the `AuthorizationManager` into the `SecurityFilterChain` bean, as follows:

```
//src/main/java/com/packtpub/springsecurity/configuration/
SecurityConfig.java

@Bean
public SecurityFilterChain filterChain(HttpSecurity
http, AuthorizationManager<RequestAuthorizationContext>
authorizationManager) throws Exception {
    http
            .authorizeHttpRequests(authorize -> authorize
                .anyRequest()
                .access(authorizationManager));

...omitted for brevity
    return http.build();
}
```

We will update the `CustomAuthorizationConfig` configuration by replacing the bean of type `DefaultWebSecurityExpressionHandler` with another one of type `DefaultHttpSecurityExpressionHandler`, as we have chosen to use `http.authorizeHttpRequest()` instead of `http.authorizeRequests()`:

```
//src/main/java/com/packtpub/springsecurity/configuration/
CustomAuthorizationConfig.java

@Bean
public DefaultHttpSecurityExpressionHandler
defaultHttpSecurityExpressionHandler(){
    return new DefaultHttpSecurityExpressionHandler();
}
```

You can remove the `FilterInvocationServiceSecurityMetadataSource` and restart your application. You should have the same results as the previous example.

> **Important note**
> Your code should now look like this: `calendar13.06-calendar`.

Summary

After reading this chapter, you should have a firm understanding of how `Spring Security` authorization works for HTTP requests and methods. With this knowledge, and the provided concrete examples, you should also know how to extend authorization to meet your needs. Specifically, in this chapter, we covered the `Spring Security` authorization architecture for both HTTP requests and methods. We also demonstrated how to configure secured URLs from a database.

We also saw how to create a custom `AuthorizationManager`, `PermissionEvaluator` object, and custom `Spring Security` expression.

In the next chapter, we will explore how `Spring Security` performs session management. We will also gain an understanding of how it can be used to restrict access to our application.

Part 5:
Advanced Security Features
and Deployment Optimization

This part commences with an explanation of session fixation attacks and Spring Security's defense mechanisms against them. It proceeds to explore methods for managing logged-in users and limiting the number of concurrent sessions per user. The association of a user to `HttpSession` by Spring Security and techniques for customizing this behavior are also detailed.

Then, we delve into common security vulnerabilities such as **Cross-Site Scripting (XSS)**, **Cross-Site Request Forgery (CSRF)**, synchronizer tokens, and clickjacking, along with strategies to mitigate these risks effectively.

Following this, we present a migration path to Spring Security 6, highlighting notable configuration changes, class and package migrations, and significant new features, including support for Java 17 and enhanced authentication mechanisms with OAuth 2.

Subsequently, we explore microservices-based architectures and examine the role of OAuth 2 with **JSON Web Tokens (JWTs)** in securing microservices within a Spring-based application. Additionally, we discuss the implementation of **Single Sign-On (SSO)** using the **Central Authentication Service (CAS)**.

Concluding this part, we delve into the process of building native images using GraalVM, offering insights into enhancing performance and security within Spring Security applications.

This part has the following chapters:

- *Chapter 14, Session Management*
- *Chapter 15, Additional Spring Security Features*
- *Chapter 16, Migration to Spring Security 6*
- *Chapter 17, Microservice Security with OAuth 2 and JSON Web Tokens*
- *Chapter 18, Single Sign-On with the Central Authentication Service*
- *Chapter 19, Build GraalVM Native Images*

14

Session Management

This chapter discusses Spring Security's session management functionality. It starts off with an example of how Spring Security defends against session fixation. We will then discuss how concurrency control can be leveraged to restrict access to software licensed on a per-user basis. We will also see how session management can be leveraged for administrative functions. Last, we will explore how `HttpSession` is used in Spring Security and how we can manage sessions:

The following is a list of topics that will be covered in this chapter:

- Session management/session fixation
- Concurrency control
- Managing logged-in users
- How `HttpSession` is used in Spring Security and how to control creation
- How to use the `DebugFilter` class to discover where `HttpSession` was created

This chapter's code in action link is here: `https://packt.link/qaJyz`.

Configuring session fixation protection

As we are using the security namespace style of configuration, session fixation protection is already configured on our behalf. If we wanted to explicitly configure it to mirror the default settings, we would do the following:

```
http.sessionManagement(session -> session.sessionFixation().
migrateSession());
```

Session fixation protection is a feature of the framework that you most likely won't even notice unless you try to act as a malicious user. We'll show you how to simulate a session-stealing attack; before we do, it's important to understand what session fixation does and the type of attack it prevents.

Understanding session fixation attacks

Session fixation is a type of attack whereby a malicious user attempts to steal the session of an unauthenticated user of your system. This can be done by using a variety of techniques that result in the attacker obtaining the unique session identifier of the user (for example, JSESSIONID). If the attacker creates a cookie or a URL parameter with the user's JSESSIONID identifier in it, they can access the user's session.

Although this is obviously a problem, typically, if a user is unauthenticated, they haven't entered any sensitive information. This becomes a more critical problem if the same session identifier continues to be used after a user has been authenticated. If the same identifier is used after authentication, the attacker may now gain access to the authenticated user's session without even having to know their username or password!

> **Important note**
>
> At this point, you may scoff in disbelief and think this is extremely unlikely to happen in the real world. In fact, session-stealing attacks happen frequently. We would suggest that you spend some time reading the very informative articles and case studies on the subject, published by the **Open Web Application Security Project (OWASP)** organization (http://www.owasp. org/). Specifically, you will want to read the OWASP top 10 lists. Attackers and malicious users are real, and they can do very real damage to your users, your application, or your company if you don't understand the techniques that they commonly use.

The following diagram illustrates how a session fixation attack works:

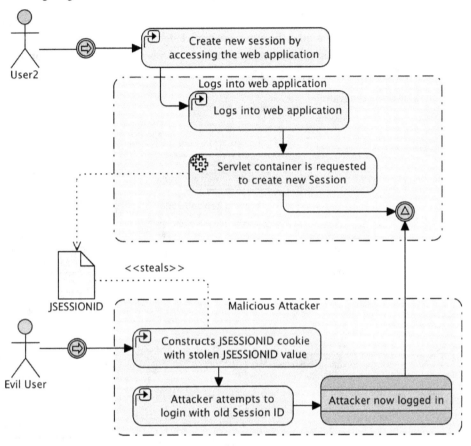

Figure 14.1 – Session fixation attack

In this diagram, we see that by fixing the session identifier to a known value, the attacker bypasses the normal authentication process and gains unauthorized access to the victim's account or session. This type of attack underscores the importance of properly managing and securing session identifiers to prevent session fixation vulnerabilities.

Now that we have seen how an attack like this works, we'll see what Spring Security can do to prevent it.

Preventing session fixation attacks with Spring Security

If we can prevent the same session that the user had prior to authentication from being used after authentication, we can effectively render the attacker's knowledge of the session ID useless. Spring Security session fixation protection solves this problem by explicitly creating a new session when a user is authenticated and invalidating their old session.

Let's look at the following diagram:

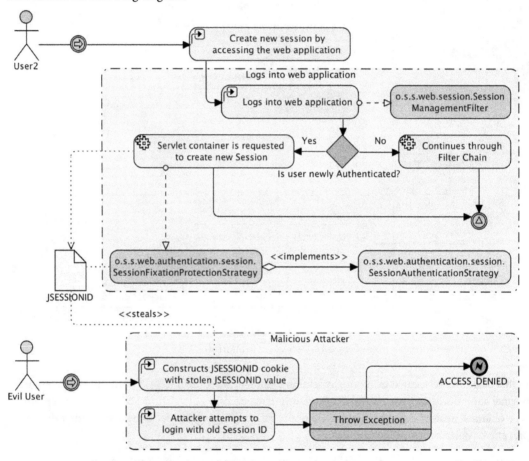

Figure 14.2 – Preventing session fixation attacks with Spring Security

We can see that a new filter, `o.s.s.web.session.SessionManagementFilter`, is responsible for evaluating if a particular user is newly authenticated. If the user is newly authenticated, a configured `o.s.s.web.authentication.session.SessionAuthenticationStrategy` interface determines what to do. `o.s.s.web.authentication.session.SessionFixation ProtectionStrategy` will create a new session (if the user already had one) and copy the contents of the existing session to the new one. That's pretty much it—seems simple. However, as we can see in the preceding diagram, it effectively prevents the malicious user from reusing the session ID after the unknown user is authenticated.

Simulating a session fixation attack

At this point, you may want to see what's involved in simulating a session fixation attack:

> **Important note**
> You should start with the code from `chapter14.00-calendar`.

1. You'll first need to disable session fixation protection in the `SecurityConfig.java` file by adding the `sessionManagement()` method as a child of the HTTP element.

    Let's take a look at the following code snippet:

    ```
    //src/main/java/com/packtpub/springsecurity/configuration/
    SecurityConfig.java
    http.sessionManagement(session -> session.sessionFixation().
    none());
    ```

> **Important note**
> Your code should now look like `chapter14.01-calendar`.

2. Next, you'll need to open two browsers. We'll initiate the session in Google Chrome, steal it from there, and our attacker will log in using the stolen session in Firefox. We will use the *Google Chrome* and *Firefox Web Developer* add-ons in order to view and manipulate cookies.

3. Open the JBCP calendar home page in Google Chrome.

4. Next:

    - **Open Developer Tools**: Right-click on the web page and select **Inspect**, or press *Ctrl + Shift + I* (Windows/Linux) or *Cmd + Opt + I* (Mac) to open the Developer Tools.

    - **Navigate to the Application Tab**: In the Developer Tools, you will see a menu at the top. Click on the **Application** tab.

    - **Locate Cookies in the Sidebar**: On the left sidebar, you should see a **Cookies** section. Expand it to see the list of domains with their associated cookies.

- **Select the Specific Domain**: Click on the domain relevant to the website you are interested in. This will display the list of cookies associated with that domain.

- **View the Cookie Values**: You can see the details of each cookie, including its name, value, domain, path, and so on. Look for the specific cookie you are interested in, and you will find its value.

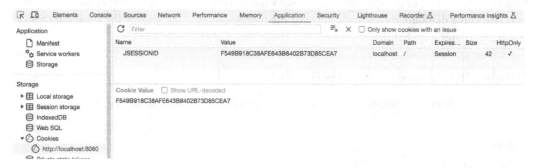

Figure 14.3 – Cookies explorer in Google Chrome

5. Select the `JSESSIONID` cookie, copy the value of **Content** to the clipboard, and log in to the JBCP calendar application. If you repeat the **View Cookie Information** command, you'll see that `JSESSIONID` did not change after you logged in, making you vulnerable to a session fixation attack!

6. In Firefox, open the JBCP calendar website. You will have been assigned a session cookie, which you can view by using *Ctrl + F2* to open the **bottom: Cookie** console. Then, type in *cookie list [enter]* to bring up cookies for the current page.

7. To complete our hack, we'll click on the **Edit Cookie** option and paste in the `JSESSIONID` cookie that we copied to the clipboard from Google Chrome, as shown in the following screenshot:

Figure 14.4 – Cookies hack in Firefox

8. Keep in mind that newer versions of Firefox include web developer tools, too. However, you will need to ensure that you are using the extension and not the built-in one, as it provides additional capabilities.

Our session fixation hack is complete! If you now reload the page in Firefox, you will see that you are logged in as the same user who was logged in using Google Chrome, but without the knowledge of the username and password. Are you scared of malicious users yet?

9.   Now, re-enable session fixation protection and try this exercise again. You'll see that, in this case, the JSESSIONID changes after the user logs in. Based on our understanding of how session fixation attacks occur, this means that we have reduced the likelihood of an unsuspecting user falling victim to this type of attack. Excellent job!

Cautious developers should note that there are many methods of stealing session cookies, some of which—such as XSS—may make even session fixation-protected sites vulnerable. Please consult the OWASP site for additional resources on preventing these types of attacks.

Comparing the session-fixation-protection options

The session-fixation-protection attribute has the following three options that allow you to alter its behavior; they are as follows:

Attribute value	Description
none()	This option disables session fixation protection and (unless other sessionManagement() attributes are non-default) does not configure SessionManagementFilter.
migrateSession()	When the user is authenticated and a new session is allocated, it ensures that all attributes of the old session are moved to the new session.
newSession()	When the user is authenticated, a new session is created and no attributes from the old (unauthenticated) session will be migrated.

Table 14.1 – session-fixation-protection options

In most cases, the default behavior of migrateSession() will be appropriate for sites that wish to retain important attributes of the user's session (such as click interest and shopping carts) after the user has been authenticated.

Restricting the number of concurrent sessions per user

In the software industry, software is often sold on a per-user basis. This means that, as software developers, we have an interest in ensuring that only a single session per user exists, to combat the sharing of accounts. Spring Security's concurrent session control ensures that a single user cannot have more than a fixed number of active sessions simultaneously (typically one). Ensuring that this maximum limit is enforced involves several components working in tandem to accurately track changes in user session activity.

Let's configure the feature, review how it works, and then test it out!

Configuring concurrent session control

Now that we have understood the different components involved in concurrent session control, setting it up should make much more sense. Let's take a look at the following steps to configure concurrent session control:

1. Firstly, you update your `SecurityConfig.java` file as follows:

    ```
    //src/main/java/com/packtpub/springsecurity/configuration/
    SecurityConfig.java



    http.sessionManagement(session -> session.maximumSessions(1));
    ```

2. Next, we need to enable `o.s.s.web.session.HttpSessionEventPublisher` in the `SecurityConfig.java` deployment descriptor so that the servlet container will notify Spring Security (through `HttpSessionEventPublisher`) of session life cycle events, as follows:

    ```
    //src/main/java/com/packtpub/springsecurity/configuration/
    SessionConfig.java
    @Configuration
    public class SessionConfig {

        @Bean
        public HttpSessionEventPublisher httpSessionEventPublisher()
    {

            return new HttpSessionEventPublisher();
        }
    }
    ```

With these two configuration bits in place, concurrent session control will now be activated. Let's see what it actually does, and then we'll demonstrate how it can be tested.

Understanding concurrent session control

Concurrent session control uses `o.s.s.core.session.SessionRegistry` to maintain a list of active HTTP sessions and the authenticated users with which they are associated. As sessions are created and expired, the registry is updated in real time based on the session life cycle events published by `HttpSessionEventPublisher` to track the number of active sessions per authenticated user.

Refer to the following diagram:

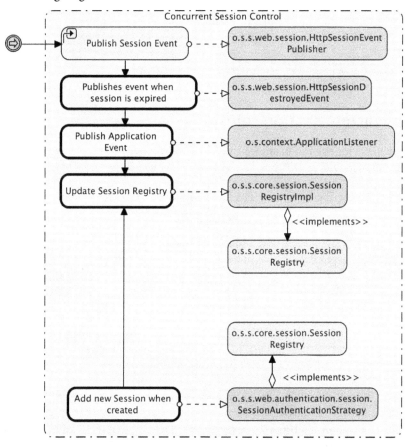

Figure 14.5 – Concurrent session control

An extension of `SessionAuthenticationStrategy`, `o.s.s.web.authentication.session.ConcurrentSessionControlStrategy` is the method by which new sessions are tracked and the method by which concurrency control is actually enforced. Each time a user accesses the secured site, `SessionManagementFilter` is used to check the active session against `SessionRegistry`. If the user's active session isn't in the list of active sessions tracked in `SessionRegistry`, the least recently used session is immediately expired.

The secondary actor in the modified concurrent session control filter chain is `o.s.s.web.session.ConcurrentSessionFilter`. This filter will recognize expired sessions (typically, sessions that have been expired either by the servlet container or forcibly by the `ConcurrentSessionControlStrategy` interface) and notify the user that their session has expired.

Now that we have understood how concurrent session control works, it should be easy for us to reproduce a scenario in which it is enforced.

> **Important note**
>
> Your code should now look like `chapter14.02-calendar`.

Testing concurrent session control

As we did when verifying session fixation protection, we will need to access two web browsers by performing the following steps:

1. In Google Chrome, log in to the site as `user1@example.com/user1`.

2. Now, in Firefox, log in to the site as the same user.

3. Finally, go back to Google Chrome and take any action. You will see a message indicating that your session has expired, as shown in the following screenshot:

This session has been expired (possibly due to multiple concurrent logins being attempted as the same user).

Figure 14.6 – Testing concurrent session control

If you were using this application and received this message, you'd probably be confused. This is because it's obviously not a friendly method of being notified that only a single user can access the application at a time. However, it does illustrate that the session has been forcibly expired by the software.

> **Important note**
>
> Concurrent session control tends to be a very difficult concept for new Spring Security users to grasp. Many users try to implement it without truly understanding how it works and what the benefits are. If you're trying to enable this powerful feature and it doesn't seem to be working as you expect, make sure you have everything configured correctly, and then review the theoretical explanations in this section—hopefully, they will help you understand what may be wrong.

When a session expiration event occurs, we should probably redirect the user to the login page and provide them with a message to indicate what went wrong.

Configuring expired session redirect

Fortunately, there is a simple method for directing users to a friendly page (typically the login page) when they are flagged by concurrent session control—simply specify the `expired-url` attribute and set it to a valid page in your application. Update your `SecurityConfig.java` file as follows:

```
//src/main/java/com/packtpub/springsecurity/configuration/
SecurityConfig.java
http.sessionManagement(session -> session.maximumSessions(1)
        .expiredUrl("/login/form?expired"));
```

In the case of our application, this will redirect the user to the standard login form. We will then use the query parameter to display a friendly message, indicating that we determined that they had multiple active sessions and should log in again. Update your `login.html` page to use this parameter to display our message:

```
//src/main/resources/templates/login.html
<div th:if="${param.expired != null}" class="alert alert-success">
    <strong>Session Expired</strong>
    <span>You have been forcibly logged out due to multiple
    sessions on the same account (only one active
        session per user is allowed).</span>
</div>
```

Go ahead and give it a try by logging in as the user `admin1@example.com/admin1` using both Google Chrome and Firefox.

> **Important note**
> Your code should now look like `chapter14.03-calendar`.

This time, you should see a login page with a custom error message:

Figure 14.7 – A concurrent session login page custom error message

After setting up redirection for expired sessions, we'll delve into typical challenges associated with concurrency control.

Common problems with concurrency control

There are a few common reasons that logging in with the same user does not trigger a logout event. The first occurs when using the custom `UserDetails` (as we did in *Chapter 3, Custom Authentication*) while the equals and `hashCode` methods are not properly implemented. This occurs because the default `SessionRegistry` implementation uses an in-memory map to store `UserDetails`. In order to resolve this, you must ensure that you have properly implemented the `hashCode` and equals methods.

The second problem occurs when restarting the application container while the user sessions are persisted to a disk. When the container has started back up, the users who were already logged in with a valid session are logged in. However, the in-memory map of `SessionRegistry` that is used to determine if the user is already logged in will be empty. This means that Spring Security will report that the user is not logged in, even though the user is. To solve this problem, either a custom `SessionRegistry` is required along with disabling session persistence within the container, or you must implement a container-specific way to ensure that the persisted sessions get populated into the in-memory map at startup.

The last common reason we will cover is that concurrency control will not work in a clustered environment with the default `SessionRegistry` implementation. The default implementation uses an in-memory map. This means that if `user1` logs in to `application server A`, the fact that they are logged in will be associated with that server. Thus, if `user1` then authenticates to `Application Server B`, the previously associated authentication will be unknown to `Application Server B`.

Preventing authentication instead of forcing logout

Spring Security can also prevent a user from being able to log in to the application if the user already has a session. This means that instead of forcing the original user to log out, Spring Security will prevent the second user from being able to log in. The configuration changes can be seen as follows:

```
//src/main/java/com/packtpub/springsecurity/configuration/
SecurityConfig.java
http.sessionManagement(session -> session.maximumSessions(1)
        .expiredUrl("/login/form?expired").
maxSessionsPreventsLogin(true));
```

Make the updates and log in to the calendar application with Google Chrome. Now, attempt to log in to the calendar application with Firefox using the same user. You should see our custom error message from our `login.html` file:

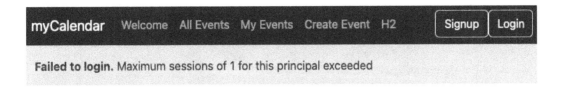

Figure 14.8 – A custom error message preventing authentication for concurrent sessions

> **Important note**
> Your code should now look like `chapter14.04-calendar`.

There is a disadvantage to this approach that may not be apparent without some thought. Try closing Google Chrome without logging out and then opening it up again. Now, attempt to log in to the application again. You will observe that you are unable to log in. This is because when the browser is closed, the `JSESSIONID` cookie is deleted. However, the application is not aware of this, so the user is still considered authenticated. You can think of this as a kind of memory leak, since `HttpSession` still exists but there is no pointer to it (the `JSESSIONID` cookie is gone). It is not until the session times out that our user will be able to authenticate again. Thankfully, once the session times out, our `SessionEventPublisher` interface will remove the user from our `SessionRegistry` interface. What we can take away from this is that if a user forgets to log out and closes the browser, they will not be able to log in to the application until the session times out.

> **Important note**
> Just as in *Chapter 7, Remember-me Services*, this experiment may not work if the browser decides to remember a session even after the browser is closed. Typically, this will happen if a plugin or the browser is configured to restore sessions. In this event, you might want to delete the `JSESSIONID` cookie manually to simulate the browser being closed.

Other benefits of concurrent session control

Another benefit of concurrent session control is that `SessionRegistry` exists to track active (and, optionally, expired) sessions. This means that we can get runtime information about what user activity exists in our system (for authenticated users, at least) by performing the following steps:

1. You can even do this if you don't want to enable concurrent session control. Simply set `maximumSessions` to `-1`, and session tracking will remain enabled, even though no maximum will be enforced. Instead, we will use the explicit bean configuration provided in the `SessionConfig.java` file of this chapter, as follows:

    ```
    //src/main/java/com/packtpub/springsecurity/configuration/
    SessionConfig.java
    ```

```
@Bean
public SessionRegistry sessionRegistry(){
    return new SessionRegistryImpl();
}
```

2. We have already added the import of the `SessionConfig.java` file to the `SecurityConfig.java` file. So, all that we need to do is reference the custom configuration in our `SecurityConfig.java` file. Go ahead and replace the current `sessionManagement` and `maximumSessions` configurations with the following code snippet:

```
//src/main/java/com/packtpub/springsecurity/configuration/
SecurityConfig.java

http.sessionManagement(session -> session.maximumSessions(-1)
        .sessionRegistry(sessionRegistry)
        .expiredUrl("/login/form?expired")
        .maxSessionsPreventsLogin(true));
```

> **Important note**
> Your code should now look like `chapter14.05-calendar`.

Now, our application will allow an unlimited number of authentications for the same user. However, we can use `SessionRegistry` to forcibly log out the users. Let's see how we can use this information to enhance the security of our users.

Displaying active sessions for a user

You've probably seen how many websites allow a user to view and forcibly log out sessions for their account. We can easily use this forcible logout functionality to do the same. We have already provided `UserSessionController`, which obtains the active sessions for the currently logged-in user. You can see the implementation as follows:

```
//src/main/java/com/packtpub/springsecurity/web/controllers/
UserSessionController.java
@Controller
public class UserSessionController {

    private final SessionRegistry sessionRegistry;

    public UserSessionController(SessionRegistry sessionRegistry) {
        this.sessionRegistry = sessionRegistry;
    }
```

```
    @GetMapping("/user/sessions/")
    public String sessions(Authentication authentication, ModelMap
model) {
        List<SessionInformation> sessions = sessionRegistry.
getAllSessions(authentication.getPrincipal(),
                false);
        model.put("sessions", sessions);
        return "user/sessions";
    }


    @PostMapping(value="/user/sessions/{sessionId}")
    public String removeSession(@PathVariable String sessionId,
RedirectAttributes redirectAttrs) {
        SessionInformation sessionInformation = sessionRegistry.
getSessionInformation(sessionId);
        if(sessionInformation != null) {
            sessionInformation.expireNow();
        }
        redirectAttrs.addFlashAttribute("message", "Session was
removed");
        return "redirect:/user/sessions/";
    }
}
```

Our sessions method will use a Spring **Model-View-Controller** (**MVC**) to automatically obtain the current Spring Security Authentication. If we were not using a Spring MVC, we could also get the current Authentication from SecurityContextHolder, as discussed in *Chapter 3, Custom Authentication*. The principal is then used to obtain all the SessionInformation objects for the current user. The information is easily displayed by iterating over the SessionInformation objects in our sessions.html file, as follows:

```
//src/main/resources/templates/user/sessions.html
...
<tr th:each="currentSession : ${sessions}">

    <td th:text="${#calendars.format(currentSession.lastReques, 'yyyy-
MM-dd HH:mm')}">lastUsed</td>
    <td th:text="${currentSession.sessionId}"></td>

    <td>
        <form action="#" th:action="@{'/user/sessions/
{id}'(id=${currentSession.sessionId})}"
            th:method="post" cssClass="form-horizon"al">
          <input type="sub"it" value="Del"te" class=""tn"/>
```

```
        </form>
      </td>
  </tr>
  ...
```

You can now safely start the JBCP calendar application and log in to it using `user1@example.com`/`user1` in Google Chrome. Now, log in using Firefox and click on the `user1@example.com` link in the upper-right corner. You will then see both sessions listed on the display, as shown in the following screenshot:

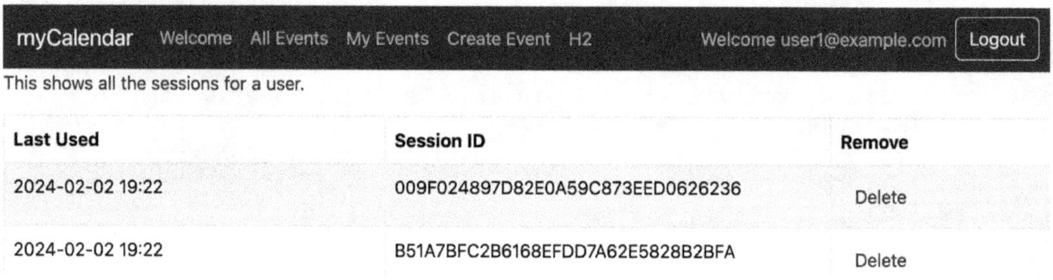

| myCalendar | Welcome | All Events | My Events | Create Event | H2 | | Welcome user1@example.com | Logout |

This shows all the sessions for a user.

Last Used	Session ID	Remove
2024-02-02 19:22	009F024897D82E0A59C873EED0626236	Delete
2024-02-02 19:22	B51A7BFC2B6168EFDD7A62E5828B2BFA	Delete

Figure 14.9 – A list of available sessions

While in Firefox, click on the **Delete** button for the first session. This sends the request to our `deleteSession` method of `UserSessionsController`. This indicates that the session should be terminated. Now, navigate to any page within Google Chrome. You will see the custom message saying that the session has been forcibly terminated. While the message could use updating, we see that this is a nice feature for users to terminate other active sessions.

Other possible uses include allowing an administrator to list and manage all active sessions, displaying the number of active users on the site, or even extending the information to include things like an IP address or location information.

How Spring Security use the HttpSession method?

We have already discussed how Spring Security uses `SecurityContextHolder` to determine the currently logged-in user. However, we have not explained how `SecurityContextHolder` gets automatically populated by Spring Security. The secret to this lies in the `o.s.s.web.context.SecurityContextPersistenceFilter` filter and the `o.s.s.web.context.SecurityContextRepository` interface. Let's take a look at the following diagram:

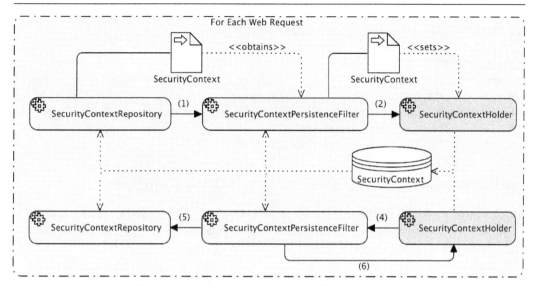

Figure 14.10 – Spring Security usage of the HttpSession

Here is an explanation for each step shown in the preceding diagram:

1. At the beginning of each web request, `SecurityContextPersistenceFilter` is responsible for obtaining the current `SecurityContext` implementation using `SecurityContextRepository`.

2. Immediately afterward, `SecurityContextPersistenceFilter` sets `SecurityContext` on `SecurityContextHolder`.

3. For the remainder of the web request, `SecurityContext` is available via `SecurityContextHolder`. For example, if a Spring MVC controller or `CalendarService` wanted to access `SecurityContext`, it could use `SecurityContextHolder` to access it.

4. Then, at the end of each request, `SecurityContextPersistenceFilter` gets the `SecurityContext` from `SecurityContextHolder`.

5. Immediately afterward, `SecurityContextPersistenceFilter` saves `SecurityContext` in `SecurityContextRepository`. This ensures that if `SecurityContext` is updated at any point during the web requests (that is, when a user creates a new account, as done in *Chapter 3, Custom Authentication*), `SecurityContext` is saved.

6. Lastly, `SecurityContextPersistenceFilter` clears `SecurityContextHolder`.

The question that now arises: How is this related to `HttpSession`? This is all tied together by the default `SecurityContextRepository` implementation, which uses `HttpSession`.

The HttpSessionSecurityContextRepository interface

The default implementation of `SecurityContextRepository`, `o.s.s.web.context.HttpSessionSecurityContextRepository`, uses `HttpSession` to retrieve and store the current `SecurityContext` implementation. There are no other `SecurityContextRepository` implementations provided out of the box. However, since the usage of `HttpSession` is abstracted behind the `SecurityContextRepository` interface, we could easily write our own implementation if we desired.

Configuring how Spring Security uses HttpSession

Spring Security has the ability to configure when the session is created by Spring Security. This can be done with the `http` element's `create-session` attribute. A summary of the options can be seen in the following table:

Attribute value	Description
ifRequired	Spring Security will create a session only if one is required (the default value).
always	Spring Security will proactively create a session if one does not exist.
never	Spring Security will never create a session but will make use of one if the application does create it. This means that if there is a `HttpSession` method, `SecurityContext` will be persisted or retrieve from it.
stateless	Spring Security will not create a session and will ignore the session for obtaining a Spring `Authentication`. In such instances, `NullSecurityContextRepository` is used, which will always state that the current `SecurityContext` is `null`.

Table 14.2 – The session-fixation-protection options

In practice, controlling session creation can be more difficult than it first appears. This is because the attributes only control a subset of Spring Security's usage of `HttpSession`. It does not apply to any other components, such as **Java Server Pages (JSPs)**, in the application. To help figure out when the `HttpSession` method was created, we can add Spring Security's `DebugFilter`.

Debugging with Spring Security's DebugFilter

Let's take a look at the following steps and learn about how to debug with `DebugFilter` of Spring Security:

1. Update your `SecurityConfig.java` file to have a session policy of `NEVER`. Also, add the debug flag to `true` on the `@EnableWebSecurity` annotation so that we can track when the session was created. The updates can be seen as follows:

    ```
    //src/main/java/com/packtpub/springsecurity/configuration/
    SecurityConfig.java
    @Configuration
    @EnableWebSecurity(debug = true)
    public class SecurityConfig {
    ...
    http.sessionManagement(session -> session
            .sessionCreationPolicy(SessionCreationPolicy.NEVER)
            .maximumSessions(-1)
            .sessionRegistry(sessionRegistry)
            .expiredUrl("/login/form?expired")
            .maxSessionsPreventsLogin(true));
    ```

2. When you start up the application, you should see something similar to the following code written to standard output. If you have not already, ensure that you have logging enabled across all levels of the Spring Security debugger category:

    ```
    *********************************************************
    ******     Security debugging is enabled.        ******
    ******     This may include sensitive information.******
    ******     Do not use in a production system!     ******
    ```

3. Now, clear out your cookies (this can be done in Firefox with *Shift* + *Ctrl* + *Delete*), start up the application, and navigate directly to `http://localhost:8080`. When we look at the cookies, as we did earlier in the chapter, we can see that `JSESSIONID` is created even though we stated that Spring Security should never create `HttpSession`. Look at the logs again, and you will see a call stack of the code that created `HttpSession` as follows:

    ```
    *********************************************************
    2024-02-02T20:17:27.859+01:00  INFO 54253 --- [nio-8080-exec-1]
    Spring Security Debugger          :
    *********************************************************
    New HTTP session created: 8C85C6E21D976ED6A1EDE2F8877EB227
    ```

There are several other uses for `DebugFilter`, which we encourage you to explore on your own, for example, determining when a request will match a particular URL, which Spring Security filters are being invoked, and so on.

Summary

After reading this chapter, you should be familiar with how Spring Security manages sessions and protects against session fixation attacks. We also know how to use Spring Security's concurrency control to prevent the same user from being authenticated multiple times.

We explored the utilization of concurrency control to allow a user to terminate sessions associated with their account. Also, we saw how to configure Spring Security's creation of sessions. We also covered how to use Spring Security's `DebugFilter` filter to troubleshoot issues related to Spring.

We also learned about security, including determining when a `HttpSession` method was created and the cause of it.

This concludes our discussion about Spring Security's session management. In the next chapter, we will discuss some specifics about integrating Spring Security with other frameworks.

15

Additional Spring Security Features

In this chapter, we will explore several additional `Spring Security` features that we have not covered so far in this book, including the following topics:

- **Cross-Site Scripting (XSS)**
- **Cross-Site Request Forgery (CSRF)**
- Synchronizer token pattern
- Clickjacking
- Testing Spring Security Applications
- Reactive Applications support

We will understand how to include various HTTP headers to protect against common security vulnerabilities, using the following methods:

- `Cache-Control`
- `Content-Type Options`
- `HTTP Strict Transport Security (HSTS)`
- `X-Frame-Options`
- `X-XSS-Protection`

Before you read this chapter, you should already understand how `Spring Security` works. This means you should already be able to set up authentication and authorization in a simple web application. If you are unable to do this, you will want to ensure you have read up to *Chapter 3, Custom Authentication*, before proceeding with this chapter.

This chapter's code in action link is here: `https://packt.link/aXvKi`.

Security vulnerabilities

In the age of the internet, there is a multitude of possible vulnerabilities that can be exploited. A great resource to learn more about web-based vulnerabilities is the **Open Web Application Security Project** (**OWASP**), which is located at `https://owasp.org`.

In addition to being a great resource to understand various vulnerabilities, OWASP categorizes the top 10 vulnerabilities based on industry trends.

Cross-Site Scripting

Cross-site scripting or **XSS** attacks involve malicious scripts that have been injected into a trusted site.

XSS attacks occur when an attacker exploits a given web application that allows unventilated input to be sent to the site, generally in the form of browser-based scripts, which are then executed by a different user of the website.

There are many forms that attackers can exploit, based on validated or unencoded information provided to websites.

XSS can be described by the following sequence diagram:

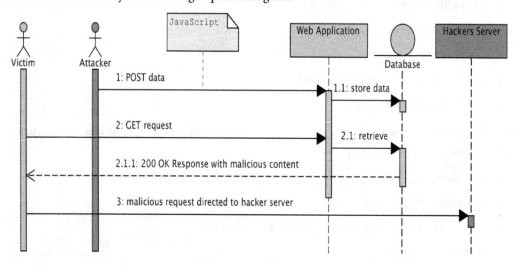

Figure 15.1 – Cross-Site Scripting (XSS)

At the core of this issue is expecting a user to trust the site's information that is being sent. The end user's browser has no way of knowing that the script should not be trusted because there is implicit trust in the website they're browsing. Because the end user thinks the script came from a trusted source, the malicious script can access any cookies, session tokens, or other sensitive information retained by the browser and used with that website.

Cross-Site Request Forgery

Cross-site request forgery (**CSRF**) is an attack that tricks the victim into submitting a malicious request. This type of attack inherits or hijacks the identity and privileges of the victim and performs unauthorized functions and gains access on the victim's behalf.

For web applications, most browsers automatically include credentials associated with the site, which includes a user session, cookie, IP address, Windows domain credentials, and so forth.

So, if a user is currently authenticated on a site, that given site will have no way to distinguish between the forged request sent by the victim and a legitimate request.

CSRF attacks target functionality that causes a state change on the server, such as changing the victim's email address or password, or engaging in a financial transaction.

This forces the victim to retrieve data that doesn't benefit an attacker because the attacker does not receive the response; the victim does. Thus, CSRF attacks target state-changing requests.

The following sequence diagram details how a CSRF attack would occur:

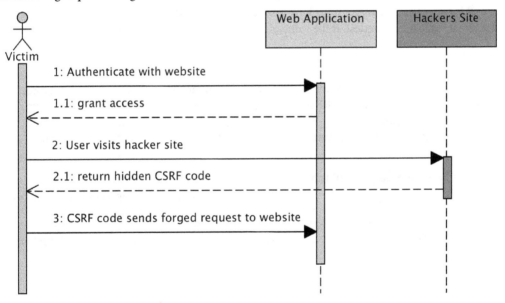

Figure 15.2 – CSRF

There are several different design measures that may be taken to attempt to prevent CSRF; however, measures such as secret cookies, `HTTP POST` requests, multistep transactions, URL rewriting, and HTTPS in no way prevent this type of attack.

> **Important note**
>
> OWASP's top 10 security vulnerabilities list details CSRF as the eighth most common attack at `https://owasp.org/www-community/attacks/csrf`.

In summary, CSRF is the general concept of an attack where a user is tricked into performing unintended actions. In the next section, we will explore the `Synchronizer Tokens pattern`, which is a specific method used to mitigate CSRF attacks by using unique tokens associated with each user session.

Synchronizer token pattern

A solution to CSRF is to use the **synchronizer token** pattern. This solution ensures that each request requires, in addition to our session cookie, a randomly generated token as an HTTP parameter. When a request is submitted, the server must look up the expected value for the parameter and compare it to the actual value in the request. If the values do not match, the request should fail.

> **Important note**
>
> The *Cross-Site Request Forgery Prevention Cheat Sheet* recommends the `synchronizer token` pattern as a viable solution for CSRF attacks: `https://cheatsheetseries.owasp.org/cheatsheets/Cross-Site_Request_Forgery_Prevention_Cheat_Sheet.html#General_Recommendation:_%20Synchronizer_Token_Pattern`.

Let's look at how our example would change. Assume the randomly generated token is present in an HTTP parameter named `_csrf`. For example, the request to transfer money would look as follows:

```
POST /transfer HTTP/1.1 Host: bank.example.com
Cookie: JSESSIONID=randomid; Domain=bank.example.com; Secure; HttpOnly
Content-Type: application/x-www-form-urlencoded
amount=100.00&routingNumber=1234&account=9876&_csrf=<secure-random
token>
```

You will notice that we added the `_csrf` parameter with a random value. Now, the malicious website will not be able to guess the correct value for the `_csrf` parameter (which must be explicitly provided on the malicious website), and the transfer will fail when the server compares the actual token to the expected token.

The following diagram shows a standard use case for a `synchronizer token` pattern:

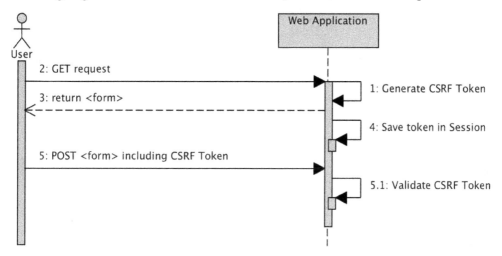

Figure 15.3 – Synchronizer token pattern

`Spring Security` provides `synchronizer token` support that is turned on by default. You might have noticed from the previous chapters that in our `SecurityConfig.java` file, we disabled CSRF protection, as shown in the following code snippet:

```
//src/main/java/com/packtpub/springsecurity/configuration/
SecurityConfig.java

    @Bean
    public SecurityFilterChain filterChain(HttpSecurity http) throws
Exception {
...

    http.csrf(AbstractHttpConfigurer::disable);
    return http.build();
}
```

Up to this point in the book, we have disabled `synchronizer token` protection so that we can focus on other security concerns.

If we start the application at this point, and there will be no `synchronizer token` support added to any of the pages.

> **Important note**
> You should start with the code from `chapter15.00-calendar`.

After exploring the `synchronizer token` pattern, we will explore when to use CSRF protection.

When to use CSRF protection

It is recommended you use CSRF protection for any request that could be processed by a browser or normal users. If you are only creating a service that is used by non-browser clients, you will most likely want to disable CSRF protection.

CSRF protection and JSON

A common question is, do I need to protect JSON requests made by JavaScript? The short answer is, it depends. However, you must be very careful, as there are CSRF exploits that can impact JSON requests. For example, a malicious user can create a CSRF with JSON using the following form:

```
<form action="https://example.com/secureTransaction" method="post"
enctype="text/plain">
    <input
name='{"amount":100,"routingNumber":"maliciousRoutingNumber",
"account":"evilsAccountNumber", "ignore_me":"' value='test"}'
type='hidden'>
    <input type="submit" value="Win Money!"/>
</form>
```

This will produce the following JSON structure:

```
{
   "amount": 100,
   "routingNumber": "maliciousRoutingNumber",
   "account": "maliciousAccountNumber",
   "ignore_me": "=test"
}
```

If an application were not validating the `Content-Type` method, then it would be exposed to this exploit. Depending on the setup, a Spring MVC application that validates the `Content- Type` method could still be exploited by updating the URL suffix to end with `.json`, as shown in the following code:

```
<form action="https://example.com/secureTransaction.json"
method="post" enctype="text/plain">
    <input
name='{"amount":100,"routingNumber":"maliciousRoutingNumber",
"account":"maliciousAccountNumber", "ignore_me":"' value='test"}'
type='hidden'>
    <input type="submit" value="Win Money!"/>
</form>
```

In this section, we have explored CSRF protection and JSON. In the next section, we will cover CSRF and stateless browser applications.

CSRF and stateless browser applications

What if your application is stateless? That doesn't necessarily mean you are protected. In fact, if a user does not need to perform any actions in the web browser for a given request, they are likely still vulnerable to CSRF attacks.

For example, consider an application using a custom cookie that contains all of the states within it for authentication instead of the JSESSIONID cookie. When the CSRF attack happens, the custom cookie will be sent with the request in the same manner that the JSESSIONID cookie was sent in our previous example.

Users using basic authentication are also vulnerable to CSRF attacks, since the browser will automatically include the username and password in any requests, in the same manner, that the JSESSIONID cookie was sent in our previous example.

Using Spring Security CSRF protection

So, what are the steps necessary to use Spring Security to protect our site against CSRF attacks? The steps for using Spring Security's CSRF protection are as follows:

1. Use proper HTTP verbs.
2. Configure CSRF protection.
3. Include the CSRF token.

Let's understand these steps better now.

Using proper HTTP verbs

The first step to protecting against CSRF attacks is to ensure that your website uses proper HTTP verbs. Specifically, before Spring Security's CSRF support can be of use, you need to be certain that your application uses PATCH, POST, PUT, and/or DELETE for anything that modifies state.

This is not a limitation of Spring Security's support but, instead, a general requirement for proper CSRF prevention. The reason is that including private information in an HTTP GET method can cause the information to be leaked.

Refer to *RFC 2616, Section 15.1.3, Encoding Sensitive Information in URI's*, for general guidance on using POST instead of GET for sensitive information (https://www.rfc-editor.org/rfc/rfc2616.html#section-15.1.3).

Configuring CSRF protection

The next step is to include Spring Security's CSRF protection within your application. Some frameworks handle invalid CSRF tokens by invaliding the user's session, but this causes its own problems. Instead, by default, Spring Security's CSRF protection will produce HTTP 403 access denied. This can be customized by configuring AccessDeniedHandler to process InvalidCsrfTokenException differently.

For passivity reasons, if you are using the XML configuration, CSRF protection must be explicitly enabled using the <csrf> element. Refer to the <csrf> element's documentation for additional customizations.

Default CSRF support

CSRF protection is enabled by default with Java configuration. Refer to the Java documentation of csrf() for additional customizations regarding how CSRF protection is configured.

Just to be verbose in this configuration, we are going to add the CSRF method to our SecurityConfig.java file, as follows:

```
//src/main/java/com/packtpub/springsecurity/configuration/
SecurityConfig.java

    @Bean
    public SecurityFilterChain filterChain(HttpSecurity http) throws
Exception {
...
    http.csrf(Customizer.withDefaults());
    return http.build();
}
```

To have access to the H2 console, CSRF needs to be disabled, as follows:

```
//src/main/java/com/packtpub/springsecurity/configuration/
SecurityConfig.java

    @Bean
    public SecurityFilterChain filterChain(HttpSecurity http) throws
Exception {
...
    http
        .csrf(csrf -> csrf
                .ignoringRequestMatchers(toH2Console())
                .disable());
    return http.build();
}
```

After configuring CSRF in the `Spring Security SecurityFilterChain` bean, we will see how we can enable the support in web forms.

Including the CSRF token in the <form> submissions

The last step is to ensure that you include the CSRF token in all `PATCH`, `POST`, `PUT`, and `DELETE` methods. One way to approach this is to use the `_csrf` request attribute to obtain the current `CsrfToken` token. An example of doing this with a **Java Server Page (JSP)** is shown as follows:

```
<c:url var="logoutUrl" value="/logout"/>
<form action="${logoutUrl}"
      method="post">
    <input type="submit"
           value="Log out" />
    <input type="hidden"
           name="${_csrf.parameterName}"
           value="${_csrf.token}"/>
</form>
```

After including the CSRF token in the `<form>`, we will explore another option by including the CSRF token based on `Spring Security` JSP tags.

Including the CSRF token using the Spring Security JSP tag library

If CSRF protection is enabled, the `Spring Security` tag inserts a hidden form field with the correct name and value for the CSRF protection token. If CSRF protection is not enabled, this tag has no output.

Normally, `Spring Security` automatically inserts a CSRF form field for any `<form:form>` tags you use, but if for some reason you cannot use `<form:form>`, `csrfInput` is a handy replacement.

You should place this tag within an HTML `<form></form>` block, where you would normally place other input fields. Do not place this tag within a Spring `<form:form></form:form>` block. `Spring Security` handles Spring forms automatically, as follows:

```
<form method="post" action="/logout">
    <sec:csrfInput />
    . . .
</form>
```

Default CSRF token support

If you are using the Spring MVC `<form:form>` tag, or `Thymeleaf 2.1+`, and you are also using `@EnableWebSecurity`, the `CsrfToken` token is automatically included for you (using the `CsrfRequestDataValue` token we have been processing).

So, for this book, we have been using *Thymeleaf* for all of our web pages. Thymeleaf has CSRF support enabled by default if we enable CSRF support in `Spring Security`.

The `logout` link will not work with CSRF support enabled and needs to be replaced by the following code:

```
//src/main/webapp/WEB-INF/templates/fragments/header.html
<form th:action="@{/logout}" method="post">
    <input type="submit" value="Logout" class="btn btn-outline-light"
/>
</form></li>
```

> **Important note**
> Your code should now look like `chapter15.01-calendar`.

If we start up the JBCP calendar application and navigate to the login page at `http://localhost:8080/login.html`, we can view the generated source for the `login.html` page, as follows:

```
<form class="form-horizontal" method="POST" action="/login">
    <input type="hidden" name="_csrf" value="eEOF9AiMgfLo353Q19oTxL
z5JFiNDUwbVnp-UiIExznGwFV9HiK2xGq_4sLFvfng4fcn9oTNCTrpay82YEhLNBBh
pV_x8DFM"/>
...
</form>
```

Ajax and JavaScript requests

If you are using JSON, then it is not possible to submit the CSRF token within an HTTP parameter. Instead, you can submit the token within an HTTP header. A typical pattern would be to include the CSRF token within your `<meta>` HTML tags. An example with a JSP is as follows:

```
<html>
<head>
    <meta name="_csrf" th:content="${_csrf.token}"/>
    <!-- default header name is X-CSRF-TOKEN -->
    <meta name="_csrf_header" th:content="${_csrf.headerName}"/>
    <!-- ... -->
</head>
```

Instead of manually creating the meta tags, you can use the simpler `csrfMetaTags` tag from the `Spring Security` JSP tag library.

The csrfMetaTags tag

If CSRF protection is enabled, this tag inserts meta tags containing the CSRF protection token form field, header names, and CSRF protection token value. These meta tags are useful for employing CSRF protection within JavaScript in your applications.

You should place the `csrfMetaTags` tag within an HTML `<head></head>` block, where you would normally place other meta tags. Once you use this tag, you can access the form field name, header name, and token value easily using JavaScript, as follows:

```
<!DOCTYPE html>
<html>
<head>
    <title>CSRF Protected JavaScript Page</title>
    <meta name="description" content="This is the description for this
page"/>
    <sec:csrfMetaTags/>
    <script type="text/javascript">

        var csrfParameter = $("meta[name='_csrf_parameter']").
attr("content");
        var csrfHeader = $("meta[name='_csrf_header']").
attr("content");
        var csrfToken = $("meta[name='_csrf']").attr("content");

        // using XMLHttpRequest directly to send an x-www-form-
urlencoded request
        var ajax = new XMLHttpRequest();
        ajax.open("POST", "https://www.example.org/do/something",
true);
        ajax.setRequestHeader("Content-Type", "application/x-www-form-
urlencoded data");
        ajax.send(csrfParameter + "=" + csrfToken + "&name=John&...");

        // using XMLHttpRequest directly to send a non-x-www-form-
urlencoded request
        var ajax = new XMLHttpRequest();
        ajax.open("POST", "https://www.example.org/do/something",
true);
        ajax.setRequestHeader(csrfHeader, csrfToken);
        ajax.send("...");
    </script>
</head>>
<body>
...
</body>
</html>
```

If CSRF protection is not enabled, `csrfMetaTags` outputs nothing.

jQuery usage

You can then include the token within all of your Ajax requests. If you were using jQuery, this could be done with the following code snippet:

```
// using JQuery to send an x-www-form-urlencoded request
var data = {};
data[csrfParameter] = csrfToken;
data["name"] = "John";
...
$.ajax({
    url: "https://www.example.org/do/something",
    type: "POST",
    data: data,
    ...
});

// using JQuery to send a non-x-www-form-urlencoded request
var headers = {};
headers[csrfHeader] = csrfToken;
$.ajax({
    url: "https://www.example.org/do/something",
    type: "POST",
    headers: headers,
    ...
});
```

After exploring the default CSRF configuration, we will conclude with some CSRF caveats.

CSRF caveats

There are a few caveats when implementing CSRF in `Spring Security` that you need to be aware of. Let us see these caveats in the next few sections.

Timeouts

One issue is that the expected CSRF token is stored in the `HttpSession` method, so as soon as the `HttpSession` method expires, your configured `AccessDeniedHandler` handler will receive `InvalidCsrfTokenException`. If you are using the default `AccessDeniedHandler` handler, the browser will get an `HTTP 403` and display a poor error message.

Another disadvantage is that by removing the state (the timeout), you lose the ability to forcibly terminate the token if something is compromised.

A simple way to mitigate an active user experiencing a timeout is to have some JavaScript that lets the user know their session is about to expire. The user can click a button to continue and refresh the session.

Alternatively, specifying a custom `AccessDeniedHandler` handler allows you to process `InvalidCsrfTokenException` any way you like, as we can see in the following code:

```
//src/main/java/com/packtpub/springsecurity/configuration/
SecurityConfig.java

@Configuration
@EnableWebSecurity
public class SecurityConfig {

    @Bean
    public SecurityFilterChain securityFilterChain(HttpSecurity http,
CustomAccessDeniedHandler accessDeniedHandler) throws Exception {
        http
            // ...
            .exceptionHandling(exceptionHandlin) -> exceptionHandling
                .accessDeniedHandler(accessDeniedHandler);
            );
        return http.build();
    }

    @Bean
    public CustomAccessDeniedHandler accessDeniedHandler(){ return new
AccessDeniedHandlerImpl();
    }

}
```

Logging in

To protect against forged login requests, the login form should be protected against CSRF attacks, too. Since the `CsrfToken` token is stored in `HttpSession`, this means an `HttpSession` method will be created as soon as the `CsrfToken` attribute is accessed.

While this sounds bad in a RESTful/stateless architecture, the reality is that the state is necessary to implement practical security. Without the state, we have nothing we can do if a token is compromised. Practically speaking, the CSRF token is quite small in size and should have a negligible impact on our architecture.

> **Important note**
> An attacker may forge a request to log the victim into a target website using the attacker's credentials; this is known as login CSRF (https://en.wikipedia.org/wiki/Cross-site_request_forgery#Forging_login_requests).

Logging out

Adding CSRF will update the `LogoutFilter` filter to only use `HTTP POST`. This ensures that logging out requires a CSRF token and that a malicious user cannot forcibly log out your users.

One approach to prevent CSRF attacks is to use a `<form>` tag for logout. If you want an HTML link, you can use JavaScript to have the link perform an `HTTP POST` (which can be in a hidden form). For browsers with JavaScript disabled, you can optionally have the link take the user to a logout confirmation page that will perform the `HTTP POST`.

If you want to use `HTTP GET` with logout, you can do so, but remember, this is generally not recommended. For example, the following Java configuration will perform logout when the logout URL pattern is requested with any `HTTP` method:

```
//src/main/java/com/packtpub/springsecurity/configuration/
SecurityConfig.java
@Configuration
@EnableWebSecurity
public class SecurityConfig {

    @Bean
    public SecurityFilterChain securityFilterChain(HttpSecurity http)
throws Exception {
        http
                // ...
                .logout(logout -> logout
                        .logoutRequestMatcher(new AntPathRequestMatcher("/
logout"))
                );
        return http.build();
    }
}
```

In this section, we've discussed CSRF now, let's delve into securing HTTP response headers.

Security HTTP response headers

The following sections discuss `Spring Security`'s support for adding various security headers to the response.

`Spring Security` allows users to easily inject default security headers to assist in protecting their applications. The following is a list of the current default security headers provided by `Spring Security`:

- Cache-Control
- Content-Type Options
- HTTP Strict Transport Security
- X-Frame-Options
- X-XSS-Protection

While each of these headers is considered best practice, it should be noted that not all clients utilize these headers, so additional testing is encouraged. For passivity reasons, if you are using `Spring Security`'s `XML namespace` support, you must explicitly enable the security headers. All of the default headers can be easily added using the `<headers>` element with no child elements.

If you are using `Spring Security`'s Java configuration, all of the default security headers are added by default. They can be disabled using Java configuration, as follows:

```
//src/main/java/com/packtpub/springsecurity/configuration/
SecurityConfig.java
@Configuration
@EnableWebSecurity
public class SecurityConfig {

    @Bean
    public SecurityFilterChain filterChain(HttpSecurity http) throws
Exception {
        http
            // ...
            .headers(headers -> headers.disable());
        return http.build();
    }
}
```

By default, `Spring Security` instructs browsers to disable the XSS Auditor by using `<headers-xss-protection,X-XSS-Protection header>`. You can disable the `X-XSS-Protection` header entirely:

```
//src/main/java/com/packtpub/springsecurity/configuration/
SecurityConfig.java
@Bean
SecurityFilterChain springSecurityFilterChain(HttpSecurity http)
throws Exception {
    http
```

```
            .headers(headers -> headers
                    .xssProtection(XXssConfig::disable)
            );
    return http.build();
}
```

You can also set the header value:

```
//src/main/java/com/packtpub/springsecurity/configuration/
SecurityConfig.java
@Bean
SecurityFilterChain springSecurityFilterChain(HttpSecurity http)
throws Exception {
    http
            .headers(headers -> headers
                    .xssProtection(xssProtection -> xssProtection.
headerValue(XXssProtectionHeaderWriter.HeaderValue.ENABLED_MODE_
BLOCK))
            );
    return http.build();
}
```

As soon as you specify any headers that should be included, then only those headers will be included. For example, the following configuration will include support for X-Frame-Options only:

```
//src/main/java/com/packtpub/springsecurity/configuration/
SecurityConfig.java
@Bean
SecurityFilterChain springSecurityFilterChain(HttpSecurity http)
throws Exception {
    http
            .headers(headers -> headers
                    .frameOptions(FrameOptionsConfig::sameOrigin));
    return http.build();
}
```

Cache-Control

In the past, `Spring Security` required you to provide your own `Cache-Control` method for your web application. This seemed reasonable at the time, but browser caches have evolved to include caches for secure connections as well. This means that a user may view an authenticated page and log out, and then a malicious user can use the browser history to view the cached page.

To help mitigate this, `Spring Security` has added `Cache-Control` support, which will insert the following headers into your response:

```
Cache-Control: no-cache, no-store, max-age=0, must-revalidate Pragma:
no-cache
Expires: 0
```

To be secure, `Spring Security` adds these headers by default. However, if your application provides its own cache control headers, `Spring Security` will rely on your own application headers only. This allows for applications to ensure that static resources (such as CSS and JavaScript) can be cached.

Content-Type Options

Historically, browsers, including Internet Explorer, would try to guess the content type of a request using content sniffing. This allowed browsers to improve the user experience by guessing the content type of resources that had not specified the content type. For example, if a browser encountered a JavaScript file that did not have the content type specified, it would be able to guess the content type and then execute it.

> **Important note**
> There are many additional things one should do, such as only displaying the document in a distinct domain, ensuring the `Content-Type` header is set, sanitizing the document, and so on, when allowing content to be uploaded. However, these measures are out of the scope of what `Spring Security` provides. It is also important to point out that when disabling content sniffing, you must specify the content type for things to work properly.

The problem with content sniffing is that this allows malicious users to use *polyglots* (a file that is valid as multiple content types) to execute XSS attacks. For example, some sites may allow users to submit a valid PostScript document to a website and view it. A malicious user might create a PostScript document that is also a valid JavaScript file and execute an XSS attack with it (`http://webblaze.cs.berkeley.edu/papers/barth-caballero-song.pdf`).

Content sniffing can be disabled by adding the following header to our response:

```
X-Content-Type-Options: nosniff
```

By default, `Spring Security` disables content sniffing by adding this header to HTTP responses.

HTTP Strict Transport Security

When you type in your bank's website, do you enter `mybank.example.com`, or do you enter `https://mybank.example.com`? If you omit the HTTPS protocol, you are potentially vulnerable to **Man-in-the-Middle (MitM)** attacks. Even if the website performs a redirect to `https://mybank.`

example.com, a malicious user could intercept the initial HTTP request and manipulate the response (redirect to https://mibank.example.com and steal your credentials).

Many users omit the HTTPS protocol, and this is why HSTS was created.

In accordance with RFC6797, the HSTS header is only injected into HTTPS responses. In order for the browser to acknowledge the header, the browser must first trust the **certificate authority** (**CA**) that signed the **Secure Sockets Layer** (**SSL**) certificate used to make the connection, not just the SSL certificate (https://datatracker.ietf.org/doc/html/rfc6797).

Once mybank.example.com is added as an HSTS host, a browser can know beforehand that any request to mybank.example.com should be interpreted as https://mybank.example.com. This greatly reduces the possibility of a **Man-in-the-Middle** (**MitM**) attack occurring.

One way for a site to be marked as an HSTS host is to have the host preloaded into the browser. Another is to add the Strict-Transport-Security header to the response. For example, the following would instruct the browser to treat the domain as an HSTS host for a year (there are approximately 31,536,000 seconds in a year):

```
Strict-Transport-Security: max-age=31536000 ; includeSubDomains ;
preload
```

The optional includeSubDomains directive instructs Spring Security that subdomains (such as secure.mybank.example.com) should also be treated as an HSTS domain.

The optional preload directive instructs the browser that the domain should be preloaded in the browser as an HSTS domain. For more details on HSTS preload, see hstspreload.org.

You can explicitly customize the results. The following example explicitly provides HSTS:

```
//src/main/java/com/packtpub/springsecurity/configuration/
SecurityConfig.java
@Configuration
@EnableWebSecurity
public class SecurityConfig {

    @Bean
    public SecurityFilterChain filterChain(HttpSecurity http) throws
Exception {
        http
            // ...
            .headers(headers -> headers
                .httpStrictTransportSecurity(hsts -> hsts
                    .includeSubDomains(true)
                    .preload(true)
                    .maxAgeInSeconds(31536000)
                )
```

```
                );
            return http.build();
        }
}
```

HTTP Public Key Pinning (HPKP)

`Spring Security` provides servlet support for HPKP (see here: `https://docs.spring.io/spring-security/reference/features/exploits/headers.html#headers-hpkp`), but it is no longer recommended (`https://docs.spring.io/spring-security/reference/features/exploits/headers.html#headers-hpkp-deprecated`).

You can enable HPKP headers with the following configuration:

```
//src/main/java/com/packtpub/springsecurity/configuration/
SecurityConfig.java
@Configuration
@EnableWebSecurity
public class SecurityConfig {

    @Bean
    public SecurityFilterChain filterChain(HttpSecurity http) throws
Exception {
        http
            // ...
            .headers(headers -> headers
                .httpPublicKeyPinning(hpkp -> hpkp
                    .includeSubDomains(true)
                    .reportUri("https://example.net/pkp-report")
                    .
addSha256Pins("d6qzRu9zOECb90Uez27xWltNsj0e1Md7GkYYkVoZWmM=",
"E9CZ9INDbd+2eRQozYqqbQ2yXLVKB9+xcprMF+44U1g=")
                )
            );
        return http.build();
    }
}
```

X-Frame-Options

Allowing your website to be added to a frame can be a security issue. For example, through the use of clever CSS styling, users could be tricked into clicking on something that they did not intend to.

For example, a user who is logged in to their bank might click a button that grants access to other users. This sort of attack is known as **clickjacking**.

Read more about Clickjacking at `https://owasp.org/www-community/attacks/Clickjacking`.

Another modern approach to dealing with Clickjacking is using a **Content Security Policy (CSP)**.

There are a number of ways to mitigate Clickjacking attacks. For example, to protect legacy browsers from Clickjacking attacks, you can use frame-breaking code. While not perfect, frame-breaking code is the best you can do for legacy browsers.

A more modern approach to address Clickjacking is to use the `X-Frame-Options` header, as follows:

```
X-Frame-Options: DENY
```

If you want to change the value for the `X-Frame-Options` header, then you can use an `XFrameOptionsHeaderWriter` instance.

Some browsers have built-in support for filtering out reflected XSS attacks. This is by no means foolproof, but it does assist with XSS protection.

Filtering is typically enabled by default, so adding the header just ensures that it is enabled and instructs the browser as to what to do when an XSS attack is detected. For example, the filter might try to change the content in the least invasive way to still render everything. At times, this type of replacement can become an XSS vulnerability. Instead, it is best to block the content, rather than attempt to fix it. To do this, we can add the following header:

```
X-XSS-Protection: 1; mode=block
```

CSP

`Spring Security` does not add a CSP (`https://docs.spring.io/spring-security/reference/features/exploits/headers.html#headers-csp`) by default, because a reasonable default is impossible to know without knowing the context of the application. The web application author must declare the security policy (or policies) to enforce or monitor them for the protected resources.

Consider the following security policy:

```
Content-Security-Policy: script-src 'self' https://trustedscripts.
example.com; object-src https://trustedplugins.example.com; report-uri
/csp-report-endpoint/
```

Given the preceding security policy, you can enable the CSP header:

```
//src/main/java/com/packtpub/springsecurity/configuration/
SecurityConfig.java
@Configuration
@EnableWebSecurity
```

```
public class SecurityConfig {

    @Bean
    public SecurityFilterChain filterChain(HttpSecurity http) throws
Exception {
        http
                // ...
                .headers(headers -> headers
                        .contentSecurityPolicy(csp -> csp
                                .policyDirectives("script-src 'self' https://
trustedscripts.example.com; object-src https://trustedplugins.example.
com; report-uri /csp-report-endpoint/")
                        )
                );
        return http.build();
    }
}
```

To enable the CSP report-only header, provide the following configuration:

```
//src/main/java/com/packtpub/springsecurity/configuration/
SecurityConfig.java
@Configuration
@EnableWebSecurity
public class SecurityConfig {

    @Bean
    public SecurityFilterChain filterChain(HttpSecurity http) throws
Exception {
        http
                // ...
                .headers(headers -> headers
                        .contentSecurityPolicy(csp -> csp
                                .policyDirectives("script-src 'self' https://
trustedscripts.example.com; object-src https://trustedplugins.example.
com; report-uri /csp-report-endpoint/")
                                .reportOnly()
                        )
                );
        return http.build();
    }
}
```

Referrer Policy

Spring Security does not add Referrer Policy (https://docs.spring.io/spring-security/reference/features/exploits/headers.html#headers-referrer) headers by default. You can enable the `ReferrerPolicy` header by using this configuration:

```
//src/main/java/com/packtpub/springsecurity/configuration/
SecurityConfig.java
@Configuration
@EnableWebSecurity
public class SecurityConfig {

    @Bean
    public SecurityFilterChain filterChain(HttpSecurity http) throws
Exception {
        http
            // ...
            .headers(headers -> headers
                .referrerPolicy(referrer -> referrer
                    .policy(ReferrerPolicy.SAME_ORIGIN)
                )
            );
        return http.build();
    }
}
```

Feature Policy

Spring Security does not add Feature Policy (https://docs.spring.io/spring-security/reference/features/exploits/headers.html#headers-feature) headers by default. Consider the following `Feature-Policy` header:

```
Feature-Policy: geolocation 'self'
```

You can enable the preceding feature policy header by using the following configuration:

```
//src/main/java/com/packtpub/springsecurity/configuration/
SecurityConfig.java
@Configuration
@EnableWebSecurity
public class SecurityConfig {

    @Bean
    public SecurityFilterChain filterChain(HttpSecurity http) throws
Exception {
```

```
        http
            // ...
            .headers(headers -> headers
                .featurePolicy("geolocation 'self'")
            );
        return http.build();
    }
}
```

Permissions Policy

`Spring Security` does not add Permissions Policy (`https://docs.spring.io/spring-security/reference/features/exploits/headers.html#headers-permissions`) headers by default. Consider the following `Permissions-Policy` header:

```
Permissions-Policy: geolocation=(self)
```

You can enable the preceding permissions policy header using the following configuration:

```
//src/main/java/com/packtpub/springsecurity/configuration/
SecurityConfig.java
@Configuration
@EnableWebSecurity
public class SecurityConfig {

    @Bean
    public SecurityFilterChain filterChain(HttpSecurity http) throws
Exception {
        http
            // ...
            .headers(headers -> headers
                .featurePolicy("geolocation 'self'")
            );
        return http.build();
    }
}
```

Clear Site Data

`Spring Security` does not add `Clear-Site-Data` (`https://docs.spring.io/spring-security/reference/features/exploits/headers.html#headers-clear-site-data`) headers by default. Consider the following `Clear-Site-Data` header:

```
Clear-Site-Data: "cache", "cookies"
```

You can send the preceding header to logout function with the following configuration:

```
//src/main/java/com/packtpub/springsecurity/configuration/
SecurityConfig.java
@Configuration
@EnableWebSecurity
public class SecurityConfig {

    @Bean
    public SecurityFilterChain filterChain(HttpSecurity http) throws
Exception {
        http
        // ...
        .logout((logout) -> logout
                .addLogoutHandler(new HeaderWriterLogoutHandler(new
ClearSiteDataHeaderWriter(CACHE, COOKIES)))
        );
        return http.build();
    }
}
```

Static headers

There may be times when you wish to inject custom security headers into your application that are not supported out of the box. For example, perhaps you wish to have early support for a CSP to ensure that resources are only loaded from the same origin. Since support for a CSP has not been finalized, browsers use one of two common extension headers to implement the feature. This means we will need to inject the policy twice. An example of the headers can be seen in the following code snippet:

```
X-Content-Security-Policy: default-src 'self' X-WebKit-CSP: default-
src 'self'
```

When using Java configuration, these headers can be added to the response using the header() method, as follows:

```
//src/main/java/com/packtpub/springsecurity/configuration/
SecurityConfig.java
@Configuration
@EnableWebSecurity
public class SecurityConfig {

@Bean
public SecurityFilterChain filterChain(HttpSecurity http) throws
Exception {
    http.headers(headers -> headers
            .addHeaderWriter(
                new StaticHeadersWriter("X-Content-Security-Policy",
```

```
"default-src 'self'"))
            .addHeaderWriter(
                new StaticHeadersWriter( "X-WebKit-CSP","default-src
'self'");
    return http.build();
  }
}
```

The HeadersWriter instance

When the namespace or Java configuration does not support the headers you want, you can create a custom `HeadersWriter` instance or even provide a custom implementation of `HeadersWriter`.

Let's look at an example of using a custom instance of `XFrameOptionsHeaderWriter`. Perhaps you want to allow the framing of content for the same origin. This is easily supported by setting the policy attribute to `SAMEORIGIN`, but let's look at a more explicit example using the `ref` attribute, as shown in the following code snippet:

```
//src/main/java/com/packtpub/springsecurity/configuration/
SecurityConfig.java
@Configuration
@EnableWebSecurity
public class SecurityConfig {

    @Bean
    public SecurityFilterChain filterChain(HttpSecurity http) throws
Exception {
        http
                // ...
                .headers(headers -> headers
                        .frameOptions(FrameOptionsConfig::sameOrigin
                        )
                );
        return http.build();
    }
}
```

The DelegatingRequestMatcherHeaderWriter class

At times, you may want to only write a header for certain requests. For example, perhaps you want to only protect your login page from being framed. You could use the `DelegatingRequestMatcherHeaderWriter` class to do so. When using Java configuration, this can be done with the following code:

```
//src/main/java/com/packtpub/springsecurity/configuration/
SecurityConfig.java
```

```
@Configuration
@EnableWebSecurity
public class SecurityConfig {

    @Bean
    public SecurityFilterChain filterChain(HttpSecurity http) throws
Exception {
        RequestMatcher matcher = new AntPathRequestMatcher("/login");
        DelegatingRequestMatcherHeaderWriter headerWriter =
            new DelegatingRequestMatcherHeaderWriter(matcher,new
XFrameOptionsHeaderWriter());
        http
            // ...
            .headers(headers -> headers
                .frameOptions(frameOptions -> frameOptions.
disable())
                .addHeaderWriter(headerWriter)
            );
        return http.build();
    }
}
```

To summarize this section, you can add CSRF protection based on DelegatingRequestMatcherHeaderWriter, explicit **HTTP Strict Transport Security (HSTS)**, a Permissions Policy, Clear-Site-Data, frame options, static headers, and DelegatingRequestMatcherHeaderWriter.

> **Important note**
>
> Your code should now look like chapter15.02-calendar.

Testing Spring Security Applications

In addition to spring-boot-starter-test, Spring Security provides an artifact tailored specifically for testing purposes. Its primary purpose is to provide utilities and classes that assist developers in writing tests for applications that use Spring Security for authentication and authorization.

Some of the key functionalities provided by org.springframework.security:spring-security-test include:

- **Mocking Authentication**: It allows you to easily mock authentication and authorization contexts during testing, enabling you to simulate different user roles and permissions.

- **Integration Testing**: It provides support for integration testing by offering utilities for setting up security configurations in your test environment, ensuring that your security configurations are correctly applied and tested.

- **Testing Security Filters**: `Spring Security` often involves configuring a chain of filters to handle authentication and authorization tasks. This module provides utilities to test these filters in isolation or as part of the filter chain.

To get started, you need to include the following dependencies in your `Spring Security` project:

```
//build.gradle
dependencies {
...
testImplementation 'org.springframework.boot:spring-boot-starter-test'
testImplementation 'org.springframework.security:spring-security-test'
}!
```

By using `spring-security-test`, you can effectively test the security features of the applications, ensuring that sensitive resources are properly protected and accessible only to authorized users.

Reactive Applications Support

Reactive programming revolves around asynchronous and non-blocking interactions, characterized by callback and declarative approaches. It incorporates a backpressure mechanism to regulate the throughput from the producer, aiding in consumer control. In Java, Reactive programming relies on `Streams`, `CompletableFuture`, and backpressure controls. There exist numerous relevant use cases where reactive programming proves beneficial, including supporting high peak workloads, microservices, contention avoidance, IoT and Big Data applications.

Spring relies on the project reactor(`https://projectreactor.io/`) that is the f foundation of `Spring Webflux` (`https://docs.spring.io/spring-framework/reference/web/webflux.html`).

To add `Spring Security` reactive support into your Spring Boot project, you need to include `spring-boot-starter-security` and `spring-boot-starter-webflux` dependencies in your projects:

```
//build.gradle
dependencies {
...
    // spring-webflux

    implementation 'org.springframework.boot:spring-boot-starter-webflux'
    // spring-security
```

```
    implementation 'org.springframework.boot:spring-boot-starter-
security'...
}!
```

This table provides an overview of key classes to facilitate the transition from the Spring Security Servlet implementation to the Spring Security Reactive implementation.

Spring Security Servlet implementation	Spring Security Reactive implementation
`o.s.s.w.SecurityFilterChain`	`o.s.s.w.s.Security WebFilterChain`
`o.s.s.c.u.UserDetailsService`	`o.s.s.c.u. ReactiveUserDetailsService`
`o.s.s.c.a.w.b.HttpSecurity`	`o.s.s.c.w.s.ServerHttpSecurity`
`o.s.s.c.c. SecurityContextHolder`	`o.s.s.c.c. ReactiveSecurityContextHolder`
`o.s.s.a.AuthenticationManager`	`o.s.s.a.ReactiveAuthentication Manager`
`o.s.s.c.a.w.c.EnableWebSecurity`	`o.s.s.c.a.m.r. EnableWeb FluxSecurity`
`o.s.s.c.a.m.c. EnableMethodSecurity`	`o.s.s.c.a.m.c.EnableReactive MethodSecurity`
`o.s.s.p.InMemoryUser DetailsManager`	`o.s.s.c.u. MapReactiveUserDetailsService`
`o.s.s.o.c.OAuth2Authorized ClientManager`	`o.s.s.o.c.ReactiveOAuth2 AuthorizedClientManager`
`o.s.s.o.c. OAuth2Authorized ClientProvider`	`o.s.s.o.c.ReactiveOAuth2 AuthorizedClientProvider`
`o.s.s.o.c.w.DefaultOAuth2Authorized ClientManager`	`o.s.s.o.c.w.DefaultReactiveO Auth2AuthorizedClientManager`
`o.s.s.o.c.r.ClientRegistration Repository`	`o.s.s.o.c.r.ReactiveClient RegistrationRepository`
`o.s.s.o.c.e. OAuth2Access TokenResponseClient`	`o.s.s.o.c.e.ReactiveOAuth2 AccessTokenResponseClient`
`o.s.s.o.j.JwtDecoder`	`o.s.s.o.j.ReactiveJwtDecoder`

Table 15.1 – Transitioning from Spring Security Servlet to Reactive Implementation

> **Important note**
>
> In this section, we provide a fully functional reactive implementation available in `chapter15.03-calendar`.

You will notice the following main changes in to enable reactive support:

1. Configure `Spring Security` to use our custom `ReactiveUserDetailsService` object, as follow:

```java
//com/packtpub/springsecurity/service/UserDetailsServiceImpl.java
@Service
public class UserDetailsServiceImpl implements
ReactiveUserDetailsService {

    private final CalendarUserRepository userRepository;

    public UserDetailsServiceImpl(CalendarUserRepository
userRepository) {
        this.userRepository = userRepository;
    }

    @Override
    public Mono<UserDetails> findByUsername(String username) {
        return userRepository.findByEmail(username)
            .flatMap(user -> {
                Set<GrantedAuthority> grantedAuthorities = new
HashSet<>();
                for (Role role : user.getRoles()) {
                    grantedAuthorities.add(new
SimpleGrantedAuthority(role.getName()));
                }
                return Mono.just(new User(user.getEmail(), user.
getPassword(), grantedAuthorities));
            });
    }

}
```

2. Use `@EnableWebFluxSecurity` annotation, to enable Spring Security reactive support:

```
//com/packtpub/springsecurity/configuration/SecurityConfig.java
@Configuration
@EnableWebFluxSecurity
public class SecurityConfig {

...

}
```

3. Define a `ReactiveAuthenticationManager` bean:

```
//com/packtpub/springsecurity/configuration/SecurityConfig.java
@Bean
public ReactiveAuthenticationManager
reactiveAuthenticationManager(final ReactiveUserDetailsService
userDetailsService,
        final PasswordEncoder passwordEncoder) {
    UserDetailsRepositoryReactiveAuthenticationManager
authenticationManager = new
UserDetailsRepositoryReactiveAuthenticationManager(
userDetailsService);
    authenticationManager.setPasswordEncoder(passwordEncoder);
    return authenticationManager;
}
```

4. Create the `SecurityWebFilterChain` bean:

```
//com/packtpub/springsecurity/configuration/SecurityConfig.java
@Bean
public SecurityWebFilterChain filterChain(ServerHttpSecurity
http) {
    http.authorizeExchange(exchanges -> exchanges
                .pathMatchers("/webjars/**").permitAll()
                .pathMatchers("/css/**").permitAll()
                .pathMatchers("/favicon.ico").permitAll()
                // H2 console:
                .pathMatchers("/admin/h2/**").permitAll()
                .pathMatchers("/").permitAll()
                .pathMatchers("/login/*").permitAll()
                .pathMatchers("/logout").permitAll()
                .pathMatchers("/signup/*").permitAll()
                .pathMatchers("/errors/**").permitAll()
                .pathMatchers("/admin/*").hasRole("ADMIN")
                .pathMatchers("/events/").hasRole("ADMIN")
                .pathMatchers("/**").hasRole("USER"));

    http.formLogin(Customizer.withDefaults());
```

```
        http.exceptionHandling(exceptions -> exceptions
            .accessDeniedHandler(new
    HttpStatusServerAccessDeniedHandler(HttpStatus.FORBIDDEN)));
        return http.build();
    }
```

Launch the application and try to access it through `http://localhost:8080`.

You should be able to log in with the configured users `admin1@example.com/admin1` or `user1@example.com/user1` and test the events creation.

Furthermore, creating a new user enables immediate login with the newly created credentials.

In summary, `Spring Security` also provide a reactive implementation. This implementation is best suited for applications that require handling a large number of concurrent connections, high scalability, and efficient resource utilization. It's particularly beneficial in scenarios involving non-blocking I/O operations, such as web applications with heavy user traffic or real-time data processing.

Summary

In this chapter, we covered several security vulnerabilities, as well as employing `Spring Security` to circumvent those vulnerabilities. After reading this chapter, you should understand the threat of CSRF and the use of *synchronizer token* pattern to prevent CSRF.

You should also know how to include various HTTP headers to protect against common security vulnerabilities using the *Cache-Control*, *Content-Type Options*, *HSTS*, *X-Frame-Options*, and *X-XSS-Protection* methods.

In the next chapter, we will discuss how to migrate to **Spring Security 6.x**.

16

Migration to Spring Security 6

In this final chapter, we will review information relating to common migration issues when moving from **Spring Security 5.x** to **Spring Security 6.x**. We'll spend much more time discussing the differences between these two versions because this is what most users will struggle with, as the updates from Spring Security 5.x to Spring Security 6.x contain a lot of non-passive refactoring.

At the end of the chapter, we will also highlight some of the new features that can be found in Spring Security 6.x. However, we do not explicitly cover changes from Spring Security 5.x to Spring Security 6.x. This is because by explaining the differences between this two versions, users should be able to update to Spring Security 6.x.

You may be planning to migrate an existing application to Spring Security 6.x or you may be trying to add functionality to a Spring Security 5.x application and are looking for guidance in the pages of this book. We'll try to address both of your concerns in this chapter.

First, we'll run through the important differences between Spring Security 5.x and 6.x—both in terms of features and configuration.

Second, we'll provide some guidance on mapping configuration or class name changes. This will better enable you to translate the examples in the book from Spring Security 6.x back to Spring Security 5.x (where applicable).

> **Important note**
>
> Spring Security 6.x mandates a migration to **Spring Framework 6** and **Java 17** or greater.
>
> Be aware that in many cases, migrating these other components may have a greater impact on your application than the upgrade of Spring Security!

During this chapter, we will cover the following topics:

- Reviewing important enhancements in `Spring Security 6.x`.

- Understanding configuration changes required in your existing Spring version.

- Reviewing `Spring Security 5.x` applications when moving them to `Spring Security 6.x`.

- Illustrating the overall movement of important classes and packages in `Spring Security 6.x`.

- Highlighting some of the new features found in `Spring Security 6.x`. Once you have completed the review of this chapter, you will be in a good position to migrate an existing application from `Spring Security 5.x` to `Spring Security 6.x`.

- Migrating from `Spring Security 5.x`.

This chapter's code in action link is here: `https://packt.link/wDOSk`.

Exploit Protection

In `Spring Security 5.8`, the default `CsrfTokenRequestHandler` responsible for providing the `CsrfToken` to the application is `CsrfTokenRequestAttributeHandler`. The default setting for the field `csrfRequestAttributeName` is `null`, leading to the loading of the CSRF token on every request.

Examples of situations where reading the session should be deemed unnecessary include endpoints explicitly marked with `permitAll()`, such as static assets, static HTML pages, and single-page applications hosted under the same domain/server.

In `Spring Security 6`, `csrfRequestAttributeName` now defaults to `_csrf`. If you had configured the following solely for the purpose of transitioning to version 6.0, you can now safely remove it:

```
requestHandler.setCsrfRequestAttributeName("_csrf");
```

Now that we have explored how to define the `CsrfToken`, we will explore how to protect against CSRF attacks.

Protecting against CSRF attacks

In `Spring Security 5.8`, the default `CsrfTokenRequestHandler` facilitating the availability of `CsrfToken` to the application is `CsrfTokenRequestAttributeHandler`. `XorCsrfTokenRequestAttributeHandler` was introduced to enable opting into CSRF attack support.

In Spring Security 6, XorCsrfTokenRequestAttributeHandler becomes the default CsrfTokenRequestHandler for providing the CsrfToken. If you had configured XorCsrfTokenRequestAttributeHandler solely for the purpose of transitioning to version 6.0, it can now be safely removed.

> **Important note**
> If you've set the csrfRequestAttributeName to null to exclude deferred tokens, or if you've established a CsrfTokenRequestHandler for any particular purpose, you can maintain the current configuration.

CSRF attack with WebSocket support

In Spring Security 5.8, the default ChannelInterceptor used to provide the CsrfToken with WebSocket security is CsrfChannelInterceptor. XorCsrfChannelInterceptor was introduced to enable opting into **CSRF attack support**.

In Spring Security 6, XorCsrfChannelInterceptor becomes the default ChannelInterceptor for providing the CsrfToken. If you had configured XorCsrfChannelInterceptor solely for the purpose of transitioning to version 6.0, it can now be safely removed.

After exploring how to protect against CSRF attacks, we will deep dive into configuration migration options.

Configuration Migrations

The subsequent sections pertain to alterations in configuring HttpSecurity, WebSecurity, and AuthenticationManager.

Adding @Configuration annotation to @Enable* annotations

In version 6.0, the annotations @EnableWebSecurity, @EnableMethodSecurity, @EnableGlobalMethodSecurity, and @EnableGlobalAuthentication no longer include @Configuration.

For instance, @EnableWebSecurity will be modified from:

```
@EnableWebSecurity
public class SecurityConfig {
    // ...
}
```

to:

```
@Configuration
@EnableWebSecurity
public class SecurityConfig {
    // ...
}
```

To adapt to this change, wherever you utilize these annotations, you might need to add @Configuration.

Using the new requestMatchers methods

In Spring Security 5.8, the methods antMatchers, mvcMatchers, and regexMatchers were deprecated in favor of the new requestMatchers methods.

The introduction of the new requestMatchers methods extended to authorizeHttpRequests, authorizeRequests, CSRF configuration, WebSecurityCustomizer, and other locations with specialized RequestMatcher methods. As of Spring Security 6, the deprecated methods have been removed.

The new methods come with more secure defaults by automatically selecting the most suitable RequestMatcher implementation for your application.

To provide a summary, these methods:

- Opt for the MvcRequestMatcher implementation if your application includes Spring MVC in the classpath.

- Fall back to the AntPathRequestMatcher implementation, in the absence of Spring MVC, aligning their behavior with the Kotlin equivalent methods.

The following table should guide you in your migration journey:

Spring Security 5	Spring Security 6
antMatchers("/api/admin/**")	requestMatchers("/api/admin/**")
mvcMatchers("/admin/**")	requestMatchers("/admin/**")
mvcMatchers("/admin").servletPath("/path")	requestMatchers(mvcMatcherBuilder.pattern("/admin"))

Table 16.1 – Migration to the new requestMatchers

If you encounter difficulties with the new requestMatchers methods, you have the option to revert to the RequestMatcher implementation you were previously using. For instance, if you prefer to continue using AntPathRequestMatcher and RegexRequestMatcher implementations, you can utilize the requestMatchers method that accepts a RequestMatcher instance:

Spring Security 5	Spring Security 6
antMatchers("/api/admin/**")	requestMatchers(antMatcher("/user/**"))

Table 16.2 – Alternatives with the new requestMatchers

> **Important note**
> Please note that the provided example utilizes static factory methods from AntPathRequestMatcher and RegexRequestMatcher to enhance readability.

When you are employing the WebSecurityCustomizer interface, you can substitute the deprecated antMatchers methods:

```
@Bean
public WebSecurityCustomizer webSecurityCustomizer() {
    return web -> web.ignoring().antMatchers("/ignore1", "/ignore2");
}
```

with their corresponding requestMatchers alternatives:

```
@Bean
public WebSecurityCustomizer webSecurityCustomizer() {
    return web -> web.ignoring().requestMatchers("/ignore1", "/
ignore2");
}
```

Similarly, if you are customizing the CSRF configuration to exclude specific paths, you can substitute the deprecated methods with the requestMatchers counterparts.

Using the new securityMatchers methods

In Spring Security 5.8, the antMatchers, mvcMatchers, and requestMatchers methods in HttpSecurity underwent deprecation in favor of the new securityMatchers methods.

It's important to note that these methods differ from the authorizeHttpRequests methods, which were deprecated in favor of the requestMatchers methods. However, the securityMatchers methods share similarities with the requestMatchers methods in that they automatically select the most suitable RequestMatcher implementation for your application.

To elaborate, the new methods:

- Opt for the `MvcRequestMatcher` implementation if your application includes Spring MVC in the classpath.
- Fall back to the `AntPathRequestMatcher` implementation, in the absence of Spring MVC, aligning their behavior with the Kotlin equivalent methods. The introduction of `securityMatchers` methods also serves to prevent confusion with the `requestMatchers` methods from `authorizeHttpRequests`.

The following table should guide you in your migration journey where `http` is of type `HttpSecurity`:

Spring Security 5	Spring Security 6
`http.antMatcher("/api/**")`	`http.securityMatcher("/api/**")`
`http.requestMatcher(new MyCustomRequestMatcher())`	`http.securityMatcher(new MyCustomRequestMatcher())`

`http`
 	
`.requestMatchers((matchers) -> matchers`
 	
`.antMatchers("/api/**", "/app/**")`
 	
`.mvcMatchers("/admin/**")`
 	
`.requestMatchers(new MyCustomRequestMatcher()))`
	`http.securityMatchers((matchers) -> matchers.requestMatchers("/api/**", "/app/**", "/admin/**")`
	
	`.requestMatchers(new MyCustomRequestMatcher()))`

Table 16.3 – Migration to the new securityMatchers

If you encounter challenges with the automatic selection of `RequestMatcher` implementation by the `securityMatchers` methods, you have the option to manually choose the `RequestMatcher` implementation yourself:

```
@Bean
public SecurityFilterChain filterChain(HttpSecurity http) throws
Exception {
    http
        .securityMatchers(matchers -> matchers
            .requestMatchers(antMatcher("/api/**"), antMatcher("/
app/**"))
        );
    return http.build();
}
```

After exploring the new securityMatchers methods, we'll now proceed to examine the process of replacing WebSecurityConfigurerAdapter in Spring Security 6.x.

Replacing the WebSecurityConfigurerAdapter class

The WebSecurityConfigurerAdapter class was deprecated and then removed in Spring Security 6.x. In the following sub-sections, we will explore the impacts of this major change.

Exposing a SecurityFilterChain Bean

In Spring Security 5.4, a new feature was introduced allowing the publication of a SecurityFilterChain bean instead of extending WebSecurityConfigurerAdapter. However, in version 6.0, WebSecurityConfigurerAdapter has been removed. To accommodate this change, you can substitute constructs resembling:

```
@Configuration
public class SecurityConfiguration extends
WebSecurityConfigurerAdapter {

    @Override
    protected void configure(HttpSecurity http) throws Exception {
       http
              .authorizeHttpRequests(authorize -> authorize
                   .anyRequest().authenticated()
              )
              .httpBasic(withDefaults());
    }

}
```

with:

```
@Configuration
public class SecurityConfiguration {

    @Bean
    public SecurityFilterChain filterChain(HttpSecurity http) throws
Exception {
       http
              .authorizeHttpRequests((authorize) -> authorize
                   .anyRequest().authenticated()
              )
              .httpBasic(withDefaults());
       return http.build();
```

```
    }

}
```

Exposing a WebSecurityCustomizer Bean

Spring Security 5.4 introduced WebSecurityCustomizer as a replacement for configure(WebSecurity web) in WebSecurityConfigurerAdapter. To prepare for its removal, you can update code similar to the following:

```
@Configuration
public class SecurityConfiguration extends
WebSecurityConfigurerAdapter {
    @Override
    public void configure(WebSecurity web) {
        web.ignoring().antMatchers("/ignore1", "/ignore2");
    }
}
```

with:

```
@Configuration
public class SecurityConfiguration {
    @Bean
    public WebSecurityCustomizer webSecurityCustomizer() {
        return (web) -> web.ignoring().antMatchers("/ignore1", "/
ignore2");
    }
}
```

Exposing an AuthenticationManager Bean

With the removal of W e b S e c u r i t y C o n f i g u r e r A d a p t e r, the configure(AuthenticationManagerBuilder) method is also eliminated.

LDAP Authentication

When using auth.ldapAuthentication() for **Lightweight Directory Access Protocol (LDAP)** authentication support, you can replace this:

```
@Configuration
public class SecurityConfig extends WebSecurityConfigurerAdapter {
    @Override
    protected void configure(AuthenticationManagerBuilder auth) throws
Exception {
        auth
```

```
            .ldapAuthentication()
            .userDetailsContextMapper(new PersonContextMapper())
            .userDnPatterns("uid={0},ou=people")
            .contextSource()
            .port(0);
    }
}
```

with:

```
@Configuration
public class SecurityConfiguration {
    @Bean
    public EmbeddedLdapServerContextSourceFactoryBean
contextSourceFactoryBean() {
        EmbeddedLdapServerContextSourceFactoryBean
contextSourceFactoryBean =
                EmbeddedLdapServerContextSourceFactoryBean.
fromEmbeddedLdapServer();
        contextSourceFactoryBean.setPort(0);
        return contextSourceFactoryBean;
    }

    @Bean
    AuthenticationManager
ldapAuthenticationManager(BaseLdapPathContextSource contextSource) {
        LdapBindAuthenticationManagerFactory factory =
                new LdapBindAuthenticationManagerFactory(contextSource);
        factory.setUserDnPatterns("uid={0},ou=people");
        factory.setUserDetailsContextMapper(new PersonContextMapper());
        return factory.createAuthenticationManager();
    }
}
```

JDBC Authentication

If you are currently utilizing auth.jdbcAuthentication() for **Java Database Connectivity (JDBC)** authentication support, you can substitute:

```
@Configuration
public class SecurityConfig extends WebSecurityConfigurerAdapter {

    private final DataSource dataSource;
    public SecurityConfig(DataSource dataSource) {
        this.dataSource = dataSource;
```

```
    }

    @Override
    protected void configure(AuthenticationManagerBuilder auth) throws
Exception {
        UserDetails user = User.withDefaultPasswordEncoder()
                .username("user")
                .password("password")
                .roles("USER")
                .build();
        auth.jdbcAuthentication()
                .withDefaultSchema()
                .dataSource(this.dataSource)
                .withUser(user);
    }
}
```

with:

```
@Configuration
public class SecurityConfig {

    private final DataSource dataSource;
    public SecurityConfig(DataSource dataSource) {
        this.dataSource = dataSource;
    }

    @Bean
    public UserDetailsManager users(DataSource dataSource) {
        UserDetails user = User.withDefaultPasswordEncoder()
                .username("user")
                .password("password")
                .roles("USER")
                .build();
        JdbcUserDetailsManager users = new
JdbcUserDetailsManager(dataSource);
        users.createUser(user);
        return users;
    }
}
```

In-Memory Authentication

If you are currently utilizing `auth.inMemoryAuthentication()` for in-memory Authentication support, you can replace:

```
@Configuration
public class SecurityConfig extends WebSecurityConfigurerAdapter {
    @Override
    protected void configure(AuthenticationManagerBuilder auth) throws
Exception {
        UserDetails user = User.withDefaultPasswordEncoder()
                .username("user")
                .password("password")
                .roles("USER")
                .build();
        auth.inMemoryAuthentication()
                .withUser(user);
    }
}
```

with:

```
@Configuration
public class SecurityConfig {
    @Bean
    public InMemoryUserDetailsManager userDetailsService() {
        UserDetails user = User.withDefaultPasswordEncoder()
                .username("user")
                .password("password")
                .roles("USER")
                .build();
        return new InMemoryUserDetailsManager(user);
    }
}
```

After exploring the impacts of `WebSecurityConfigurerAdapter` removal, we will delve into updates regarding password encoding.

Password Encoding Updates

In `Spring Security 6.0`, the minimum requirements for password encoding have been revised for **PBKDF2**, **SCrypt**, and **Argon2**.

If you use the default password encoder, there's no need to follow any preparatory steps, and you can skip this section.

Pbkdf2PasswordEncoder updates

If you are using `Pbkdf2PasswordEncoder`, the constructors have been substituted with static factories that correspond to the `Spring Security` version relevant to the provided settings:

```
@Bean
PasswordEncoder passwordEncoder2() {
    return Pbkdf2PasswordEncoder.defaultsForSpringSecurity_v5_5();
}
```

And if you have custom settings, use the constructor that specifies all settings:

```
@Bean
PasswordEncoder passwordEncoder() {
    return new Pbkdf2PasswordEncoder("secret".getBytes(UTF_8), 16,
185000, 256);
}
```

SCryptPasswordEncoder Updates

If you are employing `SCryptPasswordEncoder`, the constructors have been substituted with static factories that correspond to the `Spring Security` version associated with the provided settings.

Your initial step should be to modify the deprecated constructor:

```
@Bean
PasswordEncoder passwordEncoder() {
    return SCryptPasswordEncoder.defaultsForSpringSecurity_v4_1();
}
```

Argon2PasswordEncoder Updates

If you are using `Argon2PasswordEncoder`, the constructors have been substituted with static factories that correspond to the `Spring Security` version associated with the provided settings. For example:

```
@Bean
PasswordEncoder passwordEncoder() {
    return Argon2PasswordEncoder.defaultsForSpringSecurity_v5_2();
}
```

Delegating PasswordEncoder usage

If you haven't employed the deprecated constructor, it's essential to update your code to adhere to the latest standards. This entails configuring the `DelegatingPasswordEncoder` to identify passwords that adhere to current standards and update them to the latest ones. The following example using `Pbkdf2PasswordEncoder` can also be applied to `SCryptPasswordEncoder` or `Argon2PasswordEncoder`:

```
@Bean
PasswordEncoder passwordEncoder() {
    String prefix = "pbkdf2@5.8";
    PasswordEncoder current = Pbkdf2PasswordEncoder.
defaultsForSpringSecurity_v5_5();
    PasswordEncoder upgraded = Pbkdf2PasswordEncoder.
defaultsForSpringSecurity_v5_8();
    DelegatingPasswordEncoder delegating = new
DelegatingPasswordEncoder(prefix, Map.of(prefix, upgraded));
    delegating.setDefaultPasswordEncoderForMatches(current);
    return delegating;
}
```

Abandoning Encryptors.queryableText

The use of `Encryptors.queryableText(CharSequence, CharSequence)` is considered unsafe as identical input data will yield the same output (CVE-2020-5408 - `https://github.com/advisories/GHSA-2ppp-9496-p23q`).e

`Spring Security 6.x` no longer endorses data encryption through this method. To facilitate the upgrade, you must either re-encrypt the data using a supported mechanism or store it in a decrypted form.

Following the examination of password encoding updates, we will delve into the details of session management updates.

Session Management Updates

In the upcoming sections, we'll thoroughly examine session management updates, encompassing the primary deprecations and modifications.

Requiring Explicit Saving of SecurityContextRepository

In `Spring Security 5`, the default process involves automatically saving the `SecurityContext` to the `SecurityContextRepository` through the `SecurityContextPersistenceFilter`. This saving occurs just before the `HttpServletResponse` is committed and right before the `SecurityContextPersistenceFilter`. However, this automatic persistence can catch users off guard, especially when performed just before the request completes (i.e., prior to committing the `HttpServletResponse`). It also introduces complexity in tracking the state to determine the necessity of saving, leading to unnecessary writes to the `SecurityContextRepository` (e.g., `HttpSession`) at times.

With the advent of `Spring Security 6`, the default behavior has shifted. The `SecurityContextHolderFilter` will now solely read the `SecurityContext` from the `SecurityContextRepository` and populate it in the `SecurityContextHolder`. Users are now required to explicitly save the `SecurityContext` using the `SecurityContextRepository` if they wish for the `SecurityContext` to persist between requests. This modification eliminates ambiguity and enhances performance by mandating writes to the `SecurityContextRepository` (e.g., `HttpSession`) only when necessary.

To opt into the new `Spring Security 6` default, the following configuration can be used:

```
public SecurityFilterChain filterChain(HttpSecurity http) {
    http
            .securityContext((securityContext) -> securityContext
                .requireExplicitSave(true)
            );
    return http.build();
}
```

When using the configuration, it is crucial that any code responsible for setting the `SecurityContextHolder` with a `SecurityContext` also ensures the saving of the `SecurityContext` to the `SecurityContextRepository` if persistence between requests is required.

For example, the following code:

```
SecurityContextHolder.setContext(securityContext);
```

should be replaced with:

```
SecurityContextHolder.setContext(securityContext);
securityContextRepository.saveContext(securityContext,
httpServletRequest, httpServletResponse);
```

Changing HttpSessionSecurityContextRepository to DelegatingSecurityContextRepository

In `Spring Security 5`, the default `SecurityContextRepository` is `HttpSessionSecurityContextRepository`.

In Spring Security 6, the default SecurityContextRepository is DelegatingSecurityContextRepository. To adopt the new Spring Security 6 default, the following configuration can be utilized:

```
@Bean
public SecurityFilterChain filterChain(HttpSecurity http) throws
Exception {
    http
            // ...
        .securityContext((securityContext) -> securityContext
            .securityContextRepository(new
DelegatingSecurityContextRepository(
                new RequestAttributeSecurityContextRepository(),
                new HttpSessionSecurityContextRepository()
            ))
        );
    return http.build();
}
```

Addressing SecurityContextRepository Deprecations

In Spring Security 6, the following method in the class SecurityContextRepository has been deprecated:

```
Supplier<SecurityContext> loadContext(HttpServletRequest request)
```

The method should be replaced with the following:

```
DeferredSecurityContext loadDeferredContext(HttpServletRequest
request)
```

Improving Querying of RequestCache

In Spring Security 5, the standard procedure involves querying the saved request with every incoming request. In a typical configuration, this implies that the HttpSession is consulted on each request to utilize the RequestCache.

In Spring Security 6, the new default is such that the RequestCache will only be interrogated for a cached request if the HTTP parameter continue is explicitly defined. This approach enables Spring Security to skip unnecessary reads of the HttpSession when working with the RequestCache.

Requiring Explicit Invocation of SessionAuthenticationStrategy

In Spring Security 5, the standard configuration depends on the SessionManagementFilter to identify whether a user has recently authenticated and to trigger the SessionAuthenticationStrategy. However, this setup entails reading the HttpSession for every request in a typical scenario.

In Spring Security 6, the new default is for authentication mechanisms to directly invoke the SessionAuthenticationStrategy. Consequently, there is no requirement to identify when authentication occurs, eliminating the need to read the HttpSession for every request.

Following the investigation of session management updates, we will delve deeply into authentication updates.

Authentication Updates

We'll examine the main updates in authentication, including the adoption of SHA-256 for Remember Me functionality and enhancements related to AuthenticationServiceExceptions.

Utilizing SHA-256 for Remember Me

The TokenBasedRememberMeServices implementation in Spring Security 6 now defaults to using SHA-256 for Remember Me tokens, enhancing the default security stance. This change is motivated by the recognition of MD5 as a weak hashing algorithm susceptible to collision attacks and modular differential attacks.

The newly generated tokens include information about the algorithm used for token generation. This information is leveraged for matching purposes. If the algorithm name is absent, the matchingAlgorithm property is employed to verify the token. This design allows for a seamless transition from MD5 to SHA-256.

The following code shows how you can enable Remember Me feature, with the default implementation:

```
@Configuration
@EnableWebSecurity
public class SecurityConfig {
    @Bean
    SecurityFilterChain securityFilterChain(HttpSecurity http,
RememberMeServices rememberMeServices) throws Exception {
        http
                // ...
                .rememberMe(remember -> remember
                    .rememberMeServices(rememberMeServices)
                );
        return http.build();
    }
    @Bean
```

```
    RememberMeServices rememberMeServices(UserDetailsService
userDetailsService) {
        return new TokenBasedRememberMeServices(myKey,
userDetailsService);
    }
}
```

To embrace the new `Spring Security 6` default for encoding tokens while maintaining compatibility with MD5-encoded tokens, you can set the `encodingAlgorithm` property to `SHA-256` and the `matchingAlgorithm` property to MD5:

```
@Configuration
@EnableWebSecurity
public class SecurityConfig {

    @Bean
    SecurityFilterChain securityFilterChain(HttpSecurity http,
RememberMeServices rememberMeServices) throws Exception {
        http
                .rememberMe(remember -> remember
                        .rememberMeServices(rememberMeServices)
                );
        return http.build();
    }

    @Bean
    RememberMeServices rememberMeServices(UserDetailsService
userDetailsService) {
        RememberMeTokenAlgorithm encodingAlgorithm =
RememberMeTokenAlgorithm.SHA256;
        TokenBasedRememberMeServices rememberMe = new
TokenBasedRememberMeServices(myKey, userDetailsService,
encodingAlgorithm);
        rememberMe.setMatchingAlgorithm(RememberMeTokenAlgorithm.MD5);
        return rememberMe;
    }

}
```

Propagating AuthenticationServiceExceptions

The `AuthenticationFilter` forwards `AuthenticationServiceException` to the `AuthenticationEntryPoint`. As `AuthenticationServiceExceptions` indicate a server-side error rather than a client-side error, in version 6.0, this mechanism is adjusted to propagate them to the container.

Therefore, if you had previously enabled this behavior by setting `rethrowAuthenticationServiceException` to `true`, you can now eliminate it as follows:

```
AuthenticationFilter authenticationFilter = new
AuthenticationFilter(...);
AuthenticationEntryPointFailureHandler handler = new
AuthenticationEntryPointFailureHandler(...);
handler.setRethrowAuthenticationServiceException(true);
authenticationFilter.setAuthenticationFailureHandler(handler);
```

This can be changed to:

```
AuthenticationFilter authenticationFilter = new
AuthenticationFilter(...);
AuthenticationEntryPointFailureHandler handler = new
AuthenticationEntryPointFailureHandler(...);
authenticationFilter.setAuthenticationFailureHandler(handler);
```

Following the examination of authentication updates, we will delve deeply into authorization updates.

Authorization Updates

In this section, we will explore several key enhancements in authorization management within `Spring Security`. We'll begin by discussing how to utilize the `AuthorizationManager` for `Method Security`, enabling fine-grained control over method-level access. Next, we'll delve into leveraging the `AuthorizationManager` for message security, facilitating secure communication over messaging protocols. Additionally, we'll highlight some deprecations like `AbstractSecurityWebSocketMessageBrokerConfigurer`.

Leveraging AuthorizationManager for Method Security

Method Security is now streamlined with the `AuthorizationManager` API and direct utilization of **Spring AOP**.

In case you encounter challenges while implementing these adjustments, it's essential to note that even though `@EnableGlobalMethodSecurity` is deprecated, it has not been removed in version 6.0. This ensures you have the option to opt out by continuing to use the deprecated annotation.

Replacing Global Method Security with Method Security

`@EnableGlobalMethodSecurity` and `<global-method-security>` are now deprecated in favor of `@EnableMethodSecurity` and `<method-security>`, respectively. The updated annotation and XML element automatically activate Spring's pre-post annotations and internally utilize `AuthorizationManager`.

Changing the order value in @EnableTransactionManagement

@EnableTransactionManagement and @EnableGlobalMethodSecurity both have the same order value, Integer.MAX_VALUE. As a result, their relative order in the Spring AOP Advisor chain is undefined.

While this is generally acceptable, as most **Method Security** expressions don't rely on an open transaction to function correctly, there were historical cases where it was necessary to ensure a specific order by setting their order values.

On the contrary, @EnableMethodSecurity lacks an order value because it dispatches multiple interceptors. Unlike @EnableTransactionManagement, it cannot maintain backward compatibility, as it cannot position all interceptors within the same advisor chain location.

Instead, the order values for the @EnableMethodSecurity interceptors are based on an offset of 0. For example, the @PreFilter interceptor has an order of 100, @PostAuthorize has an order of 200, and so forth.

If, after updating, you discover that your Method Security expressions are not functioning due to a lack of an open transaction, please modify your transaction annotation definition as the following:

```
@EnableTransactionManagement(order = 0)
```

Using a Custom @Bean instead of subclassing DefaultMethodSecurityExpressionHandler

For performance optimization, a new method has been added to MethodSecurityExpressionHandler that accepts a Supplier<Authentication> instead of an Authentication.

This enhancement allows Spring Security to defer the Authentication lookup and is automatically utilized when employing @EnableMethodSecurity instead of @EnableGlobalMethodSecurity.

For instance, suppose you aim for a customized evaluation of @PostAuthorize("hasAuthority('ADMIN')"). In such a case, you can create a custom @Bean as demonstrated here:

```
class MyAuthorizer {
    boolean isAdmin(MethodSecurityExpressionOperations root) {
        boolean decision = root.hasAuthority("ADMIN");
        // custom work ...
        return decision;
    }
}
```

Subsequently, reference it in the annotation as follows:

```
@PreAuthorize("@authz.isAdmin(#root)")
```

Exposing a MethodSecurityExpressionHandler instead of a PermissionEvaluator

`@EnableMethodSecurity` does not automatically detect a `PermissionEvaluator` to keep its API straightforward.

If you have a custom `PermissionEvaluator` declared as a `@Bean`, please update it from:

```
@Bean
static PermissionEvaluator permissionEvaluator() {
    // ... your evaluator
}
```

to:

```
@Bean
static MethodSecurityExpressionHandler expressionHandler() {
    var expressionHandler = new
DefaultMethodSecurityExpressionHandler();
    expressionHandler.setPermissionEvaluator(myPermissionEvaluator);
    return expressionHandler;
}
```

Substituting any custom AccessDecisionManagers in Method Security

Your application might feature a custom `AccessDecisionManager` or `AccessDecisionVoter` configuration. The approach to adaptation will vary depending on the specific purpose of each configuration. Continue reading to identify the most suitable adjustment for your scenario.

UnanimousBased use case

If your application utilizes `UnanimousBased` with the default voters, you probably won't need to make any changes since unanimous-based is the default behavior with `@EnableMethodSecurity`.

Nevertheless, if you find that the default authorization managers are not suitable, you can utilize `AuthorizationManagers.allOf` to construct your custom configuration.

AffirmativeBased use case

If your application relies on `AffirmativeBased`, you can create an equivalent `AuthorizationManager` as follows:

```
AuthorizationManager<MethodInvocation> authorization =
AuthorizationManagers.anyOf(
        // ... your list of authorization managers
)
```

ConsensusBased use case

For `ConsensusBased`, there is no built-in equivalent provided by the framework. In this scenario, you should implement a composite `AuthorizationManager` that considers the set of delegate `AuthorizationManager` instances.

AccessDecisionVoter use case

You can either modify the class to implement `AuthorizationManager` or create an adapter as follows:

```
public final class PreAuthorizeAuthorizationManagerAdapter implements
AuthorizationManager<MethodInvocation> {
    private final SecurityMetadataSource metadata;
    private final AccessDecisionVoter voter;

    public PreAuthorizeAuthorizationManagerAdapter
(MethodSecurityExpressionHandler expressionHandler) {
        ExpressionBasedAnnotationAttributeFactory attributeFactory =
            new
ExpressionBasedAnnotationAttributeFactory(expressionHandler);
        this.metadata = new
PrePostAnnotationSecurityMetadataSource(attributeFactory);
        ExpressionBasedPreInvocationAdvice expressionAdvice = new
ExpressionBasedPreInvocationAdvice();
        expressionAdvice.setExpressionHandler(expressionHandler);
        this.voter = new
PreInvocationAuthorizationAdviceVoter(expressionAdvice);
    }

    public AuthorizationDecision check(Supplier<Authentication>
authentication, MethodInvocation invocation) {
        List<ConfigAttribute> attributes = this.metadata.
getAttributes(invocation, AopUtils.getTargetClass(invocation.
getThis())));
        int decision = this.voter.vote(authentication.get(),
invocation, attributes);
        if (decision == ACCESS_GRANTED) {
```

```
        return new AuthorizationDecision(true);
    }
    if (decision == ACCESS_DENIED) {
        return new AuthorizationDecision(false);
    }
    return null; // abstain
    }
}
```

AfterInvocationManager or AfterInvocationProvider use case

`AfterInvocationManager` and `AfterInvocationProvider` are responsible for making an authorization decision regarding the result of an invocation. For instance, in the context of method invocation, they determine the authorization of a method's return value.

In `Spring Security 3.0`, the decision-making process for authorization was standardized through the `@PostAuthorize` and `@PostFilter` annotations. `@PostAuthorize` is used to determine whether the entire return value is allowed to be returned. On the other hand, `@PostFilter` is employed to filter individual entries from a returned collection, array, or stream.

These two annotations should fulfill most requirements, and there is encouragement to transition to one or both of them as `AfterInvocationProvider` and `AfterInvocationManager` are now deprecated.

RunAsManager use case

At present, there is no direct substitute for `RunAsManager`, although the possibility of introducing one is under consideration.

However, if required, it is relatively simple to modify a `RunAsManager` to align with the `AuthorizationManager` API.

Here is some pseudocode to assist you in getting started:

```
public final class RunAsAuthorizationManagerAdapter<T> implements
AuthorizationManager<T> {
    private final RunAsManager runAs = new RunAsManagerImpl();
    private final SecurityMetadataSource metadata;
    private final AuthorizationManager<T> authorization;

    // ... constructor

    public AuthorizationDecision check(Supplier<Authentication>
authentication, T object) {
        Supplier<Authentication> wrapped = (auth) -> {
            List<ConfigAttribute> attributes = this.metadata.
```

```
getAttributes(object);
        return this.runAs.buildRunAs(auth, object, attributes);
    };
    return this.authorization.check(wrapped, object);
  }
}
```

Verifying for AnnotationConfigurationException

`@EnableMethodSecurity` and `<method-security>` enable more stringent enforcement of `Spring Security`'s non-repeatable or otherwise incompatible annotations. If you encounter `AnnotationConfigurationException` in your logs after transitioning to either, follow the instructions provided in the exception message to rectify your application's `Method Security` annotation usage.

Leveraging AuthorizationManager for Message Security

Message Security has been enhanced with the `AuthorizationManager` API and direct utilization of Spring AOP.

To configure the `AuthorizationManager` for Message Security, you will need to follow the steps below:

1. Ensure all messages have defined authorization rules:

    ```
    @Override
    protected void
    configureInbound(MessageSecurityMetadataSourceRegistry messages)
    {
        messages
                .simpTypeMatchers(CONNECT, DISCONNECT, UNSUBSCRIBE).
    permitAll()
                .simpDestMatchers("/user/queue/errors").permitAll()
                .simpDestMatchers("/admin/**").hasRole("ADMIN")
                .anyMessage().denyAll();
    }
    ```

2. Add `@EnableWebSocketSecurity` annotation.

3. Utilize an instance of `AuthorizationManager<Message<?>>`:

    ```
    @Bean
    AuthorizationManager<Message<?>>
    messageSecurity(MessageMatcherDelegatingAuthorizationManager.
    Builder messages) {
        messages
                .simpTypeMatchers(CONNECT, DISCONNECT, UNSUBSCRIBE).
    ```

```
permitAll()
            .simpDestMatchers("/user/queue/errors").permitAll()
            .simpDestMatchers("/admin/**").hasRole("ADMIN")
            .anyMessage().denyAll();
    return messages.build();
}
```

Now we've examined the AuthorizationManager configuration for Message Security, we'll delve into the modifications associated with AbstractSecurityWebSocketMessageBrokerConfigurer.

Deprecating AbstractSecurityWebSocketMessageBrokerConfigurer

If you are employing Java configuration, you can now directly extend WebSocketMessageBrokerConfigurer.

For instance, if your class is extending AbstractSecurityWebSocketMessageBrokerConfigurer is named WebSocketSecurityConfig, then replace it with the following:

```
@EnableWebSocketSecurity
@Configuration
public class WebSocketSecurityConfig implements
WebSocketMessageBrokerConfigurer {
    // ...
}
```

Having clarified the reasons for discontinuing the implementation of AbstractSecurityWebSocketMessageBrokerConfigurer, let's now delve into the utilization of AuthorizationManager for request security.

Employing AuthorizationManager for Request Security

HTTP Request Security has been streamlined with the AuthorizationManager API. We will explain AuthorizationManager changes for security requests in Spring Security 6.x.

Ensure that all requests have well-defined authorization rules

In Spring Security 5.8 and earlier, requests without an authorization rule are allowed by default. However, for a more robust security posture, the default approach is to deny by default in Spring Security 6.0. This means that any request lacking an explicit authorization rule will be denied by default.

If you already have an `anyRequest` rule in place that meets your requirements, you can skip this step:

```
http
        .authorizeRequests((authorize) -> authorize
                .filterSecurityInterceptorOncePerRequest(true)
                .mvcMatchers("/app/**").hasRole("APP")
                // ...
                .anyRequest().denyAll()
        )
```

If you have already transitioned to `authorizeHttpRequests`, the recommended modification remains the same.

Transitioning to AuthorizationManager

To adopt the use of `AuthorizationManager`, you can utilize `authorizeHttpRequests` for Java configuration or use `use-authorization-manager` for XML configuration:

```
http
        .authorizeHttpRequests((authorize) -> authorize
                .shouldFilterAllDispatcherTypes(false)
                .mvcMatchers("/app/**").hasRole("APP")
                // ...
                .anyRequest().denyAll()
        )
```

Migrating from hasIpAddress to access(AuthorizationManager)

To migrate from `hasIpAddress` to `access(AuthorizationManager)`, use:

```
IpAddressMatcher hasIpAddress = new IpAddressMatcher("127.0.0.1");
http
        .authorizeHttpRequests((authorize) -> authorize
                    .requestMatchers("/app/**").access((authentication,
context) ->
                            new AuthorizationDecision(hasIpAddress.
matches(context.getRequest()))
                            // ...
                            .anyRequest().denyAll()
                ))
```

> **Important note**
>
> Securing by IP address is inherently delicate. Therefore, there are no intentions to transfer this support to `authorizeHttpRequests`.

Transitioning SpEL expressions to AuthorizationManager

When it comes to authorization rules, Java is generally more straightforward to test and maintain than SpEL. Consequently, `authorizeHttpRequests` does not provide a method for declaring a String SpEL. Instead, you can create your own `AuthorizationManager` implementation or utilize `WebExpressionAuthorizationManager`.

SpEL	AuthorizationManager	WebExpressionAuthorizationManager

`mvcMatchers("/`
`complicated/**").`
`access("hasRole`
`('ADMIN') ||`
`hasAuthority`
`('SCOPE_read')")`
	`mvcMatchers("/`
	`complicated/**").`
	`access(anyOf(hasRole`
	`("ADMIN"),`
	`hasAuthority`
	`("SCOPE_read"))`
		`mvcMatchers("/`
		`complicated/**").access`
		`(new`
		`WebExpressionAuthorization`
		`Manager("hasRole('ADMIN')`
		`|| hasAuthority('SCOPE_`
		`read')"))`

Table 16.4 – SpEL migration options

Transitioning to filtering all dispatcher types

In `Spring Security 5.8` and earlier, authorization is executed only once per request. Consequently, dispatcher types like `FORWARD` and `INCLUDE` that run after `REQUEST` are not secured by default. It is advisable for `Spring Security` to secure all dispatcher types. Therefore, in version 6.0, `Spring Security` modifies this default behavior.

To do this, you should change:

```
http
    .authorizeHttpRequests((authorize) -> authorize
        .shouldFilterAllDispatcherTypes(false)
        .mvcMatchers("/app/**").hasRole("APP")
        // ...
        .anyRequest().denyAll()
    )
```

to:

```
http
    .authorizeHttpRequests((authorize) -> authorize
        .shouldFilterAllDispatcherTypes(true)
        .mvcMatchers("/app/**").hasRole("APP")
        // ...
        .anyRequest().denyAll()
    )
```

Then, set:

```
spring.security.filter.dispatcher-types=request,async,error,forward,i
nclude
```

If you're using the `AbstractSecurityWebApplicationInitializer`, it's recommended to override the `getSecurityDispatcherTypes` method and return all dispatcher types:

```
public class SecurityWebApplicationInitializer extends
AbstractSecurityWebApplicationInitializer {

    @Override
    protected EnumSet<DispatcherType> getSecurityDispatcherTypes() {
        return EnumSet.of(DispatcherType.REQUEST, DispatcherType.ERROR,
DispatcherType.ASYNC,
            DispatcherType.FORWARD, DispatcherType.INCLUDE);
    }

}
```

Allowing FORWARD when employing Spring MVC

When Spring MVC identifies a mapping between the view name and the actual views, it initiates a forward to the view. As demonstrated in the previous section, `Spring Security 6.0` will, by default, apply authorization to FORWARD requests.

Substituting any custom filter-security AccessDecisionManager

In this section, we will explore the different use cases to substitute custom filter-security based on `AccessDecisionManager`.

UnanimousBased use case

If your application relies on `UnanimousBased`, begin by adjusting or replacing any `AccessDecisionVoter` Subsequently, you can create an `AuthorizationManager` as follows:

```
@Bean
AuthorizationManager<RequestAuthorizationContext>
requestAuthorization() {
    PolicyAuthorizationManager policy = ...;
    LocalAuthorizationManager local = ...;
    return AuthorizationManagers.allOf(policy, local);
}
```

Then, integrate it into the DSL as follows:

```
http
        .authorizeHttpRequests((authorize) -> authorize.anyRequest().
access(requestAuthorization))
// ...
```

AffirmativeBased use case

If your application utilizes `AffirmativeBased`, you can create an equivalent `AuthorizationManager` as follows:

```
@Bean
AuthorizationManager<RequestAuthorizationContext>
requestAuthorization() {
    PolicyAuthorizationManager policy = ...;
    LocalAuthorizationManager local = ...;
    return AuthorizationManagers.anyOf(policy, local);
}
```

Then, integrate it into the DSL as follows:

```
http
        .authorizeHttpRequests((authorize) -> authorize.anyRequest().
access(requestAuthorization))
// ...
```

ConsensusBased use case

If your application is using `ConsensusBased`, there is no equivalent provided by the framework. In this case, you should implement a composite `AuthorizationManager` that considers the set of delegate `AuthorizationManagers`.

Custom AccessDecisionVoter use case

If your application is using `AccessDecisionVoter`, you can either modify the class to implement `AuthorizationManager` or create an adapter. Without knowledge of the specific functionality of your custom voter, it's challenging to provide a generic solution.

However, here's an illustrative example of adapting `SecurityMetadataSource` and `AccessDecisionVoter` for `anyRequest().authenticated()`:

```
public final class AnyRequestAuthenticatedAuthorizationManagerAdapter
implements AuthorizationManager<RequestAuthorizationContext> {
    private final SecurityMetadataSource metadata;
    private final AccessDecisionVoter voter;
```

```
    public
PreAuthorizeAuthorizationManagerAdapter(SecurityExpressionHandler
expressionHandler) {
        Map<RequestMatcher, List<ConfigAttribute>> requestMap =
Collections.singletonMap(
            AnyRequestMatcher.INSTANCE, Collections.singletonList(new
SecurityConfig("authenticated")));
        this.metadata = new
DefaultFilterInvocationSecurityMetadataSource(requestMap);
        WebExpressionVoter voter = new WebExpressionVoter();
        voter.setExpressionHandler(expressionHandler);
        this.voter = voter;
    }

    public AuthorizationDecision check(Supplier<Authentication>
authentication, RequestAuthorizationContext context) {
        List<ConfigAttribute> attributes = this.metadata.
getAttributes(context);
        int decision = this.voter.vote(authentication.get(),
invocation, attributes);
        if (decision == ACCESS_GRANTED) {
            return new AuthorizationDecision(true);
        }
        if (decision == ACCESS_DENIED) {
            return new AuthorizationDecision(false);
        }
        return null; // abstain
    }
}
```

Having elucidated the usage of `AuthorizationManager` for Request Security, let's now delve into updates regarding OAuth.

OAuth Updates

In this section, we'll delve into OAuth updates, specifically focusing on changes related to altering default authorities in `oauth2Login()` and deprecations concerning OAuth2 clients.

Changing Default oauth2Login() Authorities

In `Spring Security` 5, when a user authenticates with an **OAuth2** or **OpenID Connect 1.0** provider through `oauth2Login()`, the default `GrantedAuthority` assigned to them is ROLE_USER.

In `Spring Security 6`, a user authenticating with an OAuth2 provider is assigned the default authority of `OAUTH2_USER`, while a user authenticating with an `OpenID Connect 1.0` provider is given the default authority of `OIDC_USER`. These default authorities provide a more distinct categorization for users based on whether they have authenticated with an `OAuth2` or `OpenID Connect 1.0` provider.

If your application relies on authorization rules or expressions such as `hasRole("USER")` or `hasAuthority("ROLE_USER")` to grant access based on specific authorities, be aware that the updated defaults in `Spring Security 6` will impact your application.

To adopt the new defaults in `Spring Security 6`, you can use the following configuration:

```
@Bean
public SecurityFilterChain securityFilterChain(HttpSecurity http)
throws Exception {
    http
            // ...
            .oauth2Login(oauth2Login -> oauth2Login
                    .userInfoEndpoint(userInfo -> userInfo
                            .
userAuthoritiesMapper(grantedAuthoritiesMapper())
                    )
            );
    return http.build();
}

private GrantedAuthoritiesMapper grantedAuthoritiesMapper() {
    return authorities -> {
        Set<GrantedAuthority> mappedAuthorities = new HashSet<>();

        authorities.forEach(authority -> {
            GrantedAuthority mappedAuthority;

            if (authority instanceof OidcUserAuthority) {
                OidcUserAuthority userAuthority = (OidcUserAuthority)
authority;
                mappedAuthority = new OidcUserAuthority(
                        "OIDC_USER", userAuthority.getIdToken(),
userAuthority.getUserInfo());
            } else if (authority instanceof OAuth2UserAuthority) {
                OAuth2UserAuthority userAuthority = (OAuth2UserAuthority)
authority;
                mappedAuthority = new OAuth2UserAuthority(
                        "OAUTH2_USER", userAuthority.getAttributes());
            } else {
```

```
            mappedAuthority = authority;
        }

        mappedAuthorities.add(mappedAuthority);
    });

    return mappedAuthorities;
    };
}
```

Handling deprecations for OAuth2 clients

In Spring Security 6, obsolete classes and methods have been eliminated from the OAuth2 client. The deprecated items are outlined as follows, along with their respective direct replacements:

Class	Deprections list
ServletOAuth2 Authorized ClientExchange FilterFunction	The method setAccessTokenExpiresSkew(...) can be replaced with one of: • ClientCredentialsOAuth2Authorized ClientProvider#setClockSkew(...) • RefreshTokenOAuth2AuthorizedClient Provider#setClockSkew(...) • JwtBearerOAuth2Authorized ClientProvider#setClockSkew(...) The method setClientCredentials TokenResponseClient(...) can be replaced with the constructor ServletOAuth2Authorized ClientExchangeFilterFunction (OAuth2AuthorizedClientManager)
OidcUserInfo	The method phoneNumberVerified(String) can be replaced with phoneNumberVerified(Boolean)
OAuth2Authorized ClientArgument} Resolver	The method setClientCredentialsTokenResponseClient(...) can be replaced with the constructor OAuth2AuthorizedClient ArgumentResolver (OAuth2AuthorizedClientManager)
ClaimAccessor	The method containsClaim(...) can be replaced with hasClaim(...)

Class	Deprections list
`OidcClient InitiatedLogout SuccessHandler`	The method `setPostLogoutRedirectUri(URI)` can be replaced with `setPostLogoutRedirectUri(String)`
`HttpSessionOAuth2 Authorization RequestRepository`	The method `setAllowMultipleAuthorizationRequests(…)` has no direct replacement
`AuthorizationRequest Repository`	The method `removeAuthorizationRequest(HttpServletRequest)` can be replaced with `removeAuthorizationRequest(HttpServletRequest, HttpServletResponse)`
`ClientRegistration`	The method `getRedirectUriTemplate()` can be replaced with `getRedirectUri()`
`ClientRegistration .Builder`	The method `redirectUriTemplate(…)` can be replaced with `redirectUri(…)`
`AbstractOAuth2 Authorization GrantRequest`	The constructor `AbstractOAuth2Authorization GrantRequest(AuthorizationGrantType)` can be replaced with `AbstractOAuth2Authorization GrantRequest(AuthorizationGrantType, ClientRegistration)`
`ClientAuthentication Method`	The static field `BASIC` can be replaced with `CLIENT_SECRET_ BASIC` The static field `POST` can be replaced with `CLIENT_SECRET_ POST`
`OAuth2Access TokenResponse HttpMessage Converter`	The field `tokenResponseConverter` has no direct replacement The method `setTokenResponseConverter(…)` can be replaced with `setAccessTokenResponseConverter(…)` The field `tokenResponseParametersConverter` has no direct replacement The method `setTokenResponseParametersConverter(…)` can be replaced with `setAccessTokenResponse ParametersConverter(…)`

Class	Deprections list
`Nimbus AuthorizationCode TokenResponseClient`	The class `NimbusAuthorizationCode TokenResponseClient` can be replaced with `DefaultAuthorizationCode TokenResponseClient`
`NimbusJwt DecoderJwkSupport`	The class `NimbusJwtDecoderJwkSupport` can be replaced with `NimbusJwtDecoder` or `JwtDecoders`
`ImplicitGrant Configurer`	The class `ImplicitGrantConfigurer` has no direct replacement
`Authorization GrantType`	The static field `IMPLICIT` has no direct replacement
`OAuth2Authorization ResponseType`	The static field `TOKEN` has no direct replacement
`OAuth2Authorization Request`	The static method `implicit()` has no direct replacement
`JwtAuthentication Converter`	The `extractAuthorities` method will be deprecated and removed. Instead of extending `JwtAuthenticationConverter`, it is recommended to provide a custom granted authorities converter using `JwtAuthenticationConverter#set JwtGrantedAuthoritiesConverter`.

Table 16.5 – List of OAuth2 deprecations

> **Important note**
> The use of the implicit grant type is discouraged, and all associated support has been removed in `Spring Security 6`.

After covering `Spring Security 6` OAuth updates, let's now delve into updates regarding SAML.

SAML Updates

Spring Security SAML Extensions deliver support for service providers through a set of filters, which can be activated by manually adding each filter in the correct sequence to various `Spring Security` filter chains.

In the case of `Spring Security`'s SAML 2.0 service provider support, you can enable it using the `Spring Security saml2Login` and `saml2Logout` DSL methods. These methods automatically select the appropriate filters and position them in the relevant locations within the filter chain.

In the following sections, we will explore the main SAML updates.

Moving to OpenSAML 4

OpenSAML 3 has reached the end of its life cycle. Consequently, `Spring Security 6` discontinues support for `OpenSAML 3` and upgrades its baseline to `OpenSAML 4`.

To upgrade to `Spring Security 6`'s SAML support, you are required to use **OpenSAML**, version `4.1.1` or later.

Utilizing the OpenSaml4AuthenticationProvider

To simultaneously accommodate both **OpenSAML 3** and **4**, `Spring Security` introduced `OpenSamlAuthenticationProvider` and `OpenSaml4AuthenticationProvider`. However, with the removal of **OpenSAML 3** support in `Spring Security 6`, `OpenSamlAuthenticationProvider` has also been discontinued.

It's important to note that not all methods from `OpenSamlAuthenticationProvider` were directly transferred to `OpenSaml4AuthenticationProvider`. Consequently, some adjustments will be necessary to address the changes when implementing the challenge.

Avoiding use of SAML 2.0 Converter constructors

In the initial version of `Spring Security` SAML 2.0 support, the `Saml2MetadataFilter` and `Saml2AuthenticationTokenConverter` were initially equipped with constructors of the `Converter` type. This level of abstraction posed challenges in evolving the class, leading to the introduction of a dedicated interface, `RelyingPartyRegistrationResolver`, in a subsequent release.

In version 6.0, the `Converter` constructors have been eliminated. To adapt to this change, modify classes that implement `Converter<HttpServletRequest, RelyingPartyRegistration>` to instead implement `RelyingPartyRegistrationResolver`.

Transitioning to utilizing Saml2AuthenticationRequestResolver

In `Spring Security 6`, `Saml2AuthenticationContextResolver`, `Saml2AuthenticationRequestFactory`, and the associated `Saml2WebSsoAuthenticationRequestFilter` are eliminated.

They are replaced by `Saml2AuthenticationRequestResolver` and a new constructor in `Saml2WebSsoAuthenticationRequestFilter`. The revised interface removes an unnecessary transport object between these classes.

While most applications won't require significant changes, if you currently use or configure `Saml2AuthenticationRequestContextResolver` or `Saml2Authentication RequestFactory`, consider the following steps to transition using `Saml2Authentication RequestResolver`.

Replacing setAuthenticationRequestContextConverter

Instead of using `setAuthenticationRequestContextConverter`, you should move to `setAuthnRequestCustomizer` in `Spring Security 6`.

Furthermore, as `setAuthnRequestCustomizer` has direct access to the `HttpServletRequest`, there is no necessity for a `Saml2AuthenticationRequestContextResolver`. Simply utilize `setAuthnRequestCustomizer` to directly retrieve the required information from the `HttpServletRequest`.

Replacing setProtocolBinding

The following implementation using `setProtocolBinding`:

```
@Bean
Saml2AuthenticationRequestFactory authenticationRequestFactory() {
    OpenSaml4AuthenticationRequestFactory factory = new
OpenSaml4AuthenticationRequestFactory();
    factory.setProtocolBinding("urn:oasis:names:tc:SAML:2.0:bindings:
HTTP-POST")
    return factory;
}
```

can be replaced as follows:

```
@Bean
Saml2AuthenticationRequestResolver authenticationRequestResolver() {
    OpenSaml4AuthenticationRequestResolver reaolver = new
OpenSaml4AuthenticationRequestResolver(registrations);
    resolver.setAuthnRequestCustomizer((context) -> context.
getAuthnRequest()
            .setProtocolBinding("urn:oasis:names:tc:SAML:2.0:bindings:H
TTP-POST"));
    return resolver;
}
```

> **Important note**
>
> As `Spring Security` exclusively supports the `POST` binding for authentication, overriding the protocol binding at this juncture doesn't yield significant value.

Utilizing the most recent constructor for Saml2AuthenticationToken

Prior to `Spring Security 6`, the `Saml2AuthenticationToken` constructor required multiple individual settings as parameters, posing challenges when adding new parameters. Recognizing that most of these settings were inherent to `RelyingPartyRegistration`, a more stable constructor was introduced. This new constructor allows the provision of a `RelyingPartyRegistration`, aligning more closely with the design of `OAuth2LoginAuthenticationToken`.

While most applications typically do not directly instantiate this class, as it is usually handled by `Saml2WebSsoAuthenticationFilter`, if your application does instantiate it, you should update the constructor as follows:

```
new Saml2AuthenticationToken(saml2Response, registration)
```

Leveraging the updated methods in RelyingPartyRegistration

In the initial version of `Spring Security`'s **SAML** support, there was some ambiguity regarding the interpretation of certain `RelyingPartyRegistration` methods and their functionalities. To address this issue and accommodate the introduction of additional capabilities to `RelyingPartyRegistration`, it became imperative to clarify the ambiguity by renaming the methods to align with the specification language.

After examining the various configuration migration options from `Spring Security 5.x` to `Spring Security 6.x`, the subsequent section will demonstrate a practical example of migrating a JDBC application from `Spring Security 5.x` to `Spring Security 6.x`.

Applying the migration steps from Spring Security 5.x to Spring Security 6.x

In this section, we'll delve into the process of migrating a sample application from `Spring Security 5.x` to `Spring Security 6.x`. This migration aims to ensure compatibility with the latest features, improvements, and security enhancements offered by the newer version.

> **Important note**
>
> The initial state of the application written in `Spring Security 5.x` is available in the project `chapter16.00-calendar`.

Reviewing Application dependencies

The following snippet defines the initial dependencies needed for Spring Security 5.x:

```
//build.gradle
plugins {
    id 'java'
    id 'org.springframework.boot' version '2.7.18'
    id 'io.spring.dependency-management' version '1.1.4'
}
...
dependencies {
    developmentOnly 'org.springframework.boot:spring-boot-devtools'
    // JPA / ORM / Hibernate:
    implementation 'org.springframework.boot:spring-boot-starter-data-
jpa'
    implementation 'org.springframework.boot:spring-boot-starter-
security'
    implementation 'org.springframework.boot:spring-boot-starter-
thymeleaf'
    implementation 'org.springframework.boot:spring-boot-starter-web'
    implementation 'org.springframework.boot:spring-boot-starter-
validation'
    implementation 'org.thymeleaf.extras:thymeleaf-extras-
springsecurity5'
    // H2 db
    implementation 'com.h2database:h2'
    // webjars
    implementation 'org.webjars:webjars-locator:0.50'
    implementation 'org.webjars:bootstrap:5.3.2'
    //Tests
    testImplementation 'org.springframework.boot:spring-boot-starter-
test'
}
```

In the migrated version of build.gradle, we will upgrade Spring Security to version 6.x:

```
//build.gradle
plugins {
    id 'java'
    id 'org.springframework.boot' version '3.2.1
    id 'io.spring.dependency-management' version '1.1.4'
}
...
dependencies {
    developmentOnly 'org.springframework.boot:spring-boot-devtools'
```

```
    // JPA / ORM / Hibernate:
    implementation 'org.springframework.boot:spring-boot-starter-data-
jpa'
    implementation 'org.springframework.boot:spring-boot-starter-
security'
    implementation 'org.springframework.boot:spring-boot-starter-
thymeleaf'
    implementation 'org.springframework.boot:spring-boot-starter-web'
    implementation 'org.springframework.boot:spring-boot-starter-
validation'
    implementation 'org.thymeleaf.extras:thymeleaf-extras-
springsecurity6'
    // H2 db
    implementation 'com.h2database:h2'
    // webjars
    implementation 'org.webjars:webjars-locator:0.50'
    implementation 'org.webjars:bootstrap:5.3.2'
    //Tests
    testImplementation 'org.springframework.boot:spring-boot-starter-
test'
}
```

Migrating from the javax to jakarta namespace

The migration from the `javax` namespace to the `jakarta` namespace in `Spring Security` 6 is primarily driven by changes in the Java ecosystem.

This change is necessary due to the evolution of the Java **Enterprise Edition** (**EE**) specifications and the community-led Jakarta EE effort.

Replacing WebSecurityConfigurerAdapter and exposing SecurityFilterChain Bean

As explained in the previous section, `Spring Security` 6 introduces enhancements and refinements to streamline security configurations. One notable evolution in recent versions involves replacing the traditional `WebSecurityConfigurerAdapter` with a more flexible approach of exposing a `SecurityFilterChain` bean.

This paradigm shift provides developers with greater control and customization over their security configurations, facilitating finer-grained security setups tailored to specific application requirements.

Before the migration, the `SecurityConfig.java` looks like this:

```
//src/main/java/com/packtpub/springsecurity/configuration/
SecurityConfig.java
@EnableWebSecurity
```

```java
public class SecurityConfig extends WebSecurityConfigurerAdapter {

    @Description("Configure HTTP Security")
    @Override
    protected void configure(final HttpSecurity http) throws Exception
{

        http.authorizeRequests(authorizeRequests -> authorizeRequests

            .antMatchers("/webjars/**").permitAll()
            .antMatchers("/css/**").permitAll()
            .antMatchers("/favicon.ico").permitAll()

            .antMatchers("/actuator/**").permitAll()
            .antMatchers("/signup/*").permitAll()
            .antMatchers("/").permitAll()
            .antMatchers("/login/*").permitAll()
            .antMatchers("/logout/*").permitAll()
            .antMatchers("/admin/h2/**").
access("isFullyAuthenticated() and hasRole('ADMIN')")
            .antMatchers("/admin/*").hasRole("ADMIN")
            .antMatchers("/events/").hasRole("ADMIN")
            .antMatchers("/**").hasRole("USER")
        );

        // The default AccessDeniedException
        http.exceptionHandling(handler -> handler
            .accessDeniedPage("/errors/403")
        );

        // Login Configuration
        http.formLogin(form -> form
            .loginPage("/login/form")
            .loginProcessingUrl("/login")
            .failureUrl("/login/form?error")
            .usernameParameter("username") // redundant
            .passwordParameter("password") // redundant
            .defaultSuccessUrl("/default", true)
            .permitAll()
        );

        // Logout Configuration
        http.logout(form -> form
            .logoutUrl("/logout")
```

```
                    .logoutSuccessUrl("/login/form?logout")
                    .permitAll()
        );



        // Allow anonymous users
        http.anonymous();

        // CSRF is enabled by default, with Java Config
        //NOSONAR
        http.csrf().disable();

        // Cross Origin Resource Sharing
        http.cors().disable();

        // HTTP Security Headers
        http.headers().disable();

        // Enable <frameset> in order to use H2 web console
        http.headers().frameOptions().disable();
    }
...
}
```

After the migration, we remove `WebSecurityConfigurerAdapter` and expose a `SecurityFilterChain` bean as follows:

```java
//src/main/java/com/packtpub/springsecurity/configuration/
SecurityConfig.java
@Configuration
@EnableWebSecurity
public class SecurityConfig {
    @Bean
    public SecurityFilterChain filterChain(HttpSecurity http) throws
Exception {
        http.authorizeHttpRequests( authz -> authz
                    .requestMatchers("/webjars/**").permitAll()
                    .requestMatchers("/css/**").permitAll()
                    .requestMatchers("/favicon.ico").permitAll()
                    .requestMatchers("/").permitAll()
                    .requestMatchers("/login/*").permitAll()
                    .requestMatchers("/logout").permitAll()
                    .requestMatchers("/signup/*").permitAll()
                    .requestMatchers("/errors/**").permitAll()
```

```
                    // H2 console
                    .requestMatchers("/admin/h2/**")
                    .access(new
WebExpressionAuthorizationManager("isFullyAuthenticated() and
hasRole('ADMIN')"))
                    .requestMatchers("/events/").hasRole("ADMIN")
                    .requestMatchers("/**").hasRole("USER"))
                .exceptionHandling(exceptions -> exceptions
                    .accessDeniedPage("/errors/403"))
                .formLogin(form -> form
                    .loginPage("/login/form")
                    .loginProcessingUrl("/login")
                    .failureUrl("/login/form?error")
                    .usernameParameter("username")
                    .passwordParameter("password")
                    .defaultSuccessUrl("/default", true)
                    .permitAll())
                .logout(form -> form
                    .logoutUrl("/logout")
                    .logoutSuccessUrl("/login/form?logout")
                    .permitAll())
                // CSRF is enabled by default, with Java Config
                .csrf(AbstractHttpConfigurer::disable);
        // For H2 Console
        http.headers(headers -> headers.
frameOptions(FrameOptionsConfig::disable));
        return http.build();
    }
...
}
```

As @EnableWebSecurity no longer includes @Configuration in Spring Security 6.x, we declared both annotations in the migrated version of SecurityConfig.java.

> **Important note**
>
> Your code should now look like this: chapter16.01-calendar.

Summary

This chapter reviewed the major and minor changes that you will find when upgrading an existing Spring Security 5.x project to Spring Security 6.x. In this chapter, we have reviewed the significant enhancements to the framework that are likely to motivate an upgrade. We also examined upgrade requirements, dependencies, common types of code, and configuration changes that will prevent applications from working post-upgrade. We also covered the investigation (at a high level) of the overall code reorganization changes that the Spring Security authors made as part of code base restructuring.

If this is the first chapter you've read, we hope that you return to the rest of the book and use this chapter as a guide to allow your upgrade to Spring Security 6.x to proceed as smoothly as possible!

17

Microservice Security with OAuth 2 and JSON Web Tokens

In this chapter, we will look at microservices-based architectures and look at how **OAuth 2** with **JSON Web Tokens (JWT)** plays a role in securing microservices in a Spring- based application.

The following is a list of topics that will be covered in this chapter:

- The general difference between **monolithic applications** and **microservices**

- Comparing **Service-Oriented Architectures (SOA)** with microservices

- The conceptual architecture of **OAuth 2** and how it provides your services with trustworthy client access

- Types of **OAuth 2** access tokens

- Types of **OAuth 2** grant types

- Examining JWT and their general structure

- Implementing a resource server and authentication server used to grant access rights to clients in order to access **OAuth 2** resources

- Implementing a **RESTful** client to gain access to resources through an **OAuth 2** grant flow

We have quite a few items to cover in this chapter, but before we dig into the details of how to start leveraging `Spring Security` to implement **OAuth 2** and **JWT**, we first want to create a baseline of the calendar application that does not have **Thymeleaf** or any other browser- based user interface.

After removing all **Thymeleaf** configuration and resources, the various controllers have been converted to **JAX-RS REST** controllers.

This chapter's code in action link is here: `https://packt.link/zEHBU`.

> **Important note**
> You should start with the code from `chapter17.00-calendar`.

What are microservices?

Microservices are an architectural approach that allows the development of physically separated modular applications which are autonomous, enabling agility, rapid development, continuous deployment, and scaling.

An application is built as a set of services, like **SOA**, such that services communicate through standard **APIs**, for example, `JSON` or `XML`, and this allows the aggregation of language-agnostic services. Basically, a service can be written in the best language for the task the service is being created for.

Each service runs in its own process and is location neutral, thus it can be located anywhere on the access network.

In the next sections, we will explore Monoliths, Microservices, and Service-oriented architectures and discern their variances. Then, we can delve into Microservices Security using spring-security.

Monoliths

The microservices approach is the opposite of the traditional monolithic software approach, which consists of tightly integrated modules that ship infrequently and have to scale as a single unit. Traditional **Java EE** applications and the `JBCP calendar` application in this book are examples of monolithic applications. Look at the following diagram which depicts the monolithic architecture:

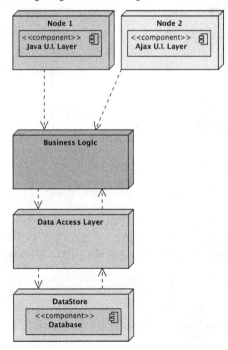

Figure 17.1 – Monolithic architecture

Although the monolithic approach fits well for some organizations and some applications, microservices is becoming popular with companies that need more options for agility and scalability in their ecosystem.

Microservices

A microservice architecture is a collection of small discrete services where each service implements a specific business capability. These services run their own process and communicate via an **HTTP API** usually using a **RESTful** service approach. These services are created to serve only one specific business function, such as user management, administrative roles, an e-commerce cart, a search engine, social media integration, and many others. Look at the following diagram which depicts the microservices architecture:

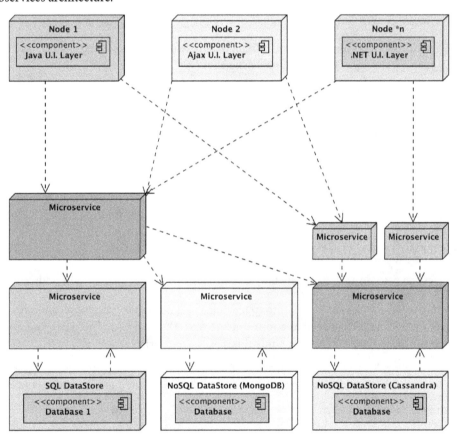

Figure 17.2 – Microservices architecture

Each service can be deployed, upgraded, scaled, restarted, and removed independently of other services in the application and other systems in the enterprise.

Because each service is created independently of the other, they can each be written in different programming languages and use different data storage. Centralized service management is virtually non-existent, and these services use lightweight **HTTP**, **REST** communicating among themselves.

Service-oriented architectures

You might be asking yourself, "Isn't this the same as SOA?" Not exactly, you could say **microservices** achieve what **SOA** promised in the first place.

An **SOA** is a style of software design where services are exposed to other components through a language-agnostic, communication protocol over a computer network.

The basic principle of **SOA** is to be independent of vendors, products, and technologies.

The definition of a service is a discrete unit of functionality that can be accessed remotely and acted upon and updated independently, such as retrieving a credit card statement online.

Although similar, **SOA** and microservices are still different types of architectures.

A typical **SOA** is often implemented inside deployment monoliths and is more platform driven, while microservices can be independently deployable and, therefore, offer more flexibility in all dimensions.

The key difference, of course, is the size; the word micro says it all. Microservices tend to be significantly smaller than regular **SOA** services. As *Martin Fowler* said:

> *We should think about SOA as a superset of microservices.*

Microservice security

Microservices can provide great flexibility but also introduce challenges that must be addressed.

- **Service communication**: Monolithic applications use in-memory communication between processes, while microservices communicate over the network. The move to network communication raises issues of not only speed but also security.

- **Tight coupling**: Microservices use many datastores rather than a few. This creates the opportunity for implicit service contracts between microservices and services that are tightly coupled.

- **Technical complexity**: Microservices can create additional complexity, which can create security gaps. If the team does not have the correct experience, then managing these complexities can quickly become unmanageable.

The OAuth 2 specification

There is sometimes a misconception that **OAuth 2** is an evolution from **OAuth 1**, but it is a completely different approach. **OAuth 1** specification requires signatures, so you would have to use cryptographic algorithms to create generate and validate those signatures that are no longer required for **OAuth 2**. The **OAuth 2** encryption is now handled by **TLS**, which is required.

> **Important note**
>
> **OAuth 2 RFC-6749**, *The OAuth 2.0 Authorization Framework* (`https://tools.ietf.org/html/rfc6749`):
>
> The **OAuth 2.0** authorization framework enables a third-party application to obtain limited access to an HTTP service, either on behalf of a resource owner by orchestrating an approval interaction between the resource owner and the HTTP service, or by allowing the third-party application to obtain access on its own behalf.
>
> This specification replaces and makes obsolete the **OAuth 1.0** protocol described in **RFC 5849**, The **OAuth 1.0** Protocol (`https://tools.ietf.org/html/rfc5849`).

To properly understand how to utilize **OAuth 2**, we need to identify certain roles and the collaboration between these roles. Let's define each of the roles that are participating in the **OAuth 2** authorization process:

- **Resource owner**: The resource owner is the entity capable of granting access to a protected resource that is located on a resource server
- **Authorization server**: The authorization server is a centralized security gateway for issuing access tokens to the client after successfully authenticating the resource owner and obtaining authorization
- **Resource server**: The resource server is the server hosting the protected resources and is capable of dissecting and responding to protected resource requests using the **OAuth 2** access token
- **Microservice client**: The client is the application making resource requests that are protected on behalf of the resource owner, but with their authorization

Access tokens

An **OAuth 2** access token, commonly referred to as `access_token` in code samples, represents a credential that can be used by a client to access an **API**. We have two types of access tokens:

- **Access token**: An access token usually has a limited lifetime and is used to enable the client to access protected resources when including this token in the HTTP request header for each request.
- **Refresh token**: A refresh token has a longer lifetime and is used to get a new access token once the access token has expired, but without the need to send credentials to the server again.

Grant types

Grant types are methods that a client can use to gain an access token that represents the permissions granted. There are different grant types that allow different types of access based on the needs of your application. Each grant type can support a different **OAuth 2** flow without worrying about the technical aspects of the implementation. We have four main grant types in **OAuth 2**:

- **Authorization code**: The **authorization code grant type**, defined in **RFC 6749, Section 4.1** (`https://tools.ietf.org/html/rfc6749`), is a redirection-based flow where the browser receives an authorization code from an authorization server and sends this to the client. The client will then interact with the authorization server and exchange the authorization code for `access_token` and, optionally, `id_token` and `refresh_token`. The client can now use this `access_token` to call the protected resource on behalf of the user.

- **Implicit**: The **implicit grant type**, defined in **RFC 6749, Section 4.1** (`https://tools.ietf.org/html/rfc6749`), is similar to the authorization code grant type, but the client application receives `access_token` directly, without the need for `authorization_code`. This happens because the client application, which is usually a JavaScript application running within a browser and is less trusted than a client application running on the server, cannot be trusted with `client_secret` (which is required in the authorization code grant type). The implicit grant type does not send a refresh token to the application due to limited trust.

- **Password credentials**: The **resource owner password grant type**, defined in **RFC 6749, Section 4.3** (`https://tools.ietf.org/html/rfc6749`), can be used directly as an authorization grant to obtain `access_token` and, optionally, `refresh_token`. This grant is used when there is a high degree of trust between the user and the client and when other authorization grant flows are not available. This grant type eliminates the need for the client to store the user credentials by exchanging the credentials with a long-lived `access_token` or `refresh_token`.

- **Client credentials**: The **Client Credentials Grant's**, defined in **RFC 6749, Section 4.4** (`https://tools.ietf.org/ html/rfc6749#section-4.4`), is for a non-interactive client (CLI), a daemon, or another service running. The client can directly ask the authorization server for `access_token` by using client-provided credentials (client id and client secret) to authenticate.

After covering the main OAuth 2 access tokens and grant types, in the next section will deep dive into the JSON Web Tokens specification.

JSON Web Tokens

JWT is an open standard, **RFC 7519** (`https://tools.ietf.org/html/rfc7519`) that defines a compact and self-contained format for securely transmitting information between parties in the form of a **JSON** object. This information can be verified and trusted because it is digitally signed. JWTs can be signed using a secret (with the **hash-based message authentication code** (HMAC) algorithm) or a public/private key pair using the **Rivest–Shamir–Adleman** (RSA) encryption algorithm.

> **Important note**
>
> **JWT RFC-7519** (`https://tools.ietf.org/html/rfc7519`):
>
> **JWT** is a compact, URL-safe means of representing claims to be transferred between two parties. The claims in a **JWT** are encoded as a **JSON** object that is used as the payload of a **JSON Web Signature** (**JWS**) structure or as the plaintext of a **JSON Web Encryption** (**JWE**) structure, enabling the claims to be digitally signed or integrity protected with a **Message Authentication Code** (**MAC**) and/or encrypted.

JWT is used to carry information related to the identity and characteristics (**claims**) of the client bearing the token. JWT is a container and is signed by the server to avoid client tampering. This token is created during the authentication process and is verified by the authorization server before any processing. It is used by a resource server to allow a client to present a token representing its *identity card* to the resource server and allows the resource server to verify the validity and integrity of the token in a stateless, secure manner.

Token structure

The structure of a JWT adheres to the following three-part structure including a header, payload, and signature:

```
[Base64Encoded(HEADER)] . [Base64Encoded (PAYLOAD)] .
[encoded(SIGNATURE)]
```

Encoded JWT

The following code snippet is the complete encoded `access_token` that is returned based on the client request:

```
eyJraWQiOiJlOTllMzEyYS0yMDJmLTRmNDItOWExNi1h
ZmE2NDA5Mzg0N2QiLCJhbGciOiJSUzI1NiJ9.
eyJzdWIiOiJqYmNwLWNhbGVuZGFyIiwiYXVkIjoiamJjcC1jYWxlbmRhciIsIm5iZiI6MT
cwODU0ODUzMCwic2NvcGUiOlsiZXZlbnRzLnJlYWQiXSwiaXNzIjoiaHR0cDovL2xvY2Fs
aG9zdDo5MDAwIiwiZXhwIjoxNzA4NTQ4ODMwLCJpYXQiOjE3MDg1NDg1MzAsImp0aSI6I
jRhMzVjZmNmLTE5YWItNDZjZC05OWI4LWQxNWM5ZmZlNjQ1MiJ9.
WNJTwQwHA4TVE1BYuizQUo88Dnf0K2by0awxVo_mSq_8n5KWkQMuKESFQwQHT3
2VExn7qHW6JoD6sfxrLK5q2o-KKIYDpL1CACtfjK0mUCWjfpLfpeyXg0FpYPw
6s4allS3zUfOSrFf53wP8k4XCNaPxU9yVQ8s2TB064Sanl7W0VwSbxoz4B-Vg-
PQwEob1cxhAXrBBy5WmM8rk7WsvPXYvMLdo
ISpkP4n66hCzdmmFiBWFhgsfRsOVG8mNmIWgeJVgLXY
BiLrbR2FuFK5KxU7Ls7IMZcWiHd95yAgA6TQ46yBiJErclNVr8Xr5M2SnzFR7HWJY
2OHCNJxnjRpbwEQ
```

Header

The encoded header for our `access_token` JWT is `base64` encoded, as shown in the following code:

```
eyJraWQiOiJlOTllMzEyYS0yMDJmLTRmNDItOWExNi1hZmE2NDA5Mzg0N2QiLCJhbGciOi
JSUzI1NiJ9
```

By decoding the encoded header, we have the following payload:

```
{
  "kid": "e99e312a-202f-4f42-9a16-afa64093847d",
  "alg": "RS256"
}
```

Payload

The encoded payload for our `access_token` JWT is `base64` encoded, as shown here:

```
eyJzdWIiOiJqYmNwLWNhbGVuZGFyIiwiYXVkIjoia
mJjcC1jYWxlbmRhciIsIm5iZiI6MTc
wODU0ODUzMCwic2NvcGUiOlsiZXZlbnRzL
nJlYWQiXSwiaXNzIjoiaHR0cDovL2xvY2FsaG
9zdDo5MDAwIiwiZXhwIjoxNzA4NTQ4ODMwL
CJpYXQiOjE3MDg1NDg1MzAsImp0aSI6IjRhM
zVjZmNmLTE5YWItNDZjZC05OWI4LWQxNWM5ZmZlNjQ1MiJ9
```

By decoding the encoded payload, we have the following payload claims:

```
{
  "sub": "jbcp-calendar",
  "aud": "jbcp-calendar",
  "nbf": 1708548530,
  "scope": [
    "events.read"
  ],
  "iss": "http://localhost:9000",
  "exp": 1708548830,
  "iat": 1708548530,
  "jti": "4a35cfcf-19ab-46cd-99b8-d15c9ffe6452"
}
```

Signature

The encoded payload for our `access_token` has been encoded with a private key by the authorization server, as seen in the following code:

```
WNJTwQwHA4TVE1BYuizQUo88Dnf0K2by0awxVo
_mSq_8n5KWkQMuKESFQwQHT32VExn7qHW
```

```
6JoD6sfxrLK5q2o-KKIYDpL1CACtfjK0mUCW
jfpLfpeyXg0FpYPw6s4allS3zUfOSrFf53
wP8k4XCNaPxU9yVQ8s2TB064Sanl7W0VwSbxoz4B-VgPQwEob1cxhAXrBBy5WmM8rk7Ws
vPXYvMLdoISpkP4n66hCzdmmFiBWFhgsfRsOV
G8mNmIWgeJVgLXYBiLrbR2FuFK5KxU7Ls
7IMZcWiHd95yAgA6TQ46yBiJErclNVr8Xr5M2SnzFR7HWJY2OHCNJxnjRpbwEQ
```

The following is pseudo code for the creation of a JWT signature:

```
var encodedString = base64UrlEncode(header) + "."; encodedString +=
base64UrlEncode(payload);
var privateKey = "[-----PRIVATE KEY
]";
var signature = SHA256withRSA(encodedString, privateKey); var JWT =
encodedString + "." + base64UrlEncode(signature);
```

JWT Authentication in Spring Security

Moving forward, let's examine the architectural elements employed by `Spring Security` to facilitate **JWT** Authentication in servlet-based applications, similar to the one we previously discussed.

The `JwtAuthenticationProvider` serves as an implementation of `AuthenticationProvider`, utilizing a `JwtDecoder` and `JwtAuthenticationConverter` to validate a **JWT** during authentication.

Now, let's delve into the workings of `JwtAuthenticationProvider` within the context of `Spring Security`. The accompanying figure elucidates the intricacies of the `AuthenticationManager`, as illustrated in the figures depicting the process of reading the **Bearer Token**.

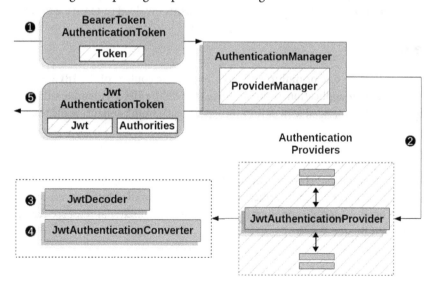

Figure 17.3 – JWT Authentication in Spring Security

The JWT authentication in spring-security entails the following steps:

1. The Authentication Filter, as part of the process outlined in reading the **Bearer Token**, transfers a `BearerTokenAuthenticationToken` to the `AuthenticationManager`, which is implemented by `ProviderManager`.

2. The `ProviderManager` is set up to utilize an `AuthenticationProvider` of the `JwtAuthenticationProvider` type.

3. `JwtAuthenticationProvider` undertakes the tasks of decoding, verifying, and validating the `Jwt` through a `JwtDecoder`.

4. Subsequently, `JwtAuthenticationProvider` employs the `JwtAuthentication Converter` to transform the **JWT** into a Collection of granted authorities.

5. Upon successful authentication, the returned Authentication takes the form of `JwtAuthenticationToken`, featuring a principal representing the **JWT** obtained from the configured `JwtDecoder`. Eventually, the `JwtAuthenticationToken` returned will be placed in the `SecurityContextHolder` by the Authentication Filter.

After covering the **OAuth 2** and **JWT** specifications we'll take a closer look at how they're implemented within spring-security.

OAuth 2 support in Spring Security

`Spring Security` provides **OAuth 2** authorization support following `Spring Framework` programming models and configuration idioms.

In the following section we will identify the main components involved in **OAuth 2** flow.

Resource owner

The resource owner can be one or multiple sources, and in the context of the **JBCP calendar**, it is going to have the calendar application as the resource owner. The **JBCP calendar** will not have any specific configuration that is needed to denote its ownership aside from configuring the resource server.

Resource server

Most of the resource server support is consolidated in `spring-security-oauth2-resource-server`. However, the decoding and verification of **JWT**s are handled by `spring-security-oauth2-jose`. Therefore, both components are essential for a functional resource server that can handle **JWT-encoded Bearer Tokens**.

In Spring Boot, setting up an application as a resource server involves two fundamental steps:

1. Firstly, include the necessary dependencies

2. Secondly, specify the location of the authorization server.

Authorization server

To enable the authorization server capability, we will use `Spring Authorization Server` that offers implementations of the **OAuth 2.1** and **OpenID Connect 1.0** specifications, along with other related specifications. Leveraging Spring Security, it establishes a secure, nimble, and adaptable foundation for constructing Identity Providers compliant with **OpenID Connect 1.0** and **OAuth2** `Authorization Server` products.

To initiate the utilization of `Spring Authorization Server`, the most straightforward approach is to build a **Spring Boot**-based application. You can employ `start.spring.io` to generate a foundational project or refer to the default authorization server sample for guidance. Subsequently, include Spring Boot's starter for `Spring Authorization Server` as a dependency as follow.

```
//build.gradle
dependencies {
...
    // Spring Authorization Server
    implementation org.springframework.boot:spring-boot-starter-
oauth2-authorization-server'
...
}
```

> **Important note**
> Your code should now look like `chapter17.00-authorization-server`.

OAuth 2 resource minimal configuration properties

Note that in the project `chapter17.00-calendar`, we already have added `spring-boot-starter-oauth2-resource-server` dependency, in addition to spring-security. This is important for our application in order to behave as a resource server.

```
//build.gradle
dependencies {
...
    // Spring Authorization Server
    implementation "org.springframework.boot:spring-boot-starter-
oauth2-resource-server"
...
}
```

In a Spring Boot application, you can easily designate the authorization server to be used by following these steps:

```
//src/main/resources/application.yml
spring:
  security:
    oauth2:
      resourceserver:
        jwt:
          issuer-uri: http://localhost:9000
```

> **Important note**
>
> The `issuer-uri` property to be effective, it's essential that one of the following endpoints is a supported endpoint for the authorization server: `idp.example.com/issuer/.well-known/openid-configuration`, `idp.example.com/.well-known/openid-configuration/issuer`, or `idp.example.com/.well-known/oauth-authorization-server/issuer`. This endpoint is commonly referred to as either a Provider Configuration endpoint or an Authorization Server Metadata endpoint.

Expectations at startup

When employing this property and its associated dependencies, the Resource Server will autonomously set up its configuration to validate Bearer Tokens encoded in JWT format.

This is accomplished through a predictable startup sequence:

1. Interrogate the Provider Configuration or Authorization Server Metadata endpoint for the `jwks_url` property.

2. Examine the `jwks_url` endpoint for the supported algorithms.

3. Configure the validation strategy to query the `jwks_url` for valid **public keys** corresponding to the identified algorithms.

4. Configure the validation strategy to verify the `"iss"` claim of each **JWT** against `idp.example.com`.

An implication of this process is that the authorization server must be operational and capable of receiving requests for the Resource Server to initialize successfully.

Expectations during Runtime

After the application has been initiated, the `Resource Server` will endeavor to handle any request that includes an `Authorization: Bearer` header.

```
GET / HTTP/1.1
Authorization: Bearer some-token-value # Resource Server will process
this
```

As long as this scheme is specified, the resource server will strive to handle the request in accordance with the Bearer Token specification.

For a properly structured **JWT**, the resource server will:

1. Validate its signature using a public key obtained from the `jwks_url` endpoint during startup, ensuring a match with the **JWT**.

2. Verify the **JWT**'s **exp** and **nbf** timestamps, along with the **JWT**'s **iss** claim.

3. Associate each scope with an authority, using the prefix **SCOPE_**.

Spring Security will automatically update and rotate the keys used to validate **JWT**s as the authorization server introduces new keys.

> **Important note**
> Your code should now look like `chapter17.01-calendar`.

By default, the resulting `Authentication#getPrincipal` is a `Spring Security Jwt` object, and if available, `Authentication#getName` corresponds to the sub property of the **JWT**.

After defining **OAuth 2** minimal configuration properties, we'll elevate our configuration by examining explicit methods for defining the `jwk-set-uri`.

Defining the JWK Set URI for the authorization server

If the **authorization server** lacks support for configuration endpoints or if the **resource server** needs the capability to initiate startup independently of the authorization server, you can provide the `jwk-set-uri` as follows:

```
//src/main/resources/application.yml
spring:
  security:
    oauth2:
      resourceserver:
        jwt:
          issuer-uri: http://localhost:9000
          jwk-set-uri: http://localhost:9000/.well-known/jwks.json
```

As a result, the **resource server** will refrain from contacting the authorization server during startup. Nevertheless, we continue to specify the `issuer-uri` to ensure that the resource server validates the `"iss"` claim within incoming JWTs.

Providing Audience Information

As demonstrated earlier, the `issuer-uri` property verifies the `"iss"` claim, identifying the entity that sent the JWT.

Additionally, **Spring Boot** includes the `audiences` property for validating the `"aud"` claim, determining the intended recipients of the JWT.

You can specify a resource server's audience as follows:

```
//src/main/resources/application.yml
spring:
  security:
    oauth2:
      resourceserver:
        jwt:
          issuer-uri: http://localhost:9000
          audiences: https://my-org.audience
```

The outcome will be that if the `"iss"` claim of the JWT is not idp.example.com, and its `"aud"` claim does not include `my-resource-server.example.com` in its list, the validation will not succeed.

Configuring Authorization using SecurityFilterChain

A JWT generated by an **OAuth 2.0** `Authorization Server` will usually contain a `scope` or `scp` attribute, signifying the granted scopes or authorities.

In such instances, the resource server will endeavor to transform these scopes into a roster of authorized authorities, adding the `SCOPE_` string as a prefix to each scope. Consequently, to secure an endpoint or method with a scope obtained from a JWT, the associated expressions should incorporate this prefix. For example:

```
//src/main/java/com/packtpub/springsecurity/configuration/
SecurityConfig.java
@Bean
SecurityFilterChain securityFilterChain(HttpSecurity http) throws
Exception {
    http
            .securityMatcher("/events/**")
            .authorizeHttpRequests(authorize -> authorize
                .requestMatchers("/events/**").hasAuthority("SCOPE_
events.read"))
```

```
        .oauth2ResourceServer(oauth2 -> oauth2.jwt(Customizer.
withDefaults())));
    return http.build();
}
```

> **Important note**
>
> Your code should now look like `chapter17.02-calendar`.

At this point, we can start the `chapter17.00-authorization-server` and `chapter17.02-calendar` and we will be ready to send **OAuth 2** requests.

Token requests

When we make the initial token request, we should get a successful response similar to the following:

```
curl -i -X POST \
  http://localhost:9000/oauth2/token \
  -H 'Content-Type: application/x-www-form-urlencoded' \
  -d 'grant_type=client_credentials&client_id=jbcp-calendar&&client_
secret=secret&scope=events.read'
```

And this is a sample response:

```
HTTP/1.1 200
Vary: Origin
Vary: Access-Control-Request-Method
Vary: Access-Control-Request-Headers
X-Content-Type-Options: nosniff
X-XSS-Protection: 0
Cache-Control: no-cache, no-store, max-age=0, must-revalidate
Pragma: no-cache
Expires: 0
X-Frame-Options: DENY
Content-Type: application/json;charset=UTF-8
Transfer-Encoding: chunked
Date: Wed, 21 Feb 2024 17:51:02 GMT
{
  "access_token": "eyJraWQiOiJjMzJjNmVlNy0yYTM5LTQ0NDY
tOWQzZS02NzA2ZWJjMWM5MGUiLCJhbGciOiJSUzI1NiJ9.eyJzdWI
iOiJqYmNwLWNhbGVuZGFyIiwiYXVkIjoiamJjcC1j
YWxlbmRhciIsIm5iZiI6MTcwODUzNzg2Miwic2NvcGUiOlsiZX
ZlbnRzLnJlYWQiXSwiaXNzIjoiaHR0cDovL2xvY2FsaG9zdDo5MD
AwIiwiZXhwIjoxNzA4NTM4MTYyLCJpYXQiOjE3MDg1Mzc4NjIsImp
0aSI6ImVkNjc1YzcwLTg4MGItNDYxYy1hMDk0LTFmMTA1ZTk3OTk0
NCJ9.OWMHZC_cRqUsshwTCIIdo6oGK_KU39hY25U5YhTUU7QTi-Sm
```

```
F7wy9QdDxJnl9brIXgjq7NpIeC9zZyi
l8lS4p7HwFP3_3iCN1NQA54vTZ0-UBfT8q6H1aEQzeEdZUDnhoYK2c
oOihbYcNH_Dfn13POMcEwBhFwIsul6tJHN_lLVFBA-CTMxSHoBWBDNq
NvU-gIdadOxFPDpWV86No8DfYgDGWKLP18k3KggLC37ebMbNkIMgK2
4gYxM_5f_g2nR_ueiV6ZQO5fyGq960nYWzePoQtdYVcvHwkQk_FG_
B75rcSrITuTTgDrcA8FWrZrOoitvEOnglHmieUguoYVG2BA",
  "scope": "events.read",
  "token_type": "Bearer",
  "expires_in": 299
}
```

Specifically, we have been granted an access token that can be used in subsequent requests. The `access_token` that will be used as our bearer.

Endpoints requests

Now we will take the `access_token` and use that token to initiate additional requests to the server with the following format:

```
curl -k -i http://localhost:8080/events/  \
-H "Authorization: Bearer eyJraWQiOiJjMzJjNmVl
Ny0yYTM5LTQ0NDYtOWQzZS02NzA2ZWJjMWM5MGUiLCJhbG
ciOiJSUzI1NiJ9.eyJzdWIiOiJqYmNwLWNhbGVuZGFyIiwiYXVkIjoiamJjC
1jYWxlbmRhciIsIm5iZiI6MTcwODUzNzg2Miwic2NvcGUi
OlsiZXZlbnRzLnJlYWQiXSwiaXNzIjoiaHR0cDovL2xvY2
FsaG9zdDo5MDAwIiwiZXhwIjoxNzA4NTM4MTYyLCJpYXQi
OjE3MDg1Mzc4NjIsImp0aSI6ImVkNjc1YzcwLTg4MGItNDY
xYy1hMDk0LTFmMTA1ZTk3OTk0NCJ9.OWMHZC_cRqUsshwT
CIIdo6oGK_KU39hY25U5YhTUU7QTi-SmF7wy9QdDxJnl9br
IXgjq7NpIeC9zZyil8lS4p7HwFP3_3iCN1NQA54vTZ0-UBf
T8q6H1aEQzeEdZUDnhoYK2coOihbYcNH_Dfn13POMcEwBhF
wIsul6tJHN_lLVFBA-CTMxSHoBWBDNqNvU-gIdadOxFPDpW
V86No8DfYgDGWKLP18k3KggLC37ebMbNkIMgK24gYxM_5f_
g2nR_ueiV6ZQO5fyGq960nYWzePoQtdYVcvHwkQk_FG_B75
rcSrITuTTgDrcA8FWrZrOoitvEOnglHmieUguoYVG2BA"
```

We should get the following response:

```
HTTP/1.1 200
Vary: Origin
Vary: Access-Control-Request-Method
Vary: Access-Control-Request-Headers
X-Content-Type-Options: nosniff
X-XSS-Protection: 0
Cache-Control: no-cache, no-store, max-age=0, must-revalidate
Pragma: no-cache
Expires: 0
Strict-Transport-Security: max-age=31536000 ; includeSubDomains
X-Frame-Options: DENY
```

```
Content-Type: application/json
Transfer-Encoding: chunked
Date: Wed, 21 Feb 2024 17:55:32 GMT
[
    {
        "id": 100,
        "summary": "Birthday Party",
        "description": "This is going to be a great birthday",
        "dateWhen": "2023-07-03T18:30:00.000+00:00"
    },
    {
        "id": 101,
        "summary": "Conference Call",
        "description": "Call with the client",
        "dateWhen": "2023-12-23T12:00:00.000+00:00"
    },
    {
        "id": 102,
        "summary": "Vacation",
        "description": "Paragliding in Greece",
        "dateWhen": "2023-09-14T09:30:00.000+00:00"
    }
]
```

Configuring Authorization using @PreAuthorize annotation

Another way to configure authorization can be done using @PreAuthorize annotation.

First step is to enable method security in SecurityConfig.java:

```
//src/main/java/com/packtpub/springsecurity/configuration/
SecurityConfig.java
@Configuration
@EnableWebSecurity
@EnableMethodSecurity
public class SecurityConfig {

    @Bean
    SecurityFilterChain securityFilterChain(HttpSecurity http) throws
Exception {
        http
                .securityMatcher("/events/**")
                .oauth2ResourceServer(oauth2 -> oauth2.jwt(Customizer.
withDefaults()));
        return http.build();
```

```
        }

    }
```

The next step is to secure `CalendarService.getEvents()` that is invoked by the `/events` endpoint.

```
//src/main/java/com/packtpub/springsecurity/service/ CalendarService.
java

public interface CalendarService {

...omitted for brevity
    @PreAuthorize("hasAuthority('SCOPE_events.read')")
    List<Event> getEvents();
...
}
```

> **Important note**
>
> Your code should now look like `chapter17.03-calendar`.

At this point, we can start the `chapter17.00-authorization-server` and `chapter17.03-calendar` and we will be ready to send **OAuth 2** requests.

You can try again, the same previous steps for the `/token` and `/events` endpoints requests.

Now that we have our `OAuth 2 server` ready to issue `access_tokens` for clients, we now can create a microservices client to interact with our system.

Configuring the OAuth 2 client

Now that we have configured our resource server, you can create REST client to consume the **OAuth2** protected resources.

1. You can use `https://start.spring.io/` to initialize you project by choosing the following dependencies:

```
//build.gradle
dependencies {
    ...
    implementation 'org.springframework.boot:spring-boot-starter-web'
    implementation 'org.springframework.boot:spring-boot-starter-
oauth2-client'
    ...
}
```

2.   Next, you will need to configure your application with the client configuration as follow:

```
//src/main/resources/application.yml
jbcp-calendar:
  events:
    api: http://localhost:8080/events/
spring:
## Chapter 17 Authorization Server
  security:
    oauth2:
      client:
        registration:
          calendar-client:
            client-id: jbcp-calendar
            client-secret: secret
            scope: events.read
            authorization-grant-type: client_credentials
            client-name: Calendar Client
        provider:
          calendar-client:
            token-uri: http://localhost:9000/oauth2/token
server:
  port: 8888
```

3.   For this example we will use `RestTemplate` and `ClientHttpRequestInterceptor` to bind the `OAuth2AccessToken` for our REST Client.

Making requests to a third-party **API** secured with **OAuth2** is a fundamental scenario for **OAuth2 Client** usage. This involves authorizing a client, represented by the `OAuth2AuthorizedClient` class in Spring Security, and gaining access to protected resources by inserting a Bearer token into the Authorization header of an outgoing request.

The provided example sets up the application to function as an **OAuth2** Client with the ability to request protected resources from a third-party **API**.

```
//src/main/java/com/packtpub/springsecurity/config/SecurityConfig.java
@Configuration
public class SecurityConfig {

    @Bean
    SecurityFilterChain securityFilterChain(HttpSecurity http) throws
Exception {
        http
                .oauth2Client(withDefaults());
        return http.build();
    }
```

```
    @Bean
    public RestTemplate
oauth2RestTemplate(OAuth2HttpRequestInterceptor
oAuth2HttpRequestInterceptor) {
        RestTemplate restTemplate = new RestTemplate();
        List<ClientHttpRequestInterceptor> interceptors = restTemplate.
getInterceptors();
        if (CollectionUtils.isEmpty(interceptors)) {
            interceptors = new ArrayList<>();
        }
        interceptors.add(oAuth2HttpRequestInterceptor);
        restTemplate.setInterceptors(interceptors);
        return restTemplate;
    }
}
```

4. The following OAuth2HttpRequestInterceptor can be defined as described in the sample code:

```
//src/main/java/com/packtpub/springsecurity/config/
OAuth2HttpRequestInterceptor.java
@Component
public class OAuth2HttpRequestInterceptor implements
ClientHttpRequestInterceptor {

    private final OAuth2AuthorizedClientManager
authorizedClientManager;
    private final ClientRegistrationRepository
clientRegistrationRepository;

    public OAuth2HttpRequestInterceptor(OAuth2AuthorizedClientManager
authorizedClientManager, ClientRegistrationRepository
clientRegistrationRepository) {
        this.authorizedClientManager = authorizedClientManager;
        this.clientRegistrationRepository =
clientRegistrationRepository;
    }

    @Override
    public ClientHttpResponse intercept(HttpRequest request, byte[]
body, ClientHttpRequestExecution execution) throws IOException {
        ClientRegistration clientRegistration =
clientRegistrationRepository.findByRegistrationId("calendar-client");
        OAuth2AuthorizeRequest oAuth2AuthorizeRequest =
OAuth2AuthorizeRequest
            .withClientRegistrationId(clientRegistration.
```

```
getRegistrationId())
            .principal(clientRegistration.getClientId())
            .build();
        OAuth2AuthorizedClient client = authorizedClientManager.
authorize(oAuth2AuthorizeRequest);
        String accessToken = client.getAccessToken().getTokenValue();
        request.getHeaders().setBearerAuth(accessToken);
        return execution.execute(request, body);
    }


}
```

5.   The last step is to make the **REST API** call using the configured **OAuth2** RestTemplate Bean:

```java
//src/main/java/com/packtpub/springsecurity/web/controllers/
OAuth2RestClient.java
@RestController
public class OAuth2RestClient {

    private final RestTemplate oauth2RestTemplate  ;

    public OAuth2RestClient(RestTemplate oauth2RestTemplate) {
        this.oauth2RestTemplate = oauth2RestTemplate;
    }

    @Value("${jbcp-calendar.events.api}")
    private String eventsApi;

    @GetMapping("/")
    public  String apiCheck() {
        return oauth2RestTemplate.getForObject(eventsApi, String.
class);
    }

}
```

We now should have the same codebase for a client application.

Important note

Your code should now look like chapter17.03-calendar-client.

We need to ensure that the chapter17.03-calendar and chapter17.00-authorization-server applications are running and ready to take **OAuth 2** requests from clients.

We can then start the `chapter17.03-calendar-client` application, which will expose a RESTful endpoint that will call our resource server to access the configured events located at `/events` on the remote resource, and will return the following result by running `http://localhost:8888/`:

```
[
  {
    "id":100,
    "summary":"Birthday Party",
    "description":"This is going to be a great birthday",
    "dateWhen":"2023-07-03T18:30:00.000+00:00"
  },
  {
    "id":101,
    "summary":"Conference Call",
    "description":"Call with the client",
    "dateWhen":"2023-12-23T12:00:00.000+00:00"
  },
  {
    "id":102,
    "summary":"Vacation",
    "description":"Paragliding in Greece",
    "dateWhen":"2023-09-14T09:30:00.000+00:00"
  }
]
```

Summary

In this chapter, you learned the general difference between monolithic applications and **microservices** and compared **SOA** with **microservices**. You also learned the conceptual architecture of **OAuth 2** and how it provides your services with trustworthy client access, and learned about the types of **OAuth 2** access tokens and the types of **OAuth 2 client credentials** types.

We examined the **JWT** and their general structure, implemented a **resource server** and **authorization server** used to grant access rights to clients to access **OAuth 2** resources, and implemented a **RESTful** client to gain access to resources through an **OAuth 2 grant flow**.

We've concluded by demonstrating a practical **OAuth 2** example implementation using spring-security. Moving forward, the next chapter will explore the integration with **Central Authentication Service (CAS)** to enable **Single Sign-On (SSO)** and **Single Logout (SLO)** functionalities for your Spring Security-enabled applications.

18

Single Sign-On with the Central Authentication Service

In this chapter, we'll examine the use of the **Central Authentication Service (CAS)** as a **Single Sign-On (SSO)** portal for Spring Security-based applications.

During the course of this chapter, we'll cover the following topics:

- Learning about **CAS**, its architecture, and how it benefits system administrators and organizations of any size

- Understanding how Spring Security can be reconfigured to handle the interception of authentication requests and redirect them to **CAS**

- Configuring the **JBCP Calendar** application to utilize CAS SSO

- Gaining an understanding of how the **Single Logout** feature can be implemented, and configuring our application to support it

- Discussing how to use **CAS** proxy ticket authentication for services, and configuring our application to utilize proxy ticket authentication

- Discussing how to customize the out-of-the-box `JA-SIG CAS` server using the recommended war overlay approach

- Integrating the **CAS** server with **LDAP**, and passing data from **LDAP** to `Spring Security` via **CAS**

This chapter's code in action link is here: `https://packt.link/1FJjp`.

Introducing the Central Authentication Service

CAS is an open source, SSO server that provides centralized access control and authentication to web-based resources within an organization. The benefits of **CAS** for administrators are numerous, and it supports many applications and diverse user communities. These benefits are as follows:

- Individual or group access to resources (applications) can be configured in one location

- Broad support for a wide variety of authentication stores (to centralize user management) provides a single point of authentication and control in widely distributed cross-machine environments

- Broad authentication support is provided for web-based and non-web-based Java applications through **CAS** client libraries

- A single point of reference for user credentials (via **CAS**) is provided so that **CAS** client applications are not required to have any knowledge of the user's credentials, or knowledge of how to verify them

In this chapter, we'll not focus much on the management of **CAS**, but on authentication and how **CAS** can act as an authentication point for the users of our site. Although **CAS** is commonly seen in intranet environments for enterprises or educational institutions, it can also be found in use at high-profile locations such as Sony Online Entertainment's public-facing site.

High-level CAS authentication flow

At a high level, **CAS** is composed of a **CAS** server, which is the central web application for determining authentication, and one or more **CAS** services, which are distinct web applications that use the CAS server to get authenticated. The basic authentication flow of **CAS** proceeds via the following actions:

1. The user attempts to access a protected resource on the website.

2. The user is redirected through the browser from the **CAS** service to the **CAS** server to request a login.

3. The **CAS** server is responsible for user authentication. If the user is not already authenticated to the **CAS** server, then the latter requests credentials from the user. As shown in the following diagram, the user is presented with a login page.

4. The user submits their credentials (that is, the username and password).

5. If the user's credentials are valid, the **CAS** server responds with a redirect through the browser with a service ticket. A service ticket is a one-time use token used to identify a user.

6. The **CAS** service calls the **CAS** server back to verify that the ticket is valid, has not expired, and so on. Note that this step does not occur through the browser.

7. The **CAS** server responds with an assertion indicating that trust has been established. If the ticket is acceptable, trust has been established and the user may proceed via normal authorization checking.

This behavior is illustrated visually in the following diagram:

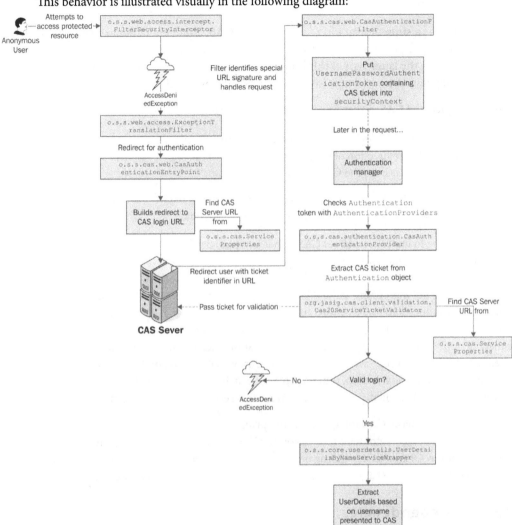

Figure 18.1 – High-level CAS authentication flow

We can see that there is a high level of interaction between the **CAS** server and the secured application, with several data-exchange handshakes required before trust in the user can be established. We assume other network security precautions, such as the use of **Secure Sockets Layer** (**SSL**) and network monitoring, are in place.

The result of this complexity is a SSO protocol that is quite hard to spoof through common techniques.

Now that we understand how **CAS** authentication works in general, let's see how it applies to Spring Security.

Spring Security and CAS

Spring Security has a strong integration capability with **CAS**, although it's not as tightly integrated into the security namespace style of configuration as the **OAuth2** and **LDAP** integrations that we've explored thus far in the latter part of this book. Instead, much of the configuration relies on bean wiring and configuration by reference from the security namespace elements to bean declarations.

The two basic pieces of **CAS** authentication when using Spring Security involve the following:

- Replacement of the standard `AuthenticationEntryPoint` implementation, which typically handles redirection of unauthenticated users to the login page with an implementation that redirects the user to the CAS server instead

- Processing the service ticket when the user is redirected back from the **CAS** server to the protected resource, through the use of a custom servlet filter

An important thing to understand about **CAS** is that in typical deployments, **CAS** is intended to replace all the alternative login mechanisms in our application. As such, once we configure **CAS** for Spring Security, our users must use **CAS** exclusively as the authentication mechanism for our application. In most cases, this is not a problem; as we discussed in the previous section, **CAS** is designed to proxy authentication requests to one or more authentication stores (just as Spring Security does when delegating to a database or **LDAP** for authentication). From the previous diagram (*Figure 18.1*), we can see that our application is no longer checking its own authentication store to validate users. Instead, it authenticates the user with service tickets. However, as we will discuss in the *Getting the UserDetails object from a CAS assertion* section, initially, Spring Security still needs a data store to determine the user's authorization. We will discuss how to remove this restriction later in the chapter.

After completing the basic **CAS** integration with Spring Security, we can remove the login link from the home page and enjoy automatic redirection to **CAS**'s login screen, where we attempt to access a protected resource. Of course, depending on the application, it can also be beneficial to still allow the user to explicitly log in (so that they can see customized content, among other things).

Required dependencies

Before we go too far, we should ensure that our dependencies are updated. The following is a list of the dependencies that we have added with comments on when they are needed:

```
//build.gradle
//Spring CAS Support
implementation " org.springframework.security:spring-security-cas"
```

Installing and configuring CAS

CAS has the benefit of having an extremely dedicated team behind it that has done an excellent job of developing both quality software and accurate, straightforward documentation on how to use it. Should you choose to follow along with the examples in this chapter, you are encouraged to read the appropriate getting started manual for your **CAS** platform. You can find this manual at `https://apereo.github.io/cas/`.

In order to make integration as simple as possible, we have included a **CAS** server application for this chapter that can be deployed in Eclipse or IntelliJ, along with the calendar application.

For the examples in this chapter, we will assume that **CAS** is deployed at `https://localhost:9443/cas/` and the calendar application is deployed at `https://localhost:8443/`. To work, **CAS** requires the use of HTTPS.

> **Important note**
>
> The examples in this chapter were written using the most recent available version of the CAS server, `7.0.1` at the time of writing, which requires Java 21. So, if you are on an earlier version of the server, these instructions may be slightly or significantly different for your environment.

Let's go ahead and configure the components required for CAS authentication.

> **Important note**
>
> You should start the chapter off with the source in `chapter18.00-calendar` and `chapter18.00-cas-server`.

To start the CAS server, run the following command from the `chapter18.00-cas-server` project:

```
./gradlew build run
```

We use the following default CAS login/password for this example: `casuser/Mellon`.

In the **JBCP Calendar** application, we should be able to use the same credentials to log in. Note that the user has admin rights.

For the next steps, we need to do the following:

- Import the CAS SSL certificate inside the **Java Runtime Environment** (**JRE**) keystore of your **JBCP Calendar** application by running the following command from the following location: `chapter18.00-cas-server/src/main/resources/etc/cas`:

  ```
  keytool -importcert -file cas.crt -alias cas-server -keystore
  $JBCP_JAVA_HOME/lib/security/cacerts
  ```

  The `$JBCP_JAVA_HOME` variable is the JVM used by the **JBCP Calendar** application.

- To check the import is done successfully, run the following command. If you are asked for the password the keystore password, the default one is *change it*.

```
keytool -list -keystore $JBCP_JAVA_HOME/lib/security/cacerts
-alias cas-server
```

The output should be similar to the following:

```
cas-server, May 6, 2024, trustedCertEntry,
Certificate fingerprint (SHA-256):
4C:E5:A1:42:58:78:69:7B:94:05:23:05:23:46:EA:DF:FB:D5:2E:10:4F:
C8:90:2D:16:A2:2A:FB:26:99:40:1D
```

- Import the **JBCP Calendar** application certificate inside the JRE keystore of your CAS server by running the following command from the following location: `chapter18.00-calendar/src/main/resources/keys`. If you are asked for the password the keystore password, the default one is *change it*:

```
keytool -importkeystore -srckeystore tomcat.jks -destkeystore
$CAS_JAVA_HOME/lib/security/cacerts  -deststoretype JKS -alias
jbcpcalendar
```

The `$CAS_JAVA_HOME` variable is the JVM used by the **CAS server**.

- To check the import is done successfully, you can run the following command. If you are asked for the password the keystore password, the default one is *change it*:

```
keytool -list -keystore $CAS_JAVA_HOME/lib/security/cacerts
-alias jbcpcalendar
```

The output should be similar to following:

```
jbcpcalendar, May 6, 2024, PrivateKeyEntry,
Certificate fingerprint (SHA-256):
79:0D:62:D7:E7:A1:25:1D:A3:C7:93:F6:03:A8:E4:B8:20:BA:FA:2B:03:
9F:5C:E3:5D:6C:61:A5:6F:CD:83:57
```

> **Important note**
>
> $JBCP_JAVA_HOME represents the path of *Java* used by the **JBCP Calendar** application. The location of the `cacerts` file is the following by default: $$JBCP_JAVA_HOME/lib/security/cacerts.
>
> $CAS_JAVA_HOME represents the path of *Java* used by the **CAS** server. The location of the `cacerts` file is the following by default: $CAS_JAVA_HOME/lib/security/cacerts.
>
> This path should be adapted to your current `cacerts` file location if you are not relying on the default JDK `cacerts` file.
>
> To adapt the command to Windows, you need to replace the $JBCP_JAVA_HOME Unix/Linux environment variable syntax with the %JBCP_JAVA_HOME% Windows syntax. In this command, %JBCP_JAVA_HOME% is assumed to be a Windows environment variable pointing to the Java installation directory. Make sure to replace it with the actual path in your system.
>
> If you didn't import both the *CAS server SSL certificate* into the *JBCP Calendar JVM*, and *JBCP Calendar SSL Certificate* into the *CAS server JVM*, you will get the following error in the logs:
>
> `javax.net.ssl.SSLHandshakeException: PKIX path building failed: sun.security.provider.certpath.SunCertPathBuilderException: unable to find valid certification path to requested target`

Configuring basic CAS integration

Since the Spring Security namespace does not support CAS configuration, there are quite a few more steps that we need to implement to get a basic working setup.

Configuring the CAS properties

The Spring Security setup relies on an `o.s.s.cas.ServiceProperties` bean in order to store common information about **CAS**. The `ServiceProperties` object plays a role in coordinating the data exchange between the various **CAS** components—it is used as a data object to store CAS configuration settings that are shared (and are expected to match) by the varying participants in the **Spring CAS** stack. You can view the configuration included in the following code snippet:

```
//src/main/java/com/packtpub/springsecurity/configuration/CasConfig.
java

@Configuration
public class CasConfig {

    @Value("${cas.base.url}")
    private String casBaseUrl;

    @Value("${cas.login.url}")
```

```
    private String casLoginUrl;

    @Value("${service.base.url}")
    private String serviceBaseUrl;

    @Bean
    public ServiceProperties serviceProperties() {
        ServiceProperties serviceProperties = new ServiceProperties();
        serviceProperties.setService(serviceBaseUrl+ "/login/cas");
        return serviceProperties;
    }
}
```

You probably noticed that we leveraged system properties to use variables named `${cas.base.url}` and `${service.base.url}`. Both values can be included in your application, and Spring will automatically replace them with the values provided in the `PropertySources` configuration. This is a common strategy when deploying a **CAS** service, since the **CAS** server will likely change as we progress from development to production. In this instance, we use `https://localhost:9443/cas` by default for the **CAS** server and `https://localhost:8443` for the calendar application.

This configuration can be overridden using a system argument when the application is taken to production. Alternatively, the configuration can be externalized into a Java properties file. Either mechanism allows us to externalize our configuration properly.

You can make the following updates to your `application.yml` file:

```
cas:
  base:
    url: https://localhost:9443/cas
  login:
    url: ${cas.base.url}/login
service:
  base:
    url: https://localhost:8443
```

Adding the CasAuthenticationEntryPoint object

As we briefly mentioned in the *Spring Security and CAS* section, `Spring Security` uses an `o.s.s.web.AuthenticationEntryPoint` interface to request credentials from the user. Typically, this involves redirecting the user to the login page. With **CAS**, we will need to redirect the **CAS** server to request a login. When we redirect to the **CAS** server, Spring Security must include a `service` parameter that indicates where the **CAS** server should send the service ticket. Fortunately, Spring Security provides the `o.s.s.cas.web.CasAuthenticationEntryPoint` object, which is specifically designed for this purpose. The configuration that is included in the example application is as follows:

```
//src/main/java/com/packtpub/springsecurity/configuration/CasConfig.
java
@Bean
public CasAuthenticationEntryPoint
casAuthenticationEntryPoint(ServiceProperties serviceProperties) {
    CasAuthenticationEntryPoint casAuthenticationEntryPoint = new
CasAuthenticationEntryPoint();
    casAuthenticationEntryPoint.setLoginUrl(this.casLoginUrl);
    casAuthenticationEntryPoint.
setServiceProperties(serviceProperties);
    return casAuthenticationEntryPoint;
}
```

The `CasAuthenticationEntryPoint` object uses the `ServiceProperties` class to specify where to send the service ticket once the user is authenticated. CAS allows for the selective granting of access per user, per application, based on configuration. We'll examine the particulars of this URL in a moment when we configure the servlet filter that is expected to process it.

Next, we will need to update Spring Security to utilize the bean with the `casAuthentication EntryPoint` ID. Make the following update to our `SecurityConfig.java` file:

```
//src/main/java/com/packtpub/springsecurity/configuration/
SecurityConfig.java
@Bean
public SecurityFilterChain filterChain(HttpSecurity http,
        CasAuthenticationEntryPoint casAuthenticationEntryPoint) throws
Exception {
...omitted for brevity

        // Exception Handling
        http.exceptionHandling(exceptions -> exceptions
            .authenticationEntryPoint(casAuthenticationEntryPoint)
            .accessDeniedPage("/errors/403"));
...
    return http.build();
}
```

Enabling CAS ticket verification

Referring to the diagram that we saw earlier (*Figure 18.1*), we can see that Spring Security is responsible for identifying an unauthenticated request and redirecting the user to CAS via the `FilterSecurityInterceptor` class. Adding the `CasAuthenticationEntryPoint` object has overridden the standard redirect to the login page functionality and provided the expected redirection from the application to the CAS server. Now, we need to configure things so that, once authenticated to CAS, the user is properly authenticated to the application.

If you remember from *Chapter 9, Opening up to OAuth2*, **OAuth2** uses a similar redirection approach by redirecting unauthenticated users to the **OAuth2** provider for authentication, and then back to the application with verifiable credentials. CAS differs from **OAuth2**. In the **CAS** protocol, upon the user's return to the application, the application is expected to call back the **CAS** server to explicitly validate that the credentials provided are valid and accurate. Compare this with **OAuth2**, which uses the presence of a date-based nonce and key-based signature so that the credentials passed by the **OAuth2** provider can be independently verified.

The benefit of the **CAS** approach is that the information passed on from the **CAS** server to authenticate the user is much simpler—only a single URL parameter is returned to the application by the **CAS** server. Additionally, the application itself need not track the active or valid tickets, and instead can wholly rely on **CAS** to verify this information. Much as we saw with **OAuth2**, a servlet filter is responsible for recognizing a redirect from CAS and processing it as an authentication request. We can see how this is configured in our `CasConfig.java` file, as follows:

```java
//src/main/java/com/packtpub/springsecurity/configuration/CasConfig.
java
@Bean
public CasAuthenticationFilter
casAuthenticationFilter(CasAuthenticationProvider
casAuthenticationProvider) {
    CasAuthenticationFilter filter = new CasAuthenticationFilter();
    filter.setAuthenticationManager(new
ProviderManager(casAuthenticationProvider));
    return filter;
}


@Bean
public CasAuthenticationProvider
casAuthenticationProvider(UserDetailsService userDetailsService,
      ServiceProperties serviceProperties, TicketValidator
cas30ServiceTicketValidator) {
    CasAuthenticationProvider provider = new
CasAuthenticationProvider();
    provider.setAuthenticationUserDetailsService(new
UserDetailsByNameServiceWrapper<>(userDetailsService));
    provider.setServiceProperties(serviceProperties);
```

```
        provider.setTicketValidator(cas30ServiceTicketValidator);
        provider.setKey("key");
        return provider;
    }


    @Bean
    public TicketValidator cas30ServiceTicketValidator() {
        return new Cas30ServiceTicketValidator(this.casBaseUrl);
    }
```

Next, we need to update Spring Security to utilize the bean with the CasAuthenticationFilter bean. Make the following update to our SecurityConfig.java file:

```
//src/main/java/com/packtpub/springsecurity/configuration/
SecurityConfig.java
public SecurityFilterChain filterChain(HttpSecurity http,
        CasAuthenticationEntryPoint casAuthenticationEntryPoint,
        CasAuthenticationFilter casAuthenticationFilter) throws
Exception {
...omitted for brevity
    // CAS Filter
    http.addFilterAt(casAuthenticationFilter, CasAuthenticationFilter.
class);
    // Exception Handling
    http.exceptionHandling(exceptions -> exceptions
                .authenticationEntryPoint(casAuthenticationEntryPoint)
                .accessDeniedPage("/errors/403"));
...
    return http.build();
}
```

The last thing to do is to remove the existing formLogin Spring Security definition in the SecurityFilterChain Bean, as we will rely on CAS login forms for user authentication.

> **Important note**
> Your code should look like that in chapter18.01-calendar.

At this point, we should be able to start both the **CAS** server and **JBCP Calendar** application. Then, visit https://localhost:8443/ and select **All Events**, which will redirect you to the **CAS** server. You can then log in using the username casuser and the password Mellon. Upon successful authentication, you will be redirected back to the **JBCP Calendar** application. Excellent job!

> **Important note**
>
> If you are experiencing issues, it is most likely due to **an improper SSL configuration**. Ensure that you have import the CAS SSL certificate inside the JRE keystore of your **JBCP Calendar** application.

Now that we have covered the basics of CAS configuration in the introduction, we will proceed to delve deeper into **Single Logout**.

Single Logout

You may notice that if you log out of the application, you get a logout confirmation page. However, if you click on a protected page, such as the **My Events** page, you are still authenticated. The problem is that the logout is only occurs locally. So, when you request another protected resource in the **JBCP Calendar** application, a login is requested from the **CAS** server. Since the user is still logged in to the **CAS** server, it immediately returns a service ticket and logs the user back into the **JBCP Calendar** application.

This also means that if the user had signed in to other applications using the **CAS** server, they would still be authenticated to those applications, since our calendar application does not know anything about the other applications. Fortunately, **CAS** and Spring Security offer a solution to this problem. Just as we can request a login from the **CAS** server, we can also request a logout.

You can see a high-level diagram of how logging out works within **CAS**, as follows:

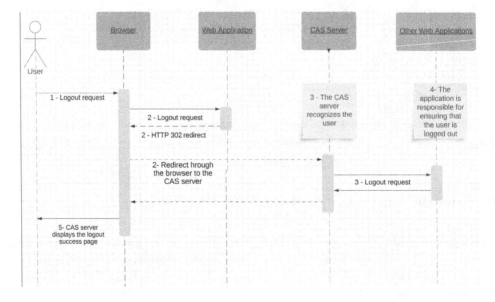

Figure 18.2 – CAS Single Logout

The following steps explain how the **Single Logout** functionality works:

1. The user requests to log out of the web application.

2. The web application then requests to log out of CAS by sending a redirect through the browser to the CAS server.

3. The CAS server recognizes the user and then sends a logout request to each CAS service that was authenticated. Note that these logout requests do not occur through the browser.

4. The CAS server indicates which user should log out by providing the original service ticket that was used to log the user in. The application is then responsible for ensuring that the user is logged out.

5. The CAS server displays the logout success page to the user.

Configuring Single Logout

The configuration for a **Single Logout** is relatively simple:

1. First, make the following updates to your `application.yml` file, adding the **CAS** logout URL:

```
cas:
  base:
    url: https://localhost:9443/cas
  login:
    url: ${cas.base.url}/login
  logout:
    url: ${cas.base.url}/logout
service:
  base:
    url: https://localhost:8443
```

2. The first step is to specify a `logout-success-url` attribute to be the logout URL of the **CAS** server in our `SecurityConfig.java` file. This means that after we log out locally, we will automatically redirect the user to the **CAS** server's logout page:

```
//src/main/java/com/packtpub/springsecurity/configuration/
SecurityConfig.java
@Configuration
@EnableWebSecurity
public class SecurityConfig {

    @Value("${cas.logout.url}")
    private String casLogoutUrl;

    @Bean
```

```
    public SecurityFilterChain filterChain(HttpSecurity http,
            CasAuthenticationEntryPoint
casAuthenticationEntryPoint) throws Exception {
...omitted for brevity
        // Logout
        http.logout(form -> form
                    .logoutUrl("/logout")
                    .logoutSuccessUrl(casLogoutUrl))

    ...

        return http.build();
    }
}
```

Since we only have one application, this is all we need to make it appear as though a single logout is occurring. This is because we log out of our calendar application before redirecting to the **CAS** server logout page. This means that by the time the **CAS** server sends the logout request to the calendar application, the user has already been logged out.

3. We will also add our `LogoutFilter` to `CasConfig.java`, as follows:

```
//src/main/java/com/packtpub/springsecurity/configuration/
CasConfig.java
@Bean
public LogoutFilter logoutFilter() {
    LogoutFilter logoutFilter = new LogoutFilter(casLogoutUrl,
new SecurityContextLogoutHandler());
    logoutFilter.setFilterProcessesUrl("/logout/cas");
    return logoutFilter;
}
```

4. If there were multiple applications and the user logged out of another application, the **CAS** server would send a logout request to our calendar application and the **CAS** server will not process the logout event. This is because our application is not listening to these logout events. The solution is simple; we must create the `SingleSignoutFilter` object. Then we need to make Spring Security aware of the `singleLogoutFilter` object in our `SecurityConfig.java`.

Place the **Single Logout** filter before the regular logout to ensure that it receives the logout events, as follows:

```
//src/main/java/com/packtpub/springsecurity/configuration/
SecurityConfig.java
@Bean
public SecurityFilterChain filterChain(HttpSecurity http,
        CasAuthenticationEntryPoint casAuthenticationEntryPoint,
        CasAuthenticationFilter casAuthenticationFilter,
        LogoutFilter logoutFilter) throws Exception {
```

```
...omitted for brevity

    // Logout Filter
    http
        .addFilterBefore(new SingleSignOutFilter(),
CasAuthenticationFilter.class)
        .addFilterBefore(logoutFilter, LogoutFilter.class);

    // Logout
    http.logout(form -> form
                .logoutUrl("/logout")
                .logoutSuccessUrl(casLogoutUrl));
...
    return http.build();
}
```

5. Go ahead and start up the application and try logging out now. You will observe that you are logged out.

6. Now, try logging back in and visiting the **CAS** server's logout URL directly. For our setup, the URL is `https://localhost:9443/cas/logout`.

7. Now, try to visit the **JBCP Calendar** application. You will observe that you are unable to access the application without authenticating again. This demonstrates that the **Single Logout** functionality is operational.

> **Important note**
> Your code should look like that in `chapter18.02-calendar`.

In this section, we covered the **Single Logout** implementation with **CAS**. Next, we will move on to discuss clustered environments.

Clustered environments

One of the things that we failed to mention in our initial diagram of **Single Logout** was how the logout is performed. Unfortunately, it is implemented by storing a mapping of the service ticket to `HttpSession` as an in-memory map. This means that **Single Logout** will not work properly within a clustered environment:

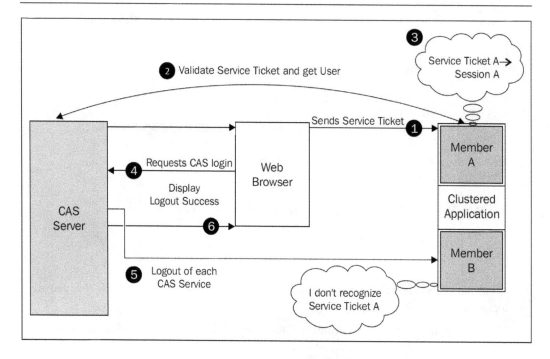

Figure 18.3 – CAS authentication in a clustered environment

Consider the following situation in the context of the preceding diagram:

1. The user logs in to **Cluster Member A**.
2. **Cluster Member A** validates the service ticket.
3. It then stores in memory, the mapping of the service ticket to the user's session.
4. The user requests to log out from the **CAS** server.

The **CAS** server sends a logout request to the **CAS** service, but **Cluster Member B** receives the logout request. It looks in its memory but does not find a session for **Service Ticket A**, because it only exists in **Cluster Member A**. This means the user will not be logged out successfully.

Proxy ticket authentication for stateless services

Centralizing our authentication using **CAS** seems to work rather well for web applications, but what if we want to call a web service using **CAS**? In order to support this, CAS has a notion of **proxy tickets** (**PT**). The following is a diagram of how it works:

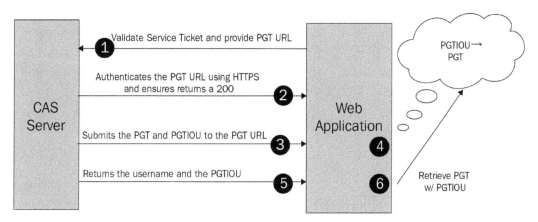

Figure 18.4 – CAS proxy ticket authentication

The flow is the same as standard **CAS** authentication until the following things take place:

1. The **Service Ticket** is validated when an additional parameter is included called the proxy ticket callback URL (**PGT URL**).

2. The **CAS** server calls the **PGT URL** over **HTTPS** to validate that the **PGT URL** is what it claims to be. Like most of the processes in a CAS, this is done by performing an SSL handshake to the appropriate URL.

3. The **CAS** server submits the **Proxy Granting Ticket (PGT)** and the **Proxy Granting Ticket I Owe You (PGTIOU)** to the **PGT URL** over **HTTPS** to ensure that the tickets are submitted to the source they claim to be.

4. The **PGT URL** receives the two tickets and must store an association of the **PGTIOU** to the **PGT**.

5. The **CAS** server finally returns a response to the request in *step 1* that includes the username and the **PGTIOU**.

6. The **CAS** service can look up the **PGT** using the **PGTIOU**.

Configuring proxy ticket authentication

Now that we know how **PT** authentication works, we will update our current configuration to obtain a **PGT** by performing the following steps:

1. The first step is to add a reference to a ProxyGrantingTicketStorage implementation. Go ahead and add the following code to our CasConfig.java file:

```
//src/main/java/com/packtpub/springsecurity/configuration/
CasConfig.java
@Bean
public ProxyGrantingTicketStorage pgtStorage() {
    return new ProxyGrantingTicketStorageImpl();
```

```
    }

    @Scheduled(fixedRate = 300_000)
    public void proxyGrantingTicketStorageCleaner(){
        logger.info("Running ProxyGrantingTicketStorage#cleanup() at
    {}",
                LocalDateTime.now());
        pgtStorage().cleanUp();
    }
```

2. The `ProxyGrantingTicketStorageImpl` implementation is an in-memory mapping of the **PGTIOU** to a **PGT**. Just as with logging out, this means we would have problems in a clustered environment using this implementation. Refer to the JA-SIG documentation to determine how to set this up in a clustered environment: `https://apereo.github.io/cas/7.0.x/high_availability/High-Availability-Guide.html`.

3. We also need to periodically clean `ProxyGrantingTicketStorage` by invoking its `cleanUp()` method. As you can see, Spring's task abstraction makes this very simple. You may consider tweaking the configuration to clear the tickets in a separate thread pool that makes sense for your environment. For more information, refer to the *Task Execution and Scheduling* section of the Spring Framework Reference documentation at `https://docs.spring.io/spring-framework/reference/integration/scheduling.html`.

4. Now we need to use `ProxyGrantingTicketStorage`, which we have just created. We just need to update the `ticketValidator` method to refer to our storage and to know the **PGT URL**. Make the following updates to `CasConfig.java`:

```
//src/main/java/com/packtpub/springsecurity/configuration/
CasConfig.java
@Value("${service.proxy.callback-url}")
private String calendarServiceProxyCallbackUrl;

@Bean
public TicketValidator cas30ServiceTicketValidator() {
    Cas30ProxyTicketValidator tv = new
Cas30ProxyTicketValidator(this.casBaseUrl);
    tv.setProxyCallbackUrl(calendarServiceProxyCallbackUrl);
    tv.setProxyGrantingTicketStorage(pgtStorage());
    return tv;
}
```

5. Then, we need to update the `application.yml` file by adding the proxy callback URL:

```
service:
  base:
    url: https://localhost:8443
  proxy:
    callback-url: ${service.base.url}/callback
```

6. The last update we need to make is to our `CasAuthenticationFilter` object, to store the **PGTIOU**-to-**PGT** mapping in our `ProxyGrantingTicketStorage` implementation when the **PGT URL** is called. It is critical to ensure that the `proxyReceptorUrl` attribute matches the `proxyCallbackUrl` attribute of the `Cas20ProxyTicketValidator` object to ensure that the **CAS** server sends the ticket to the URL that our application is listing to. Make the following changes to `CasConfig.java`:

```
//src/main/java/com/packtpub/springsecurity/configuration/
CasConfig.java
@Bean
public CasAuthenticationFilter
casAuthenticationFilter(CasAuthenticationProvider
casAuthenticationProvider,
        ProxyGrantingTicketStorage pgtStorage) {
    CasAuthenticationFilter filter = new
CasAuthenticationFilter();
    filter.setAuthenticationManager(new
ProviderManager(casAuthenticationProvider));
    filter.setProxyGrantingTicketStorage(pgtStorage);
    filter.setProxyReceptorUrl("/pgtUrl");
    return filter;
}
```

Now that we have a **PGT**, what do we do with it? A service ticket is a single-use token. However, a **PGT** can be used to produce a **PT**. Let's see how we can create a **PT** using a **PGT**.

Important note

In the `application.yml` configuration file, we can observe that the `proxyCallBackUrl` attribute matches the absolute path of our context-relative `proxyReceptorUrl` attribute path. Since we are deploying our base application to `${service.base.url}`, the full path of our `proxyReceptor` URL will be `${service.base.url}/pgtUrl`.

Following this examination of the configuration on a **CAS** Server clustered environments, we will next delve into the detailed utilization of CAS proxy tickets.

Using proxy tickets

We can now use our **PGT** to create a **PT** to authenticate it to a service. The code to do this is quite trivially demonstrated in the `EchoController` class that we have included in the code for this chapter. You can see the relevant portions of it in the following code snippet. For additional details, refer to the example source code:

```
//src/main/java/com/packtpub/springsecurity/web/controllers/
EchoController.java
@GetMapping("/echo")
    public String echo()  {
        final CasAuthenticationToken token = (CasAuthenticationToken)
SecurityContextHolder
                .getContext()
                .getAuthentication();
    // The proxyTicket could be cached in session and reused if we
wanted to
        final String proxyTicket = token.getAssertion().getPrincipal().
getProxyTicketFor(targetUrl);
        // Make a remote call using the proxy ticket
        return restTemplate.getForObject(targetUrl+"?ticket={pt}",
String.class, proxyTicket);
    }
```

This controller is a contrived example that will obtain a **PT** that will be used to authenticate a RESTful call to obtain all the events for the currently logged-in user. It then writes the JSON response to the page. The thing that may confuse some users is that the `EchoController` object is actually making a RESTful call to the `MessagesController` object that is in the same application. This means that the calendar application makes a RESTful call to itself.

Go ahead and visit `https://localhost:8443/echo` to see it in action. The page looks a lot like the **CAS** login page (minus the CSS). This is because the controller attempts to echo our **My Events** page, and our application does not yet know how to authenticate a **PT**. This means it is redirected to the **CAS** login page. Let's see how we can authenticate proxy tickets.

> **Important note**
> Your code should look like that in `chapter18.03-calendar`.

Authenticating proxy tickets

Let's look at the following steps to learn about authenticating proxy tickets:

1. We first need to tell the `ServiceProperties` object that we want to authenticate all of the tickets and not just those submitted to the `filterProcessesUrl` attribute. Make the following updates to `CasConfig.java`:

```
//src/main/java/com/packtpub/springsecurity/configuration/
CasConfig.java
@Bean
public ServiceProperties serviceProperties() {
    ServiceProperties serviceProperties = new
ServiceProperties();
    serviceProperties.setService(serviceBaseUrl+ "/login/cas");
    serviceProperties.setAuthenticateAllArtifacts(true);
    return serviceProperties;
}
```

2. We then need to update our `CasAuthenticationFilter` object for it to know that we want to authenticate all artifacts (that is, tickets) instead of only listening to a specific URL. We also need to use an `AuthenticationDetailsSource` interface that can dynamically provide the CAS service URL when validating proxy tickets on arbitrary URLs. This is important because when a CAS service asks whether a ticket is valid or not, it must also provide the CAS service URL that was used to create the ticket. Since proxy tickets can occur for any URL, we must be able to dynamically discover this URL. This is done by leveraging the `ServiceAuthenticationDetailsSource` object, which will provide the current URL from the HTTP request:

```
//src/main/java/com/packtpub/springsecurity/configuration/
CasConfig.java

@Bean
public CasAuthenticationFilter
casAuthenticationFilter(CasAuthenticationProvider
casAuthenticationProvider,
        ProxyGrantingTicketStorage pgtStorage, ServiceProperties
serviceProperties) {
    CasAuthenticationFilter filter = new
CasAuthenticationFilter();
    filter.setAuthenticationManager(new
ProviderManager(casAuthenticationProvider));
    filter.setProxyGrantingTicketStorage(pgtStorage);
    filter.setProxyReceptorUrl("/pgtUrl");
    filter.setServiceProperties(serviceProperties);
    filter.setAuthenticationDetailsSource(new
ServiceAuthenticationDetailsSource(serviceProperties));
```

```
        return filter;
    }
```

3. We will also need to ensure that we are using the `Cas30ProxyTicketValidator` object and not the `Cas30ServiceTicketValidator` implementation, and indicate which proxy tickets we will accept. We will configure our `Cas30ProxyTicketValidator` to accept a proxy ticket from any CAS service. In a production environment, you will want to consider restricting yourself to only those CAS services that are trusted:

```java
//src/main/java/com/packtpub/springsecurity/configuration/
CasConfig.java
@Bean
public TicketValidator cas30ServiceTicketValidator() {
    Cas30ProxyTicketValidator tv = new
Cas30ProxyTicketValidator(this.casBaseUrl);
    tv.setProxyCallbackUrl(calendarServiceProxyCallbackUrl);
    tv.setProxyGrantingTicketStorage(pgtStorage());
    tv.setAcceptAnyProxy(true);
    return tv;
}
```

4. Lastly, we want to provide a cache for our `CasAuthenticationProvider` object so that we do not need to hit the CAS service for every call made to our service. To do this, we need to configure a `StatelessTicketCache` as follows:

```java
//src/main/java/com/packtpub/springsecurity/configuration/
CasConfig.java
@Bean
public CasAuthenticationProvider
casAuthenticationProvider(UserDetailsService userDetailsService,
        ServiceProperties serviceProperties, TicketValidator
cas30ServiceTicketValidator, SpringCacheBasedTicketCache
springCacheBasedTicketCache) {
    CasAuthenticationProvider provider = new
CasAuthenticationProvider();
    provider.setAuthenticationUserDetailsService(new
UserDetailsByNameServiceWrapper<>(userDetailsService));
    provider.setServiceProperties(serviceProperties);
    provider.setTicketValidator(cas30ServiceTicketValidator);
    provider.setKey("key");
    provider.
setStatelessTicketCache(springCacheBasedTicketCache);
    return provider;
}
@Bean
public SpringCacheBasedTicketCache
springCacheBasedTicketCache(CacheManager cacheManager) {
```

```
        return new SpringCacheBasedTicketCache(cacheManager.
getCache("castickets"));
    }
```

> **Important note**
>
> Don't forget the `@EnableCaching` in the **CAS** configuration class-level `CasConfig.java`, so that caching can automatically enabled by Spring.

5. As you might have suspected, the Spring cache can rely on external implementations including `EhCache`. Go ahead and start the application back up and visit `https://localhost:8443/echo` again. This time, you should see a response to calling our **My Events** page:

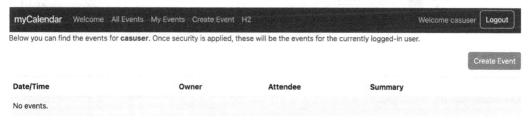

Figure 18.5 – Authenticating proxy tickets response

> **Important note**
>
> Your code should look like that in `chapter18.04-calendar`.

After using proxy tickets, we'll next explore the process of customizing the **CAS** server.

Customizing the CAS server

All the changes in this section will be made to the **CAS** server and not the calendar application. This section is only meant to be an introduction to configuring the **CAS** server, as a detailed setup is certainly beyond the scope of this book. Just as with the changes made to the calendar application, we encourage you to follow along with the changes in this chapter. For more information, you can refer to the **CAS Aperero** documentation: `https://apereo.github.io/cas`.

CAS WAR overlay

The preferred way to customize **CAS** is to use a **Maven** or **Gradle WAR overlay**. With this mechanism, you can change everything from the UI to the method in which you authenticate with the **CAS** server. The concept of a WAR overlay is simple. You add a **WAR overlay**, `cas-server-webapp`, as a dependency, and then provide additional files that will be merged with the existing **WAR overlay**.

How does the CAS internal authentication work?

Before we jump into CAS configuration, we'll briefly illustrate the standard behavior of CAS authentication processing. The following diagram should help you follow the configuration steps required to allow CAS to talk to our embedded LDAP server:

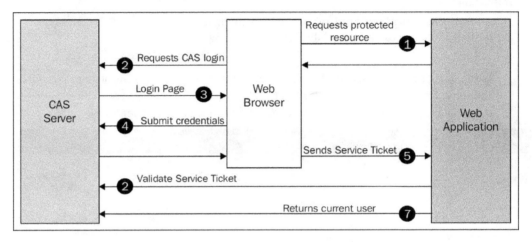

Figure 18.6 – CAS internal authentication flow

While the preceding diagram describes the internal flow of authentication within the **CAS** server itself, it is likely that if you are implementing an integration between Spring Security and **CAS**, you will need to adjust the configuration of the **CAS** server as well. It's important, therefore, that you understand how **CAS** authentication works at a high level.

The **CAS** server's `org.apereo.cas.authentication.AuthenticationManager` interface (not to be confused with the Spring Security interface of the same name) is responsible for authenticating the user based on the provided credentials. In Spring Security, the actual processing of the credentials is delegated to one (or more) processing class implementing the `org.apereo.cas.authentication.AuthenticationHandler` interface. We recognize that the analogous interface in `Spring Security` would be `AuthenticationProvider`.

While not a full review of the behind-the-scenes functionality of the **CAS** server, this should help you understand the configuration steps in the next several exercises. We encourage you to read the source code for **CAS** and consult the web-based documentation.

Configuring CAS to connect to our embedded LDAP server

CAS verifies a username/password by authenticating against an LDAP directory, such as Active Directory or OpenLDAP. There are various directory architectures, and we offer configuration options for four common scenarios.

It's important to note that **CAS** automatically generates the necessary components internally based on the specified settings. If you want to authenticate against multiple LDAP servers, you can increment the index and specify the settings for the next LDAP server.

Additionally, please be aware that attributes obtained during **LDAP** authentication are combined with attributes from other attribute repository sources, if applicable. Attributes obtained directly through **LDAP** authentication take precedence over all other attributes.

You can find the available settings and properties in the CAS documentation (`https://apereo.github.io/cas/`).

Let's take a look at the following steps to configure an embedded LDAP server with **CAS**:

1. First, we update the `build.gradle` dependencies of the `chapter18.00-cas-server` project. We enable the CAS LDAP support and add the `spring-security` built-in support for the embedded LDAP server:

    ```
    //CAS LDAP Support
    implementation "org.apereo.cas:cas-server-support-ldap"

    // Spring Security LDAP
    implementation("org.springframework.security:spring-security-
    ldap")
    implementation("org.springframework.security:spring-security-
    config")
    // Embedded LDAP Server
    implementation("com.unboundid:unboundid-ldapsdk")
    ```

2. Then we need to add the following sections to our `application.yml` file with the connection parameters for the embedded LDAP server:

    ```
    // Embedded LDAP
    spring:
      ldap:
        embedded:
          ldif: classpath:/ldif/calendar.ldif
          baseDn: dc=jbcpcalendar,dc=com
          port: 33389
          credential:
            username: uid=admin,ou=system
            password: secret
    // CAS configuration for LDAP
    cas:
      authn:
        ldap[0]:
          useStartTls: false
    ```

```
search-filter: uid={user}
type: AUTHENTICATED
ldapUrl: ldap://localhost:${spring.ldap.embedded.port}
baseDn: ${spring.ldap.embedded.baseDn}
bindDn: ${spring.ldap.embedded.credential.username}
bindCredential: ${spring.ldap.embedded.credential.
password}
```

3. For this exercise, we will be using an LDIF file created for this book, intended to capture many of the common configuration scenarios with LDAP and Spring Security (just as we did in *Chapter 6, LDAP Directory Services*). Copy the provided LDIF file to the following project location: `src/main/resources/ldif/calendar.ldif`.

4. Finally, we need to configure the `CasOverlayOverrideConfiguration.java` as follows:

```
//src/main/java/org/apereo/cas/config/
CasOverlayOverrideConfiguration.java

@Lazy(false)
@AutoConfiguration
public class CasOverlayOverrideConfiguration {

    private final BaseLdapPathContextSource contextSource;

    public
CasOverlayOverrideConfiguration(BaseLdapPathContextSource
contextSource) {
        this.contextSource = contextSource;
    }
}
```

> **Important note**
>
> Your code should look like that in `chapter18.05-cas-server` and `chapter18.05-calendar`.

So, now we've configured basic **LDAP** authentication in **CAS**. At this point, you should be able to restart **CAS**, start the **JBCP Calendar** application (if it's not already running), and authenticate it using `admin1@example.com/admin` or `user1@example.com/user1`. Go ahead and try it to check whether it works. If it does not work, try checking the logs and comparing your configuration with the example configuration.

Following the customization of the **CAS** server, we'll proceed to explore how to extract the `UserDetails` object from a **CAS** assertion.

Getting the UserDetails object from a CAS assertion

Up until this point, we have been authenticating with **CAS** by obtaining the roles from our `InMemoryUserDetailsManager` object. However, we can create the `UserDetails` object from the **CAS** assertion just as we did with **OAuth2**. The first step is to configure the **CAS** server to return the additional attributes.

Returning LDAP attributes in the CAS response

We know that **CAS** can return the username in the **CAS** response, but it can also return arbitrary attributes in the **CAS** response. Let's see how we can update the CAS server to return additional attributes. Again, all the changes in this section are in the *CAS server and not in the calendar application*.

Mapping LDAP attributes to CAS attributes

The first step requires us to map **LDAP** attributes to attributes in the **CAS** assertion (including the role attribute, which we're expecting to contain the user's `GrantedAuthority`).

We'll add another bit of configuration to the CAS `application.yml` file. This new bit of configuration is required to instruct **CAS** as to how to map attributes from the **CAS** `Principal` object to the **CAS** `principalAttributeList`, which will ultimately be serialized as part of ticket validation. The configuration should be declared as follows in the `chapter18.06-cas-server` project:

```
//src/main/resources/application.yml
cas:
  service-registry:
    core:
      init-from-json: true
    json:
      location: classpath:/etc/cas/services
  authn:
    ldap[0]:
      principalAttributeList: sn:lastName,cn:fullName,description:role
      useStartTls: false
      search-filter: uid={user}
      type: AUTHENTICATED
      ldapUrl: ldap://localhost:${spring.ldap.embedded.port}
      baseDn: ${spring.ldap.embedded.baseDn}
      bindDn: ${spring.ldap.embedded.credential.username}
      bindCredential: ${spring.ldap.embedded.credential.password}
```

The functionality behind the scenes here is confusing—essentially, the purpose of this class is to map `org.apereo.cas.authentication.principal.Principal` back to the LDAP directory. The provided `baseDN` Java bean property is searched using the LDAP query (`uid=user1@example.com`), and attributes are read from the matching entry. The attributes are mapped back to `org.apereo.cas.authentication.principal.Principal` using the key/value pairs in the `principalAttributeList` property. We recognize that LDAP's `cn` and `sn` attributes are mapped to meaningful names, and the `description` attribute is mapped to the role that will be used for determining the authorization of our user.

Last, we would love to set up the same type of query in **LDAP** as we used with Spring Security LDAP in *Chapter 6, LDAP Directory Services*, to be able to map `Principal` to a full LDAP-distinguished name, and then use that **Distinguished Name** (**DN**) to look up group membership by matching based on the `uniqueMember` attribute of a `groupOfUniqueNames` entry. Unfortunately, the CAS LDAP code doesn't have this flexibility yet, leading to the conclusion that more advanced **LDAP** mapping will require extensions to base classes in **CAS**.

Acquiring a UserDetails from CAS

When we first set up **CAS** integration with Spring Security, we configured `UserDetailsByNameServiceWrapper`, which simply translated the username presented to **CAS** into a `UserDetails` object from `UserDetailsService`, which we had referenced. Now that **CAS** is referencing the LDAP server, we can set up `LdapUserDetailsService` as we discussed at the tail end of *Chapter 6, LDAP Directory Services*, and things will work just fine.

Note that in the next section, we will switch back to modifying the *calendar application and not the CAS server*.

The GrantedAuthorityFromAssertionAttributesUser object

Now that we have modified the **CAS** server to return custom attributes, we'll experiment with another capability of the Spring Security **CAS** integration—the ability to populate `UserDetails` from the **CAS** assertion itself! This is actually as simple as switching the `AuthenticationUserDetailsService` implementation to the

`o.s.s.cas.userdetails.GrantedAuthorityFromAssertionAttributesUserDetailsService` object, whose job it is to read the **CAS** assertion, look for a certain attribute, and map the value of that attribute directly to the `GrantedAuthority` object for the user. Let's assume that there is an attribute-entitled role that will be returned with the assertion. We simply configure a new `authenticationUserDetailsService` bean (be sure to replace the previously defined `authenticationUserDetailsService` bean) in the `CaseConfig.java` file:

```
//src/main/java/com/packtpub/springsecurity/configuration/CasConfig.
java
```

```
@Bean
public AuthenticationUserDetailsService userDetailsService() {
    return new
GrantedAuthorityFromAssertionAttributesUserDetailsService(new String[]
{ "role" });
}
```

> **Important note**
> Your code should look like that in chapter18.06-cas-server and chapter18.06-calendar.

How is attribute retrieval useful?

Remember that **CAS** provides a layer of abstraction for our application, removing the ability for our application to directly access the user repository, and instead forcing all such access to be performed through **CAS** as a proxy.

This is extremely powerful! It means that our application no longer cares what kind of repository the users are stored in, nor does it have to worry about the details of how to access them. This confirms that authentication with **CAS** is sufficient to prove that a user should be able to access our application. For system administrators, this means that should an LDAP server be renamed, moved, or otherwise adjusted, they only need to reconfigure it in a single location—**CAS**. Centralizing access through **CAS** allows for a high level of flexibility and adaptability in the overall security architecture of the organization.

Now all applications authenticated through **CAS** have the same view of a user and can consistently display information across any **CAS**-enabled environment.

Be aware that, once authenticated, Spring Security does not require the **CAS** server unless the user is required to reauthenticate. This means that attributes and other user information stored locally in the application in the user's Authentication object may become stale over time, and possibly out of sync with the source **CAS** server. Take care to set session timeouts appropriately to avoid this potential issue!

Additional CAS capabilities

CAS offers additional advanced configuration capabilities outside of those that are exposed through the Spring Security **CAS** wrappers. Some of these include the following capabilities:

- Providing transparent SSO for users who are accessing multiple **CAS**- secured applications within a configurable time window on the **CAS** server.

- Applications can force users to authenticate to **CAS** by setting the renew property to true on TicketValidator; you may want to conditionally set this property in some custom code in the event that the user attempts to access a highly secured area of the application.

- A **RESTful API** for obtaining service tickets.

- The Aperero CAS server can also act as an **OAuth2** server. If you think about it, this makes sense, since **CAS** is very similar to **OAuth2**.

- Providing **OAuth** support for the **CAS** server so that it can obtain access tokens to a delegate OAuth provider (that is, Google), or so the CAS server can be the OAuth server itself.

We'd encourage you to explore the full capabilities of the CAS client and server as well as ask any questions to the helpful folks in the **CAS Aperero** community forums!

Summary

In this chapter, we learned about the **CAS** SSO portal and how it can be integrated with Spring Security, and we also covered the **CAS** architecture and communication paths between actors in a **CAS**-enabled environment. We also saw the benefits of **CAS**-enabled applications for application developers and system administrators. We also learned about configuring the **JBCP Calendar** application to interact with a basic **CAS** installation. We also covered the use of CAS's **Single Logout** support.

We also saw how proxy ticket authentication works and how to leverage it to authenticate stateless services.

We also covered the tasks of updating **CAS** to interact with **LDAP** and sharing **LDAP** data with our **CAS**-enabled application. We even learned about implementing attribute exchange with the industry-standard **SAML** protocol.

We hope this chapter was an interesting introduction to the world of SSO. There are many other SSO systems on the market, mostly commercial, but **CAS** is one of the leaders of the open source SSO world and offers an excellent platform to build out **SSO** capability in any organization.

For the last chapter, we'll learn more on building GraalVM native images.

19

Build GraalVM Native Images

This chapter is dedicated to honing skills concerning native images and `GraalVM`. Throughout this chapter, you'll receive guidance on crafting a Spring Security application that makes use of native functionalities.

`Spring Boot 3` introduces native image generation support through `GraalVM`, and Spring Security seamlessly integrates with this capability, making its features compatible with native images. This integration can be a great way to improve the performance and security of your Spring Security applications.

We'll delve into the essential steps for efficiently updating build configurations to leverage `GraalVM` tools, enabling seamless integration of native capabilities into your applications.

Additionally, we will explore `GraalVM` Native Image support with Spring Security features that we have not covered so far in this book, including the following topics:

- Introducing `GraalVM`
- Creating `GraalVM` images using `Buildpacks`
- Building a native image using Native Build Tools
- Handling method security in `GraalVM` Native Image

The following section is dedicated to offering guidance on specific `Spring Security` features that may necessitate additional hints to be supplied by the application.

This chapter's code in action link is here: `https://packt.link/fQ5AM`.

Introducing GraalVM

GraalVM is a high-performance runtime that provides significant improvements in application performance and efficiency. **GraalVM** represents a high-performance runtime **Java Development Kit (JDK)**.

Besides supporting **just-in-time** (**JIT**) compilation, **GraalVM** also enables ahead-of-time compilation of Java applications. This feature facilitates quicker initialization, enhanced runtime performance, and reduced resource consumption. However, the resulting executable is limited to running on the platform for which it was compiled. **GraalVM** extends its functionality by offering additional programming languages and execution modes. The first production-ready version, **GraalVM** 19.0, was introduced in May 2019.

In the following section, we will delve deeper into the concept of native images to gain a better understanding.

What are native images?

Native images in **GraalVM** refer to executables compiled **Ahead Of Time** (**AOT**) from Java applications. Unlike traditional Java applications, which are typically compiled to bytecode and run on the **Java Virtual Machine** (**JVM**), native images are compiled directly into machine code for a specific platform.

GraalVM's native image leverages the **GraalVM** compiler to analyze and optimize the Java application's code during compilation, resulting in a standalone native executable tailored for the target environment. These native images can be deployed independently without requiring a separate JVM installation, making them particularly useful for deploying lightweight, fast-starting applications in containerized environments or serverless platforms.

Key features of GraalVM

GraalVM distinguishes itself from the base JDK through several key features:

- The **Graal compiler**, serving as a JIT compiler.

- **GraalVM Native Image**, a technology facilitating ahead-of-time compilation of Java applications.

- **Truffle** is the language implementation framework and `GraalVM` SDK, comprising a Java-based framework and a suite of APIs tailored for developing high-performance language runtimes.

- **GraalVM** offers runtimes for `JavaScript`, `Ruby`, `Python`, and several other widely-used languages. With GraalVM's polyglot capabilities, developers can seamlessly blend multiple programming languages within a single application without incurring additional overhead from foreign language calls.

- **JavaScript runtime**, featuring an `ECMAScript 2023`-compliant JavaScript runtime, alongside Node.js.

- **Low-Level Virtual Machine** (**LLVM**) **runtime**, designed to compile languages transformable into `LLVM` bitcode.

Security benefits of GraalVM

In terms of security, **GraalVM** offers some notable advantages:

- It stops new, unfamiliar code from being loaded while the program is running.

- It only incorporates paths that the application has proven reachable within its image.

- Reflection is turned off by default and requires a specific inclusion list to enable.

- Deserialization is allowed only for a predefined list of classes.

- Issues related to the just-in-time compiler, such as crashes, incorrect compilations, or the possibility of creating machine code gadgets via techniques like **JIT spraying** (security exploit) are eliminated.

After covering the key features and security benefits of **GraalVM**, we'll proceed to explore practical examples of building **GraalVM** images. This will involve utilizing Buildpacks and applying them to our JBCP Calendar application.

GraalVM images using Buildpacks

Ensure Docker is installed by referring to the details in the Docker documentation (`https://docs.docker.com/get-docker/`). If you are using Linux, configure Docker to permit non-root users.

Important note

For macOS users, it is advisable to enhance the allocated Docker memory to a minimum of **8 GB** and consider adding more CPUs. On Microsoft Windows, ensure optimal performance by enabling the Docker WSL 2 backend (`https://docs.docker.com/desktop/wsl/`).

In the upcoming sections, we will construct **GraalVM** images employing `Buildpacks` and `Gradle`, as well as crafting **GraalVM** images utilizing `Buildpacks` and `Maven`.

Building GraalVM images using Buildpacks and Gradle

The AOT tasks are automatically configured by the **Spring Boot Gradle plugin** when the **GraalVM Native Image plugin** is applied. Ensure that your Gradle build includes a `plugins` block with `org.graalvm.buildtools.native`.

For our sample application, we need to add the plugin declaration to the `build.gradle` file:

```
//build.gradle
plugins {
    id 'java'
    id 'org.springframework.boot' version '3.2.4'
```

```
    id 'io.spring.dependency-management' version '1.1.4'
    id 'org.graalvm.buildtools.native' version '0.10.1'
}
```

Once the `org.graalvm.buildtools.native` plugin is applied, the `bootBuildImage` task will produce a native image instead of a JVM one. Execute the task using:

```
./gradlew bootBuildImage
```

Building GraalVM images using Buildpacks and Maven

To create a native image container using **Maven**, make sure that your `pom.xml` file adopts the `spring-boot-starter-parent` and includes the `org.graalvm.buildtools:native-maven-plugin`. Ensure that the `<parent>` section in your `pom.xml` resembles the following:

```
<parent>
    <groupId>org.springframework.boot</groupId>
    <artifactId>spring-boot-starter-parent</artifactId>
    <version>3.2.4</version>
</parent>
```

Additionally, you should include the following `native-maven-plugin` in the `<plugins>` section of your `pom.xml`:

```
<build>
    <plugins>
        <plugin>
            <groupId>org.graalvm.buildtools</groupId>
            <artifactId>native-maven-plugin</artifactId>
        </plugin>
    </plugins>
</build>
```

The `spring-boot-starter-parent` includes a **native profile** that sets up the necessary executions for creating a native image. You can activate profiles by using -P flag on the command line.

```
mvn -Pnative spring-boot:build-image
```

> **Important note**
> Your code should now look like that in `chapter19.01-calendar`.

Running GraalVM images from Buildpacks

After executing the relevant build command for **Maven** or **Gradle**, a Docker image should be accessible. Initiate your application by utilizing the docker run command:

```
docker run --rm -p 8080:8080 docker.io/library/chapter19.01-
calendar:0.0.1-SNAPSHOT
```

Your application should be reachable via `http://localhost:8080`.

To gracefully shut down the application, press *Ctrl + C*.

Following the construction of **GraalVM** images with `Buildpacks`, we will delve into a detailed exploration of building a Native Image using **Native Build Tools**.

Building a native image using Native Build Tools

If you prefer generating a native executable without relying on **Docker**, **GraalVM Native Build Tools** come in handy. These tools are provided as plugins by **GraalVM** for both **Maven** and **Gradle**, offering a range of **GraalVM** tasks, including the generation of a native image.

In the following sections, we will understand the process of building and running **GraalVM** images using **Native Build Tools** with either **Maven** or **Gradle**.

Prerequisites

To generate a native image with the **Native Build Tools**, ensure that you have a **GraalVM** distribution installed on your system.

For our examples, we will use `bellsoft-liberica-vm-openjdk17-23.0.3` that is available in the Liberica Native Image Kit Download Center (`https://bell-sw.com/pages/downloads/native-image-kit/#nik-23-(jdk-17)`).

Building GraalVM images using Native Build Tools and Maven

Just like with the buildpack support, it's essential to ensure the usage of `spring-boot-starter-parent` to inherit the native profile. Additionally, make sure to incorporate the `org.graalvm.buildtools:native-maven-plugin` plugin.

Once the native profile is enabled, you can initiate the `native:compile` goal to commence the native-image compilation process.

```
mvn -Pnative native:compile
```

You can locate the executable of the native image in the `target` directory.

Building GraalVM images using Native Build Tools and Gradle

For our sample application, we need to add the **Native Build Tools Gradle plugin**.

When you incorporate the Native Build Tools Gradle plugin into your project, the Spring Boot Gradle plugin will promptly activate the Spring AOT engine. Task dependencies are prearranged, enabling you to simply execute the standard `nativeCompile` task for generating a native image.

```
./gradlew nativeCompile
```

You can locate the native image executable within the directory named `build/native/nativeCompile`.

Running GraalVM images from Native Build Tools

At this stage, your application should be functional. The startup time varies across different machines, but it is expected to be considerably faster than a Spring Boot application running on a JVM.

You can run the application by executing it directly:

```
target/chapter19.01-calendar
```

By navigating to `http://localhost:8080` in your web browser, you should have access to the **JBCP Calendar** application.

To gracefully shut down the application, press *Ctrl + C*.

Upon grasping the fundamental configurations for building **GraalVM** images, we will delve deeper into specific use cases associated with **Spring Security** native images.

Method Security in GraalVM Native Image

While **GraalVM** Native Image does support **Method Security**, certain use cases may require the application to supply additional hints for proper functionality.

If you're employing custom implementations of `UserDetails` or authentication classes and using `@PreAuthorize` and `@PostAuthorize` annotations, you may need supplementary indications. Consider a scenario where you've crafted a custom implementation of the `UserDetails` class, which is returned by your `UserDetailsService`:

1. First, create a custom `UserDetails` implementation as follows:

```
//src/main/java/com/packtpub/springsecurity/service/
CalendarUserDetails.java
public class CalendarUserDetails extends CalendarUser implements
UserDetails {
    CalendarUserDetails(CalendarUser user) {
```

```java
        setId(user.getId());
        setEmail(user.getEmail());
        setFirstName(user.getFirstName());
        setLastName(user.getLastName());
        setPassword(user.getPassword());
    }

    @Override
    public Collection<? extends GrantedAuthority> getAuthorities() {
        return CalendarUserAuthorityUtils.createAuthorities(this);
    }

    public boolean hasAdminRole() {
        return getUsername().startsWith("admin");
    }

    @Override
    public String getUsername() {
        return getEmail();
    }

    @Override
    public boolean isAccountNonExpired() {
        return true;
    }

    @Override
    public boolean isAccountNonLocked() {
        return true;
    }

    @Override
    public boolean isCredentialsNonExpired() {
        return true;
    }

    @Override
    public boolean isEnabled() {
        return true;
    }

    private static final long serialVersionUID = 3384436451564509032L;
}
```

> **Important note**
>
> The provided implementation of hasAdminRole() is merely presented as an illustration of how method security can be managed with GraalVM native images. However, for production environments, it's advisable to consider a safer implementation of hasAdminRole().

2. We want to use hasAdminRole() method within a @PreAuthorize annotation in the CalendarService interface:

```
//src/main/java/com/packtpub/springsecurity/service/
CalendarUserDetails.java

public interface CalendarService{
... omitted for brevity
    @PreAuthorize("principal?.hasAdminRole()")
    List<Event> getEvents();
}
```

3. Ensure that the @EnableMethodSecurity annotation is present in SecurityConfig.java to activate method security annotations.

> **Important note**
>
> Your code should now look like that in chapter19.02-calendar.

4. Sign in to the application with admin1@example.com/admin1 and try accessing http://localhost:8080/events. You will notice all the events can be displayed.

Figure 19.1 – All the events page

5. When you sign in to the application with user1@example.com/user1 and try accessing http://localhost:8080/events, You will get the following access denied page:

Exception during execution of Spring Security application! Access Denied

Error page
Back to Home Page

Figure 19.2 – Access denied page for unauthorized user

6. If you execute your application's native image with the provided configuration, attempting to invoke the hasAdminRole() method will result in an error resembling the following:

```
Caused by: org.springframework.expression.spel.
SpelEvaluationException: EL1004E: Method call: Method
hasAdminRole() cannot be found on type com.packtpub.
springsecurity.service.CalendarUserDetails

at org.springframework.expression.spel.ast.MethodReference.
findAccessorForMethod(MethodReference.java:237) ~[na:na]

at org.springframework.expression.spel.ast.MethodReference.
getValueInternal(MethodReference.java:147) ~[na:na]

at org.springframework.expression.spel.ast.
MethodReference$MethodValueRef.getValue(MethodReference.
java:400) ~[na:na]

at org.springframework.expression.spel.ast.CompoundExpression.
getValueInternal(CompoundExpression.java:98) ~[na:na]

at org.springframework.expression.spel.ast.SpelNodeImpl.
getTypedValue(SpelNodeImpl.java:119) ~[chapter19.02-
calendar:6.1.5]

at org.springframework.expression.spel.standard.SpelExpression.
getValue(SpelExpression.java:309) ~[na:na]

at org.springframework.security.access.expression.
ExpressionUtils.evaluateAsBoolean(ExpressionUtils.java:30)
~[na:na]

... 113 common frames omitted
```

The error above indicates that the hasAdminRole() method cannot be located on the CalendarUserDetails.class.

Spring Security relies on reflection to invoke the hasAdminRole() method, and GraalVM Native Image does not inherently support reflection.

To resolve this problem, you must follow the subsequent steps:

1. Provide hints to GraalVM Native Image to enable reflection on the `CalendarUserDetails#hasAdminRole()` method. This can be achieved by supplying a custom hint, as demonstrated in the following example:

```
//src/main/java/com/packtpub/springsecurity/configuration/
SpringSecurityHints.java
@Configuration(proxyBeanMethods = false)
@ImportRuntimeHints(SpringRuntimeHints.class)
public class SpringSecurityHints {

    static class SpringRuntimeHints implements RuntimeHintsRegistrar {
        @Override
        public void registerHints(RuntimeHints hints, ClassLoader
classLoader) {
            hints.reflection()
                    .registerType(CalendarUserDetails.class,
                        MemberCategory.INVOKE_DECLARED_METHODS);
        }
    }
}
```

2. Now you can build the native image of your application, and it should function as intended.

> **Important note**
> Your code should now look like that in `chapter19.03-calendar`.

Summary

In this chapter, we delved into the support for **GraalVM** Native Image in conjunction with Spring Security features, which were not previously covered in this book. We introduced the concept of native images, including the key features and security benefits of **GraalVM**.

The key topics explored included `Spring Boot 3`, which introduces native image generation support via **GraalVM**, seamlessly integrating with Spring Security features and making them compatible with native images. We have seen that there are some cases where we need to provide hints to be used by **GraalVM**.

Moreover, we've successfully improved our application's performance and security, aligning with contemporary software development practices.

In concluding this chapter, I want to extend my heartfelt congratulations to you for reaching this milestone of completing the book. Your dedication and commitment to learning about `Spring Security 6` features are commendable.

As you reflect on the concepts covered, remember that every step you take toward mastering these technologies brings you closer to becoming an adept developer. Embrace the challenges ahead with confidence, knowing that each hurdle is an opportunity for growth. Keep exploring, keep experimenting, and never underestimate the power of continuous learning.

Wishing you continued success on your software development journey!

Appendix – Additional Reference Material

In this appendix, we will cover some reference material that we feel is helpful (and largely undocumented) but too comprehensive to insert in the chapters.

During this section, we will cover the following topics:

- Using build tools (Gradle or Maven) and IDE (**IntelliJ IDEA** or **Eclipse**)
- Getting started with the JBCP calendar sample code
- Generating a server certificate
- Additional supporting resource

This section is designed to provide you with guidance for further depth and clarity on the provided code covered in this book.

Build tools

The book's code is executable via **Gradle** or **Maven**.

In subsequent sections, we'll elaborate on executing projects utilizing either Gradle or Maven as the build tool. The code from the book operates independently of any **Integrated Development Environment (IDE)**.

However, we'll provide examples of implementations using **Eclipse** and **IntelliJ IDEA**.

In the following section, we demonstrate the usage of build tools such as Gradle or Maven.

Gradle build tool

All the code in this book can be built using the **Gradle build tool** and is organized in a chapter-by-chapter multimodule build. You can find instructions and options for getting Gradle locally at https://gradle.org/install/.

A local installation of `Gradle` is not required as the root of the source code already has the `Gradle` wrapper installed. The `Gradle` wrapper can be installed in any submodule. You can find additional information about the `Gradle` wrapper at `https://docs.gradle.org/current/userguide/gradle_wrapper.html`.

Maven build tool

Throughout the chapters ahead, you'll find a wealth of code examples, practical tips, and hands-on exercises designed for those who are comfortable with **Maven**.

Maven is a powerful build automation tool primarily used for Java projects, though it can be adapted to other languages and frameworks. It simplifies the process of project management, dependency resolution, and build configuration, allowing developers to focus on writing code rather than managing project infrastructure. You can find instructions and options for getting `Maven` locally at `https://maven.apache.org/download.cgi`.

> **Downloading the example code**
>
> You can download the example code files for all Packt books you have purchased from your account at `https://www.packtpub.com`. If you purchased this book from elsewhere, you could visit `https://www.packtpub.com/support` and register to have the files emailed directly to you.

In the following section, we'll delve into initiating the JBCP calendar sample code with your preferred IDE such as **IntelliJ IDEA** or **Eclipse**.

Getting started with the JBCP calendar sample code

As we described in *Chapter 1, Anatomy of an Unsafe Application*, we have assumed that you have installed `Java 17` as a minimum. You can download a JDK from Oracle's website (`https://www.oracle.com/java/technologies/downloads/`) or use any other OpenJDK version 17.

Upon the publication of this book, all the code has been validated with `Java 21`, the most recent **Long-Term Support (LTS)** version.

The upcoming section discusses the structure of sample code and its utilization within your preferred IDE.

Sample code structure

The sample code contains folders of a multimodule `Gradle` or `Maven` project. Each folder is named `ChapterNN`, where NN is the `chapter number`. Each `ChapterNN` folder has additional folders containing each milestone project with the format `chapterNN.mm-calendar`, where NN is the `chapter number` and mm is the `milestone` within that chapter.

For simplicity, we recommend that you extract the source to a path that does not contain any spaces. Each milestone is a checkpoint within the `chapter` and allows you to easily compare your code with the book's code. For example, `chapter02.03-calendar` contains milestone number `03` within *Chapter 2*, *Getting Started with Spring Security*, of the calendar application. The location of the preceding project would be `~/ Spring-Security-Fourth-Edition/Chapter02/ chapter02.03-calendar`.

Chapter 1, *Anatomy of an Unsafe Application*, and *Chapter 2*, *Getting Started with Spring Security*, have been created as `Spring projects`, not using `Spring Boot` as a project base.

Chapter 3, *Custom Authentication*, converted the calendar project to a `Spring Boot codebase`.

To keep each chapter as independent as possible, most chapters in the book are built on *Chapter 9*, *Opening up to OAuth2*, or *Chapter 15*, *Additional Spring Security Features*. This means that, in most cases, you can read through *Chapter 9*, *Opening up to OAuth2*, and then skip around to the other parts of the book. However, this also means that it is important to start each chapter with the chapter's `milestone 00` source code rather than continuing to work on the code from the previous chapter. This ensures that your code starts in the same place that the chapter does.

While you can get through the entire book without performing any of the steps, we recommend starting each chapter with `milestone 00` and implementing the steps in the book. This will ensure that you get the most out of the book. You can use the milestone versions to copy large portions of code or to compare your code if you run into problems.

Using the samples in IntelliJ IDEA

There are a few things that are necessary to run the sample applications within **IntelliJ IDEA**. In all the projects, a `Tomcat` plugin has been configured with `Gradle` or `Maven` to run the embedded instance to help you get started faster.

Importing projects in IntelliJ IDEA

Most of the diagrams used in this book were taken from **IntelliJ IDEA** from JetBrains (`https:// www.jetbrains.com/idea/`). IntelliJ IDEA has wonderful support for multimodule `Gradle` or `Maven` projects.

IntelliJ IDEA will allow you to import an existing project, or you can simply open `build.gradle` or `pom.xml` from the root of the source code base and IDEA will create the necessary project files for you.

1. Once you open IntelliJ IDEA, you can open the entire project using the **Open** option, as shown in the following screenshot:

Appendix figure 1 – JBCP calendar samples import

2.   Then, you will be prompted to select various options for how IntelliJ IDEA will execute this project, either **Gradle** or **Maven**, as shown in the following screenshot:

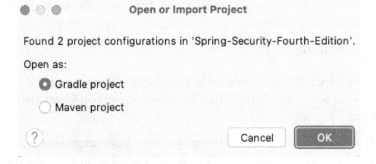

Appendix figure 2 – Gradle or Maven option

3.   Once we choose **Gradle**, for example, you will be able to work with any of the chapters, and the layout will look as shown in the following screenshot:

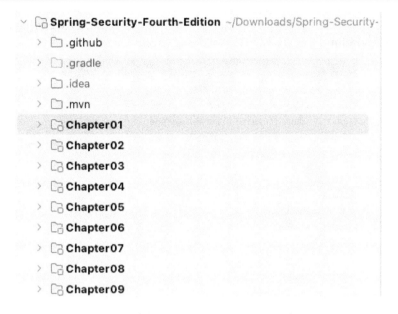

Appendix figure 3 – Chapter presentation layout with IDEA

After importing the project in **IntelliJ IDEA**, you can run your code following the guidance provided in the next section.

Running the samples within IntelliJ IDEA

Running milestone projects can be done by creating a **Run/Debug Configuration** entry for each project.

1. For each `spring-boot` project (starting from *Chapter 3, Custom Authentication*), you can simply click the green play button in the toolbar, or right-click on the main class and choose **Run CalendarApplication**. IntelliJ IDEA will start the Spring Boot application, and you will see the logs in the **Run** tool window.

```
@SpringBootApplication
public class CalendarApplication {


    public static void main(String[] args) {
        SpringApplication.run(CalendarApplication.class, args);
    }
}
```

Appendix figure 4 – Running the Spring Boot project with IntelliJ IDEA

2. For other projects, you can use the terminal or if using IntelliJ IDEA, go to **File | Run** and select **Edit Configurations…**, as shown in the following screenshot:

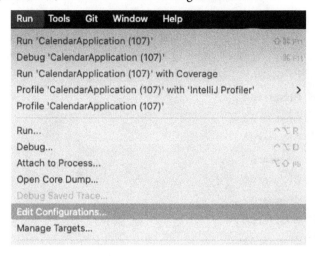

Appendix figure 5 – Custom applications, Run with IntelliJ IDEA

3. You will be presented with options to add new configurations. Select the plus (+) sign in the upper-left corner to choose a new **Gradle** configuration, as shown in the following screenshot:

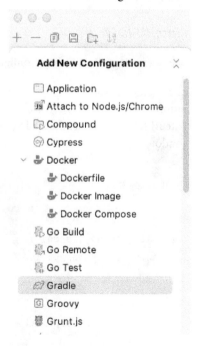

Appendix figure 6 – Custom applications, Add New Configuration with IntelliJ IDEA

4. Now, you can give it a name such as `chapter01.00` (bootRun) and select the actual milestone directory for this configuration. Finally, enter `tomcatRun` under the **Run** option to execute, as shown in the following screenshot:

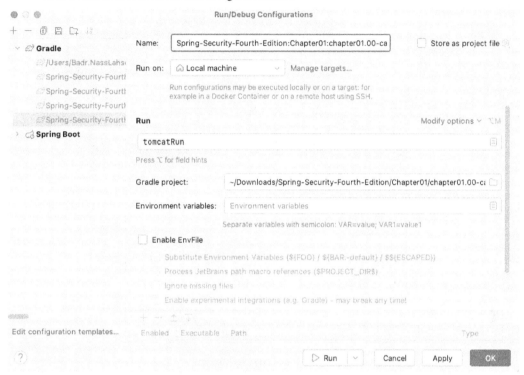

Appendix figure 7 – Custom project, Run with IntelliJ IDEA

5. Select the configuration you want to execute; click on the green **Run** button (as shown in *Appendix figure 7*).

In the next section, we will cover the usage of **Eclipse** to run the sample code.

Using the samples in Eclipse

In this section, we will cover the sample applications usage within Eclipse. Across all projects, a `Tomcat` plugin has been set up using `Gradle` or `Maven` to facilitate the rapid initiation of an embedded instance, expediting your initial setup process.

Importing projects in Eclipse

After you download and install your Eclipse IDE (https://www.eclipse.org/downloads/), launch Eclipse.

1. When you first open Eclipse, it will prompt you for the workspace location. You may need to go to **File** | **Switch Workspace** | **Other** to create a new workspace. We recommend entering a workspace location that does not contain any spaces. For example, look at the following screenshot:

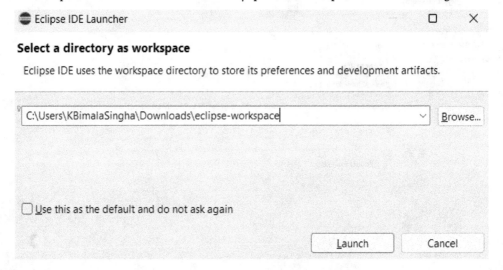

Appendix figure 8 – Eclipse workspace choice

2. Once you have created a new workspace, choose **Import Projects**.

3. This time, we will choose **Existing Maven Projects**:

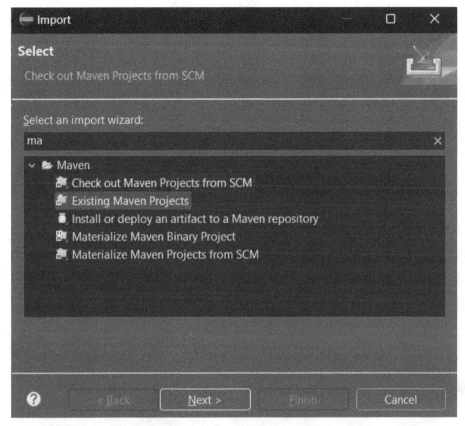

Appendix figure 9 – Eclipse project import as Maven project

4.   Browse to the location you exported the code to and select the parent folder of the code. You will see all the projects listed. You can select the projects you are interested in or leave all the projects selected. If you decide to import all the projects, you can easily focus on the current chapter since the naming conventions will ensure that the projects are sorted in the order that they are presented in the book:

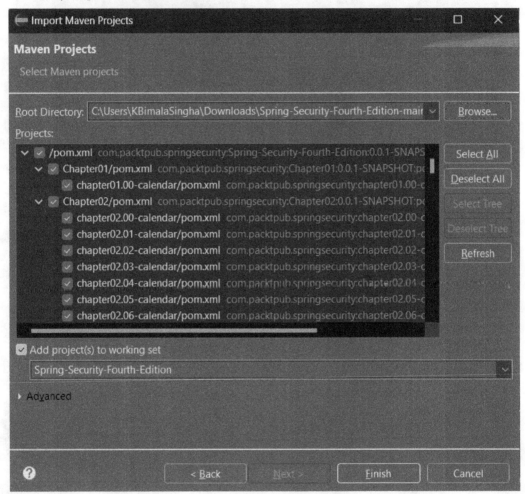

Appendix figure 10 – Eclipse Maven project import confirmation

5. You will be able to work with any of the chapters, and the layout will look as shown in the following screenshot:

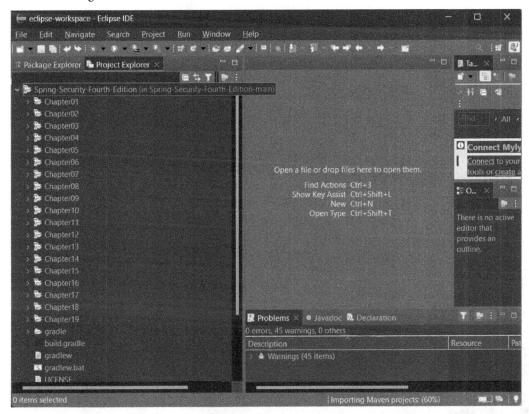

Appendix figure 11 – Chapters presentation layout with Eclipse

After importing the project in **Eclipse**, you can run your code following the guidance provided in the next section.

Running the samples within Eclipse

To execute each milestone project, you can proceed with the subsequent steps:

1. For each spring-boot project (starting from *Chapter 3, Custom Authentication*), you can simply click on the **Run** button in Eclipse, or right-click on your project and select **Run As | Java Application**.

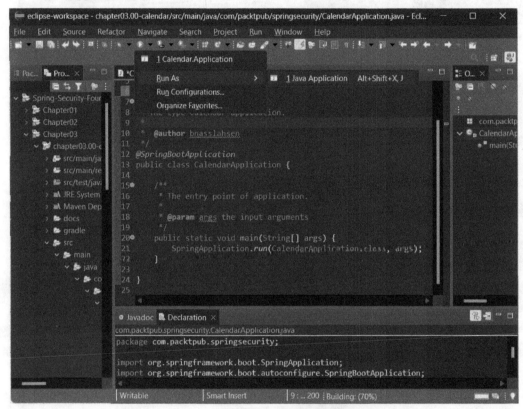

Appendix figure 12 – Running the Spring Boot project with Eclipse

2. For other projects, you can use the terminal, go to the **Run** button using **Eclipse**, or right-click on your project and select **Run Configurations**:

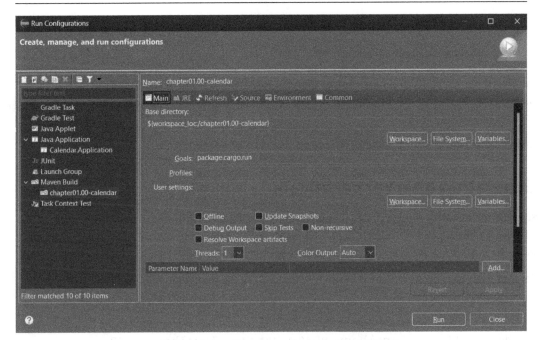

Appendix figure 13 – Custom applications, Run with Eclipse

3. In the **Goals** section, choose, for example, to run **chapter01.00-calendar**, and declare the goal: `package cargo:run`.

Starting the samples from the command line

In *Chapter 1, Anatomy of an Unsafe Application*, and *Chapter 2, Getting Started with Spring Security*, you will use different tasks to run the project.

- If you're using `Gradle`, run the following command to start the application:

  ./gradlew tomcatRun

- Alternatively, if using `Maven`, run the following command to start the application:

  ./mvnw package cargo:run

For the rest of the chapters in the book (starting from *Chapter 3, Custom Authentication*), Spring Boot has been used:

- If you're using `Gradle`, run the following command to start the application:

  ./gradlew bootRun

- Alternatively, if using Maven, run the following command to start the application:

```
./mvnw spring-boot:run
```

In general, for every section of the book, there's a README.md file in the root directory of each chapter milestone. This file includes the necessary commands to launch the application, tailored to your preferred build tool.

In the following section, we'll address the generation of a server certificate.

Generating a server certificate

Some of the chapters sample code (that is, *Chapter 8, Client Certificate Authentication with TLS, Chapter 9, Opening up to OAuth2, Chapter 10, SAML 2 Support,* and *Chapter 18, Single Sign-On with the Central Authentication Service*) requires the use of HTTPS in order for the sample code to work.

Some projects have been configured to run HTTPS; most of the configuration is managed in properties or YAML files.

Now, when you run the sample code on the embedded Tomcat server from Maven or Gradle, you can connect to http://localhost:8080 or https://localhost:8443.

If you do not already have a certificate, you must first generate one.

If you wish, you can skip this step and use the tomcat.keystore file, which contains a certificate that is located in the src/main/resources/keys directory in the book's sample source.

Enter the following command lines at the command prompt:

```
keytool -genkey -alias jbcpcalendar -keypass changeit -keyalg RSA \
-keystore tomcat.keystore
Enter keystore password: changeit
Re-enter new password: changeitWhat is your first and last name?
[Unknown]: localhost
What is the name of your organizational unit? [Unknown]: JBCP Calendar
What is the name of your organization? [Unknown]: JBCP
What is the name of your City or Locality? [Unknown]: Anywhere What is
the name of your State or Province? [Unknown]: UT
What is the two-letter country code for this unit? [Unknown]: US
Is CN=localhost, OU=JBCP Calendar, O=JBCP, L=Anywhere, ST=UT, C=US
correct? [no]: yes
```

Most of the values are self-explanatory, but you will want to ensure that the answer to **What is your first and last name?** is the host that you will be accessing your web application from. This is necessary to ensure that the SSL handshake will succeed.

You should now have a file in the current directory named `tomcat.keystore`. You can view its contents using the following command from within the same directory:

```
keytool -list -v -keystore tomcat.keystore Enter keystore password:
changeit
Keystore type: JKS Keystore provider: SUN
...
Alias name: jbcpcalendar
...
Owner: CN=localhost, OU=JBCP Calendar, O=JBCP, L=Anywhere, ST=UT, C=US
Issuer: CN=localhost, OU=JBCP Calendar, O=JBCP, L=Anywhere, ST=UT,
C=US
```

As you may have guessed, it is insecure to use `changeit` as a password, as this is the default password used with many JDK implementations. In a production environment, you should use a secure password rather than something as simple as `changeit`.

For additional information about the `keytool` command, refer to the documentation found on Oracle's website (`https://docs.oracle.com/en/java/javase/17/docs/specs/man/keytool.html`).

If you are having issues, you might also find the *CAS SSL Troubleshooting and Reference Guide* helpful (`https://apereo.github.io/cas/7.0.x/installation/Troubleshooting-Guide.html`).

Supplementary materials

This section contains a listing of additional resources to technologies and concepts that are used throughout the book:

- **Java Development Kit Downloads**: Refer to `https://www.oracle.com/java/technologies/downloads/` for downloading the JDK.

- **MVC Architecture**: Refer to `https://en.wikipedia.org/wiki/Model%E2%80%93view%E2%80%93controller`.

- **Spring Security site**: Refer to `https://spring.io/projects/spring-security`. You can find links to the Spring Security Javadoc, downloads, source code, and reference at this link.

- **Spring Framework**: Refer to `https://projects.spring.io/spring-framework/`. You can find links to the Spring Framework Javadoc, downloads, source code, and reference from this link.

- **Spring Boot**: Refer to `https://projects.spring.io/spring-boot/`. You can find links to the Spring Boot Javadoc, downloads, source code, and reference from this link.

- **Spring Data**: Refer to `https://projects.spring.io/spring-data/`. You can find links to the Spring Data Javadoc, downloads, source code, and reference from this link. In this book, we covered three of the sub-projects of Spring Data, including Spring Data JPA (`https://projects.spring.io/spring-data-jpa/`) and Spring Data MongoDB (`https://projects.spring.io/spring-data-mongodb/`).

- **Maven**: For more information about Maven, visit their site at `https://maven.apache.org`. For more information about Maven transitive dependencies, refer to the *Introduction to the Dependency Mechanism* documentation at `https://maven.apache.org/guides/introduction/introduction-to-dependency-mechanism.html#Transitive_Dependencies`.

- **Building with Gradle**: Spring Security builds with `Gradle` (`https://gradle.org`) instead of using `Maven`.

- **Object-relational mapping (ORM)**: You can find more general information on Wikipedia at `https://en.wikipedia.org/wiki/Object-relational_mapping`. If you want more hands-on instruction, you may also be interested in the Hibernate (a common Java ORM framework) documentation at `https://hibernate.org/`.

The following are UI technologies:

- **JSP**: You can find more information about JSP on Oracle's site at `https://www.oracle.com/technical-resources/articles/javase/servlets-jsp.html`.

- **Thymeleaf**: This is a modern, tempting framework that provides an excellent alternative to JSP. An additional benefit is that it provides support for both Spring and Spring Security out of the box. You can find more information about Thymeleaf at `https://www.thymeleaf.org/`.

Index

D

packtpub.com

Subscribe to our online digital library for full access to over 7,000 books and videos, as well as industry leading tools to help you plan your personal development and advance your career. For more information, please visit our website.

Why subscribe?

- Spend less time learning and more time coding with practical eBooks and Videos from over 4,000 industry professionals

- Improve your learning with Skill Plans built especially for you

- Get a free eBook or video every month

- Fully searchable for easy access to vital information

- Copy and paste, print, and bookmark content

Did you know that Packt offers eBook versions of every book published, with PDF and ePub files available? You can upgrade to the eBook version at packtpub.com and as a print book customer, you are entitled to a discount on the eBook copy. Get in touch with us at customercare@packtpub.com for more details.

At www.packtpub.com, you can also read a collection of free technical articles, sign up for a range of free newsletters, and receive exclusive discounts and offers on Packt books and eBooks.

Other Books You May Enjoy

If you enjoyed this book, you may be interested in these other books by Packt:

Modern API Development with Spring 6 and Spring Boot 3

Sourabh Sharma

ISBN: 978-1-80461-327-6

- Create enterprise-level APIs using Spring and Java
- Understand and implement REST, gRPC, GraphQL, and asynchronous APIs for various purposes
- Develop real-world web APIs and services, from design to deployment
- Expand your knowledge of API specifications and implementation best practices
- Design and implement secure APIs with authorization and authentication
- Develop microservices-based solutions with workflow and orchestration engines
- Acquire proficiency in designing and testing user interfaces for APIs
- Implement logging and tracing mechanisms in your services and APIs

Learning Spring Boot 3.0

Greg L. Turnquist

ISBN: 978-1-80323-330-7

- Create powerful, production-grade web applications with minimal fuss
- Support multiple environments with one artifact, and add production-grade support with features
- Find out how to tweak your Java apps through different properties
- Enhance the security model of your apps
- Make use of enhancing features such as native deployment and reactive programming in Spring Boot
- Build anything from lightweight unit tests to fully running embedded web container integration tests
- Get a glimpse of reactive programming and decide if it's the right approach for you

Packt is searching for authors like you

If you're interested in becoming an author for Packt, please visit `authors.packtpub.com` and apply today. We have worked with thousands of developers and tech professionals, just like you, to help them share their insight with the global tech community. You can make a general application, apply for a specific hot topic that we are recruiting an author for, or submit your own idea.

Share Your Thoughts

Now you've finished *Spring Security*, we'd love to hear your thoughts! Scan the QR code below to go straight to the Amazon review page for this book and share your feedback or leave a review on the site that you purchased it from.

`https://packt.link/r/1-835-46050-X`

Your review is important to us and the tech community and will help us make sure we're delivering excellent quality content.

Download a free PDF copy of this book

Thanks for purchasing this book!

Do you like to read on the go but are unable to carry your print books everywhere?

Is your eBook purchase not compatible with the device of your choice?

Don't worry, now with every Packt book you get a DRM-free PDF version of that book at no cost.

Read anywhere, any place, on any device. Search, copy, and paste code from your favorite technical books directly into your application.

The perks don't stop there, you can get exclusive access to discounts, newsletters, and great free content in your inbox daily

Follow these simple steps to get the benefits:

1. Scan the QR code or visit the link below

https://packt.link/free-ebook/9781835460504

2. Submit your proof of purchase
3. That's it! We'll send your free PDF and other benefits to your email directly